Ferenc Darvas, György Dormán, Volker Hessel, Steven V. Ley (Eds.)

Flow Chemistry

Also of Interest

Flow Chemistry.
Fundamentals
Ferenc Darvas, György Dormán, Volker Hessel, Steven V. Ley (Eds.), 2021
ISBN 978-3-11-069359-1, e-ISBN (PDF) 978-3-11-069367-6,
e-ISBN (EPUB) 978-3-11-069377-5
Fundamentals and *Applications*: also available as a set
Set-ISBN 978-3-11-073679-3

Chemical Reaction Engineering.
A Computer-Aided Approach
Tapio Salmi, Johan Wärnå, José Rafael Hernández Carucci, César A. de Araújo Filho, 2020
ISBN 978-3-11-061145-8, e-ISBN (PDF) 978-3-11-061160-1,
e-ISBN (EPUB) 978-3-11-061250-9

Dissipativity in Control Engineering.
Applications in Finite- and Infinite-Dimensional Systems
Alexander Schaum, 2021
ISBN 978-3-11-067793-5, e-ISBN (PDF) 978-3-11-067794-2,
e-ISBN (EPUB) 978-3-11-067809-3

Product-Driven Process Design.
From Molecule to Enterprise
Edwin Zondervan, Christhian Almeida-Rivera, Kyle Vincent Camarda, 2020
ISBN 978-3-11-057011-3, e-ISBN (PDF) 978-3-11-057013-7,
e-ISBN (EPUB) 978-3-11-057019-9

Engineering Catalysis
Dmitry Yu. Murzin, 2020
ISBN 978-3-11-061442-8, e-ISBN (PDF) 978-3-11-061443-5,
e-ISBN (EPUB) 978-3-11-061469-5

Flow Chemistry

Volume 2: Applications

Edited by
Ferenc Darvas, György Dormán, Volker Hessel,
Steven V. Ley

DE GRUYTER

Editors
Dr. Ferenc Darvas
InnoStudio Inc.
Graphisoft Park
Záhony u.7
Budapest 1031
Hungary
df.private@gmail.com

Prof. György Dormán
ThalesNano Nanotechnology Inc.
Graphisoft Park
Zahony u. 7
Budapest 1031
Hungary
gyorgy.dorman@thalesnano.com

Prof. Volker Hessel
EC&MS Research
University of Adelaide
Engineering North Building
Adelaide, SA 5005
Australia
volker.hessel@adelaide.edu.au

Prof. Steven V. Ley
Department of Chemistry
University of Cambridge
Lensfield Road
Cambridge CB2 1EW
United Kingdom
svl1000@cam.ac.uk

ISBN 978-3-11-069361-4
e-ISBN (PDF) 978-3-11-069369-0
e-ISBN (EPUB) 978-3-11-069376-8

Library of Congress Control Number: 2021940758

Bibliographic information published by the Deutsche Nationalbibliothek
The Deutsche Nationalbibliothek lists this publication in the Deutsche Nationalbibliografie; detailed bibliographic data are available on the Internet at http://dnb.dnb.de.

Cover image: Royalty Free Stock Illustration, ID: 1701258115; designer: jijomathaidesigners, Idukki, India
Typesetting: Integra Software Services Pvt. Ltd.
Printing and binding: CPI books GmbH, Leck

MIX
Papier aus verantwortungsvollen Quellen
FSC® C083411

www.degruyter.com

Preface

In the last decade, the field of flow chemistry has advanced tremendously and a plethora of applications have been reported in different fields at an unparalleled speed. The characteristics of flow reactors are their exceptionally fast heat and mass transfer. Using so-called microreactors, virtually instantaneous mixing can be achieved for all but the fastest reactions. Similarly, the accumulation of heat, formation of hot spots, and dangers of thermal runaways can be prevented. As a result of the small reactor volumes, the overall safety of the process is significantly improved, even when harsh reaction conditions are used. Thus, this technology offers a unique way to perform ultrafast, exothermic reactions, and allows the execution of reactions which proceed via highly unstable or even explosive intermediates. In addition, efficient telescoping of reaction sequences can be beneficial in terms of minimizing the number of unit operations and avoiding intermediate isolations, which are of particular interest to the pharmaceutical industry where complex multistep sequences often need to be performed. In contrast to what existed only a few years ago, the flow chemistry literature is now full of publications from not only academic groups but also from scientists working in the industry reporting the results of their many different research activities in this field.

Despite the fact that there appears to be ample literature in the flow chemistry space – including several extensive monographs, books, and highly cited review articles – there is a lack of suitable textbooks that can be used for teaching purposes and that can explain the fundamentals to newcomers to the field. A complaint often heard from companies is that there are not enough scientists with the unique training and skillsets of a flow chemist, that is, a person having been educated at the interface of synthetic chemistry and chemical engineering, with additional expertise – for example – in analytical chemistry and data-rich experimentation/machine learning. The first edition of the present *Graduate Textbook on Flow Chemistry* published in 2014 was, therefore, a highly welcome and urgently needed addition to the steadily growing flow chemistry literature! Now, several years on, the second edition of this textbook is released. The original format has been kept the same, namely, a separation into two independent volumes, one dealing with fundamentals, and a second volume, more relating to the many diverse applications that can be realized with this enabling technique. Both volumes not only discuss basic theory but also leave ample room for discussing practical considerations. The individual 22 chapters have been authored by experts in their respective fields, wisely chosen by the editors of this textbook, now – in addition to the original editorial team (Ferenc Darvas, Volker Hessel and György Dormán) – also including Steven Ley.

https://doi.org/10.1515/9783110693690-202

It is my hope and genuine expectation that the second edition of this *Graduate Textbook on Flow Chemistry* will become the standard reference work in the field, both at the university level and at other research institutions, where scientists have to get familiar with this rapidly developing field.

October, 2021

C. Oliver Kappe
Professor
University of Graz, Austria

Acknowledgments

Since the publication of the first edition of the *Graduate Textbook on Flow Chemistry* in 2014, the field has advanced tremendously. Thus, the original editors, Ferenc Darvas, György Dormán, and Volker Hessel, joined by Steven V. Ley, have decided to write a new edition, which, besides providing a broad introduction to the subject, also covers the current state of continuous-flow chemistry and also discusses practical considerations and emerging fields.

The editors would like to express their sincere gratitude to the many people who have helped to bring this book to fruition. First and foremost, the editors express their heartfelt appreciation to all authors and coauthors for their outstanding contribution, cooperation, enthusiasm, spirit, and constructive comments throughout the planning and writing of the new chapters.

A very special thanks are due to all the instrument suppliers for their contribution to the "Technology overview/Overview of the devices" chapter (AM Technology, Corning, Little Things Factory, Microinnova, Syrris, ThalesNano, Uniqsis, Vapourtec, and Zaiput).

The editors' thanks are extended to Szilvia Gilmore (Flow Chemistry Society) for her tireless efforts for coordinating and monitoring the whole project and to Réka Darvas for the great cover design, for the second time.

The editors are immensely grateful to the editorial team at De Gruyter Publishing House, especially to Nadja Schedensack, our ever-patient Project Manager, Kristin Berber-Nerlinger, for all the preliminary organization and preparation work, and Karin Sora, Vice President STEM, for the wonderful support and guidance.

Finally, the editors want to thank C. Oliver Kappe for the visionary introduction to the textbook.

https://doi.org/10.1515/9783110693690-203

Contents

About the editors

Dr. Ferenc Darvas acquired his degrees in Budapest, Hungary (medical chemistry MS, computer sciences BS, PhD in experimental biology). He has been teaching in Hungary, Spain, Austria, and the USA. Dr. Darvas has been involved in introducing microfluidics/flow chemistry methodologies for synthetizing drug candidates since the late 1990s, which led him to found ThalesNano, the inventor of H-Cube®, and the recipient of the R&D100 Award (Technical Oscar), twice.

Dr. Darvas was awarded Senator Honoris Causa by the University of Szeged, Hungary (2019), and as Fellow of the American Chemical Society (2016). Dr. Darvas is also the founder and active president of the Flow Chemistry Society, Switzerland, founder and editorial board member of the *Journal of Flow Chemistry*, founder of the Space Chemistry Consortium, organizer of the Space Chemistry Symposium series at ACS, and initiator of the world's first anti-Covid drug discovery experiments on ISS.

Prof. György Dormán obtained his PhD in organic chemistry from the Technical University of Budapest, Hungary, in 1986. Between 1982–1988 and 1996–1999, he worked at Sanofi – Chinoin in Budapest in various research positions. In 1988–1989, he spent a postdoctoral year in the UK (University of Salford). Between 1992 and 1996, he was a visiting scientist at the State University of New York, Stony Brook. Between 1999 and 2008, he served ComGenex/AMRI as chief scientific officer. In 2008, he joined ThalesNano and worked as a director of Scientific Innovation until 2015. Since 2016, he is a consultant of InnoStudio Inc. In 2011, he became honorary professor at the University of Szeged. He is an author of 116 scientific papers and book chapters. He is a member of the editorial board of Molecular Diversity and *Mini-Reviews in Medicinal Chemistry* and member of the advisory board of *Journal of Flow Chemistry*.

Prof. Volker Hessel studied chemistry at Mainz University and received his PhD in 1993. Further career steps were as follows: 1994, Institut für Mikrotechnik Mainz/D as vice director R&D and director R&D; 2005, Eindhoven University of Technology/NL as professor; 2019, at the University of Warwick/UK as part-time professor. In 2018, he was appointed as deputy dean (research) and professor at the University of Adelaide, Australia. He is research director of Adelaide's Andy Thomas Centre of Space Resources.

Prof. Hessel's research is on microfluidic and plasma processes and their application to health, chemistry, agrifood, and space. He has published 502 peer-reviewed papers (h-index: 61, Scopus) and was authority in the Parliament Enquete Commission "Future of Chemical Industry." He received the AIChE Award "Excellence in Process Development Research" and the IUPAC-ThalesNano Prize in Flow Chemistry, as well as the ERC Advanced/Proof of Concept/Synergy and FET OPEN Grants.

https://doi.org/10.1515/9783110693690-205

Prof. Steven V. Ley obtained his PhD from Loughborough University, UK, and completed postdoctoral studies at the Ohio State University, USA, and Imperial College London, UK. He was appointed to the staff of Imperial College, London, becoming professor in 1983 and head of department in 1989. He was elected to the Royal Society, London, in 1990, moved to Cambridge University to the 1702 Chair of Chemistry in 1992, and was president of the Royal Society of Chemistry 2000–02. Steve's research interests include many aspects of organic chemistry, including synthesis, products, methodology, biotransformations, enabling technologies, and, in particular, natural extensive work on flow chemistry. He has been the recipient of numerous international awards, including the IUPAC-ThalesNano Prize in Flow Chemistry and, recently, the prestigious ACS Arthur C. Cope Award.

Contributing authors

Jesús Alcázar
Discovery Chemistry
Janssen Pharmaceutical Companies of J&J
Janssen-Cilag, S.A.
45007 Toledo, Spain
jalcazar@its.jnj.com
Chapter 7

Elena Alvarez
Dept Bioquimica, Biologia Molecular e Inmunologia
Facultad de Quimica
Universidad de Murcia
Campus Reg Excelencia Int Mare Nostrum
E-30100 Murcia, Spain
Chapter 9

Aikaterini Anastasopoulou
Department of Chemical and Biomolecular Engineering
University of Delaware
Newark, DE 19716, USA
anastasopoulou.aikaterini@gmail.com
Chapter 6

Ádám Bódis
InnoStudio Inc.
Graphisoft Park
Záhony u.7
1031 Budapest, Hungary
adam.bodis@innostudio.org
Chapter 5

Balázs Buchholcz
InnoStudio Inc.
Graphisoft Park
Záhony u.7
1031 Budapest, Hungary
balazs.buchholcz@innostudio.org
Chapter 5

Ferenc Darvas
InnoStudio Inc.
Graphisoft Park
Záhony u.7
1031 Budapest, Hungary
ferenc.darvas@innostudio.org
Chapter 5

Jian Deng
Department of Chemical Engineering
State Key Laboratory of Chemical Engineering
Tsinghua University
Beijing 100084, China
jiandeng@tsinghua.edu.cn
Chapter 8

Angel Díaz-Ortiz
Facultad de Ciencias y Tecnologías Químicas
Universidad de Castilla-La Mancha
13071 Ciudad Real, Spain
Angel.Diaz@uclm.es
Chapter 7

Chencan Du
Department of Chemical Engineering
State Key Laboratory of Chemical Engineering
Tsinghua University
Beijing 100084, China
1871313378@qq.com
Chapter 8

Balázs Endrődi
Department of Physical Chemistry and Materials Science
University of Szeged
Szeged
Hungary
endrodib@chem.u-szeged.hu
Chapter 2

https://doi.org/10.1515/9783110693690-206

Francesco Ferlin
Laboratory of Green S.O.C.
Dipartimento di Chimica, Biologia e Biotecnologie
Università degli Studi di Perugia
Perugia, Italy
francesco.ferlin@unipg.it
Chapter 6

Genovéve Filipcsei
Tavanta Therapeutics Hungary Inc.
Madarász Viktor u. 47
1138 Budapest, Hungary
genoveva.filipcsei@tavanta.com
Chapter 5

Marc Escribà Gelonch
Laboratoire de Génie des Procédées Catalytiques
Centre National de la Recherche Scientifique (CNRS)
CPE-Lyon, France
escribam@hotmail.com
Chapter 6

Oliver M. Griffiths
Yusuf Hamied Department of Chemistry
University of Cambridge
Cambridge, CB2 1EW, UK
omg25@cam.ac.uk
Chapter 11

Volker Hessel
School of Chemical Engineering and Advanced Materials
University of Adelaide
Adelaide, Australia

and

School of Engineering
University of Warwick
Coventry CV4 7AL, UK
volker.hessel@adelaide.edu.au
Chapter 6

Maximilian Hielscher
Department of Chemistry
Johannes Gutenberg University Mainz
Germany
hielscher@uni-mainz.de
Chapter 2

Antonio de la Hoz
Facultad de Ciencias y Tecnologías Químicas
Universidad de Castilla-La Mancha
13071 Ciudad Real, Spain
Antonio.Hoz@uclm.es
Chapter 7

Csaba Janáky
Department of Physical Chemistry and Materials Science
University of Szeged
Szeged
Hungary
janaky@chem.u-szeged.hu
Chapter 2

Tanja Junkers
Polymer Reaction Design Group
School of Chemistry
Monash University
Clayton, VIC 3800, Australia
Tanja.Junkers@monash.edu
Chapter 4

Amol A. Kulkarni
Chem. Eng. Proc. Dev. Division
CSIR-National Chemical Laboratory
Pashan, Pune, India
aa.kulkarni@ncl.res.in
Chapter 3

Daniela Lanari
Laboratory of Green S.O.C.
Dipartimento di Chimica, Biologia e Biotecnologie
Università degli Studi di Perugia
Perugia, Italy
daniela.lanari@unipg.it
Chapter 6

Gabriele Laudadio
Flow Chemistry Van't Hoff Institute for Molecular Sciences
University of Amsterdam
Amsterdam, The Netherlands
g.laudadio@uva.nl
Chapter 1

Steven V. Ley
Yusuf Hamied Department of Chemistry
University of Cambridge
Cambridge CB2 1EW, UK
svl1000@cam.ac.uk
Chapter 11

Martin Linden
Department of Chemistry
Johannes Gutenberg University Mainz
Mainz,
Germany
malinden@uni-mainz.de
Chapter 2

Guangsheng Luo
Department of Chemical Engineering
State Key Laboratory of Chemical Engineering
Tsinghua University
Beijing 100084, China
gsluo@tsinghua.edu.cn
Chapter 8

Pedro Lozano
Dept Bioquimica, Biologia Molecular e Inmunologia
Facultad de Quimica
Universidad de Murcia
Campus Reg Excelencia Int Mare Nostrum
E-30100 Murcia, Spain
Chapter 9

Oliver S. May
Yusuf Hamied Department of Chemistry
University of Cambridge
Cambridge CB2 1EW, UK
om327@cam.ac.uk
Chapter 11

Sara Miralles-Comins
Institute of Advanced Materials (INAM)
Universitat Jaume I
Avda. Sos Baynat s/n
12071 Castellon, Spain
comins@uji.es
Chapter 9

Timothy Noël
Flow Chemistry Van't Hoff Institute for Molecular Sciences
University of Amsterdam
Amsterdam, The Netherlands
t.noel@uva.nl
Chapter 1

Zsolt Ötvös
Tavanta Therapeutics Hungary Inc.
Madarász Viktor u. 47
1138 Budapest, Hungary
zsolt.otvos@tavanta.com
Chapter 5

Francesca Paradisi
Department of Chemistry and Biochemistry
University of Bern
Bern, Switzerland
francesca.paradisi@dcb.unibe.ch
Chapter 10

Suneha Patil
Chem. Eng. Proc. Dev. Division
CSIR-National Chemical Laboratory
Pashan, Pune, India
suneha.patil@ncl.res.in
Chapter 3

László Poppe
Department for Organic Chemistry and Technology
Budapest University of Technology and Economics
Budapest, Hungary
poppe@mail.bme.hu
Chapter 10

Darbha Venkata Ravi Kumar
AMRITA Vishwa Vidyapeetham
Chennai Campus
Thiruvallur, Chennai
Tamil Nadu, India
vrk_darbha@ch.amrita.edu
Chapter 3

Antonio M. Rodríguez
Instituto Regional de Investigación Científica Aplicada (IRICA)
Universidad de Castilla-La Mancha
13071 Ciudad Real, Spain
AntonioM.Rodriguez@uclm.es
Chapter 7

Victor Sans
Institute of Advanced Materials (INAM)
Universitat Jaume I
Avda. Sos Baynat s/n
12071 Castellon, Spain
sans@uji.es
Chapter 9

Karin Sowa
Faculty of Chemistry and Pharmacy
Institute for Organic Chemistry
University of Münster
Münster, Germany
karin.sowa@wwu.de
Chapter 11

Iván Torres-Moya
Facultad de Ciencias y Tecnologías Químicas
Universidad de Castilla-La Mancha
13071 Ciudad Real, Spain
Ivan.TorresMoya@uclm.es
Chapter 7

Nam Nghiep Tran
School of Chemical Engineering and Advanced Materials
University of Adelaide
Adelaide, Australia

and

Department of Chemical Engineering
Can Tho University
Can Tho, Vietnam
namngiep.tran@adelaide.edu.au
Chapter 6

Luigi Vaccaro
Laboratory of Green S.O.C.
Dipartimento di Chimica, Biologia e Biotecnologie
Università degli Studi di Perugia
Perugia, Italy
luigi.vaccaro@unipg.it
Chapter 6

Federica Valentini
Laboratory of Green S.O.C.
Dipartimento di Chimica, Biologia e Biotecnologie
Università degli Studi di Perugia
Perugia, Italy
federicavalentinimail@gmail.com
Chapter 6

Siegfried R. Waldvogel
Department of Chemistry
Johannes Gutenberg University Mainz
Mainz, Germany
waldvogel@uni-mainz.de
Chapter 2

Kai Wang
Department of Chemical Engineering
State Key Laboratory of Chemical Engineering
Tsinghua University
Beijing 100084, China
kaiwang@tsinghua.edu.cn
Chapter 8

Gabriele Laudadio and Timothy Noël

1 Photochemical transformations in continuous-flow reactors

1.1 Introduction

The use of photons to overcome kinetic and thermodynamic barriers has provided diverse opportunities to access novel and unique synthetic pathways to organic chemists for the construction of organic molecules [1]. In the past decade, photocatalysis has become a vibrant research field with many researchers from both academia and industry implementing this mode of molecule activation into their scientific programs [2]. Despite the rapid progress in terms of synthetic chemistry, some technological issues were encountered. Most of these issues have been associated with the Bouguer–Lambert–Beer law (vide infra), which dictates that photons will be absorbed as they travel through the reaction medium. This means that the light intensity will rapidly diminish with increasing reactor diameters. Hence, photochemistry is perceived as inherently not scalable. However, this statement can be regarded as false when using continuous-flow reactors with small internal dimensions, for example, micro- or millireactor technology [3]. In such reactors, reactants can be continuously introduced into the narrow channels (i.e., <1 mm to several millimeter inner diameter) that are homogeneously irradiated [4]. As a consequence of the high photon flux, reduced reaction times are observed, which allow keeping the selectivity of the selected transformations high [5].

In this chapter, we will provide an up-to-date insight into continuous-flow photochemistry and establish guidelines on how to recognize and solve potential issues associated with this highly valuable activation mode.

1.2 Photochemical versus thermochemical activation of molecules

Classical thermochemical pathways utilize elevated temperatures to increase the reaction rate, as shown by the Arrhenius equation (Equation (1.1); Figure 1.1). The higher the reaction temperature, the more the molecules that have the minimal required energy to transform into products. However, in many cases, the temperature needed for reaction would be so high that the molecules would first decompose. Hence, additional reagents for activation, such as catalysts, acids/bases, or reductants/oxidants, are often required:

$$k = A^{\left(-\frac{E_a}{RT}\right)} \tag{1.1}$$

https://doi.org/10.1515/9783110693690-001

where k is the rate constant [unit depends on the order of the reaction], T is the absolute temperature [K], A is the pre-exponential factor, which is a constant for each chemical reaction, E_a is the activation energy for the reaction [$J \cdot mol^{-1}$], and R is the universal gas constant [8.314 $J \cdot K^{-1} \cdot mol^{-1}$].

Alternatively, the selective absorption of photons allows production of complex organic molecules that cannot be easily constructed using thermochemical pathways (Figure 1.1). Notable examples are [2 + 2] cycloadditions yielding strained cyclobutanes in a single step [6] and Norrish-type photoreactions that allow homolytic cleavage of C–C bonds [7]. Notably, the kinetics of such photochemical transformations is strongly dependent on the photon flux as follows:

$$k = \alpha \cdot I^{\beta} \tag{1.2}$$

where k is the rate constant, α is a constant depending on the type of photochemistry, I is the light intensity, and β is a constant depending on the photon flux.For lower light intensities (<around 200–250 $W \cdot m^{-2}$), β is 1.0 [8]. This means that the rate constant increases linearly with increasing photon fluxes (Figure 1.2); such a situation allows one to tune the reaction rate whilst keeping the reaction temperature constant. In other words, the reaction is purely governed by the flux of photons. Consequently, the vast majority of photochemical reactions can be easily quenched by simply switching off the light, which is important in terms of process safety. Furthermore, it can be easily understood that the higher the photocatalyst concentration, the longer the linear regime.

For intermediate light intensities, β is around 0.5 (i.e., square root behavior), while for higher light intensities, β becomes 0 and, consequently, the reaction rate becomes independent of the light intensity. In the latter situation, all catalyst molecules are permanently active, and not all photons are being absorbed. Such a situation is to be avoided because of the increased energy losses and higher costs. However, at such

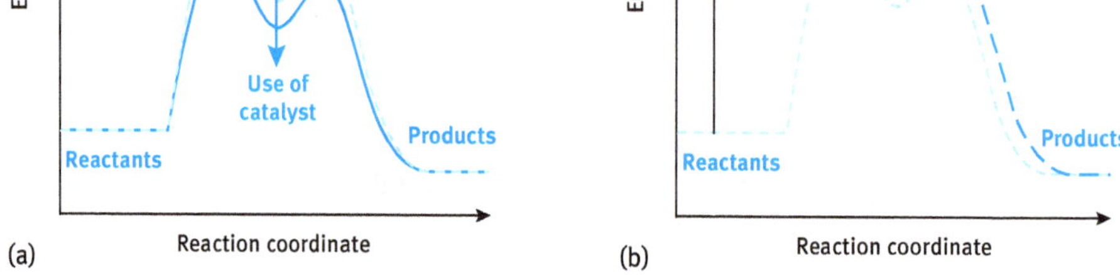

Figure 1.1: Thermochemical (A) versus photochemical activation (B).

high photon fluxes, higher catalyst loadings can be used, allowing further reduction in reaction times and thus increasing productivity.

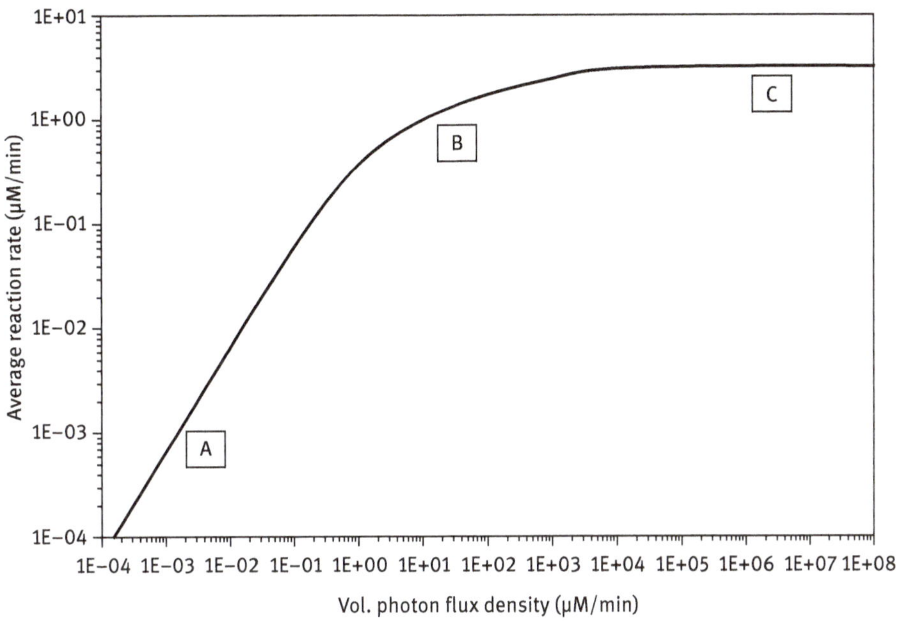

Figure 1.2: The average reaction rate in function of the photon flux. (A) Linear regime with β is 1.0 can be observed at lower light intensities. The reaction is photon limited in the whole reactor. (B) For intermediate light intensities, β is around 0.5 and kinetic limitations appear in parts of the reactor. (C) For high photon fluxes, β becomes 0 and, consequently, the reaction rate becomes independent of the light intensity. Kinetic limitations are observed in the entire reactor.

(For further details on this issue, please see Volume 1, Chapter 1, Title: Fundamentals of Flow Chemistry)

1.3 Important considerations when performing photochemistry in microreactors

The importance of light in photochemical transformations is exemplified by several important laws [9]. The first law of photochemistry (also called the Grotthüs-Draper law) details that, in order to be effective, light needs to be absorbed by the reacting system. Since photocatalysis is chromoselective, it is also important to match the emission wavelength of the light source with the absorption characteristics of the photochemical transformation. Typically, an LED with a narrow emission wavelength is selected to cover the absorption maximum of the photocatalyst, which is the wavelength range at which the substance shows maximum absorbance. The matching of light source and photochemical needs is also important to avoid undesired byproduct formation due to the absorption of other wavelengths by the starting material or product (Figure 1.3) [10]. Ideally, only the photocatalyst or the targeted functional group absorbs, while the other components in the reaction mixture are transparent to the irradiation. In other words, light *quality* is more important than light *quantity* for photochemical transformations: a light source, which is high in intensity but has a broad spectral emission (e.g., CFLs or medium/high-pressure mercury lamps have multiple emission bands; also the sun has a broad spectral distribution, see Figure 1.3), is less energy-effective and less desired compared to monochromatic LEDs. Consequently, it is important to characterize the purchased light sources, as essentially there are no two light sources that are identical (even LEDs differ slightly in color and intensity) [11]. This can be done with so-called integrating spheres that enable the precise determination of the luminous flux and ensure complete integration of all wavelengths present in the light source.

The first law of photochemistry is also important with regard to tuning the lamp positioning to the reactor; it should be noted that the light decreases with distance from the light source (so-called inverse-square law of light, Figure 1.4). To avoid lost irradiation and to maximize radiation transfer to the reactor, the distance between the light source and the reactor needs to be optimized, as shown in Figure 1.4A–C. In addition, to prevent dilution of energy, light can be focused onto the reactor by using suitable mirrors and/or waveguides.

The second law of photochemistry (i.e., the Einstein–Stark law) denotes that one photon can only activate a single molecule. However, activation does not mean that the reaction will occur due to potential energy losses, such as fluorescence, phosphorescence, or thermal decay. The efficiency of this activation process can be described by the so-called quantum efficiency, which is equal to the number of molecules that react over the number of photons absorbed. For photocatalytic processes, the quantum efficiency is typically lower than 1 [12]. However, it is also possible that the quantum efficiency is higher than 1, indicating that radical chain processes are present. The latter is actually advantageous from the vantage point of energy efficiency, as light is only used as an initiator (e.g., radical chain polymerizations, where quantum yields of 10,000 are often noted).

The absorption of photons is described by the Bouguer–Lambert–Beer law (Equation (1.3)) and explains the attenuation of light in the reaction mixture. As can be observed in Figure 1.5, the light is rapidly absorbed by the photocatalyst, resulting in short penetration depths. This is the main reason why photochemical transformations are so difficult to scale, and in larger diameter reactors, no light is present at the center of the vessel. It is also important to realize that declining light intensities in a reaction mixture will result in variable local reaction rates (see also Equation (1.2));

the reaction rate is high at the reactor wall, while it is lower at the center of the reactor. This leads to rate differences that can be two to three orders of magnitude different, depending on the position in the reactor [8]. It is generally accepted that microreactors provide a more uniform irradiation of the reaction medium, resulting in equal reaction conditions across the diameter of the microchannel.

$$A = \log_{10} T = \log_{10} \frac{I_0}{I} = \varepsilon c l \tag{1.3}$$

where A is the light absorption, T is the transmittance, I_0 is the light intensity received by the light-absorbing medium, I is the light intensity after passing through the light-absorbing medium, ε is the molar attenuation coefficient or absorptivity of the light-absorbing species, c is the concentration of the light-absorbing species, and l is the optical path length.

Finally, the Bunsen-Roscoe Law of Reciprocity expresses that the photochemical effect is directly proportional to the total energy dose. This is shown as follows:

$$I \cdot t = \text{constant} \tag{1.4}$$

where I is the light intensity and t is the exposure time.

This law is also important for the scaling of photochemical transformations where the total amount of photons that are absorbed by the reaction mixture needs to be kept the same in order to ensure similar outcomes. These so-called photon equivalents, which are defined as the ratio of absorbed photons to the substrate for a given time of irradiation, need to be kept equal at each reaction scale. In addition, the validity of the Bunsen–Roscoe law would indicate that essentially any reactor can be used, as long as one keeps the photon equivalents the same. Recently, such a strategy has been employed successfully in the scaling of a photocatalytic transformation from milligram scale in batch to multi-kilogram scale in flow [13]. However,

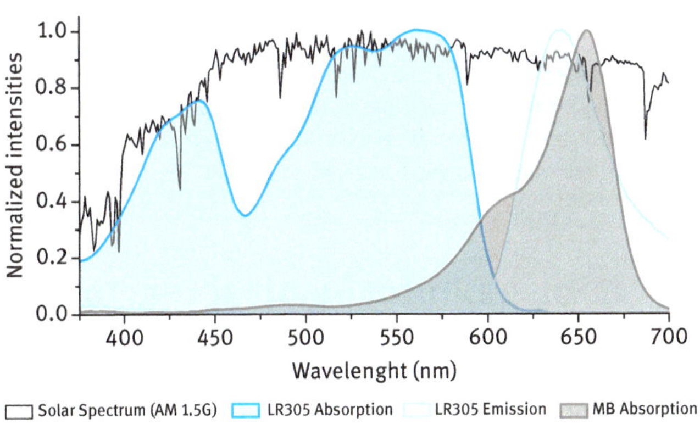

Figure 1.3: Converting solar irradiation (gray area) using a LR305 dye (red area) embedded in a PDMS matrix, which emits fluorescent light (green area) to match with the absorption of methylene blue (MB, blue area). This principle is used to harvest solar energy in so-called luminescent solar concentrator-based photomicroreactors (vide infra).

one should realize that this law is only valid when there are no follow-up reactions possible. If follow-up reactions are possible, larger diameter photochemical reactors would result in local regions at the reactor wall with high photon fluxes, where by-product formation will become problematic due to over-irradiation.

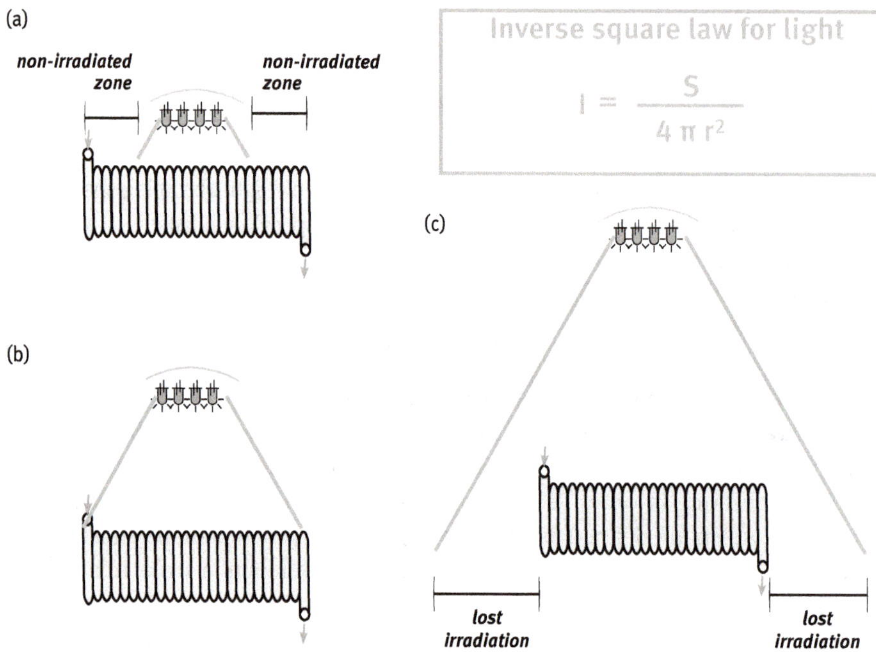

Figure 1.4: Maximizing the photon capture efficiency by minimizing the wasted radiation through optimization of the light source-reactor distance and through use of refractors. The inverse-square law for light describes the light intensity passing through a unit area (S) and is inversely proportional to the square of the distance (r) from the light source, which is treated as a point source. (A) Light source is not matched with the reactor dimensions and some regions are not irradiated. This means that the photochemical reactor is actually smaller than anticipated. (B) Optimal positioning of the reactor and the light source. (C) When the light source is positioned too far away from the reactor, the amount of wasted irradiation increases following the inverse square law for light.

1.4 How to build your own photochemical reactor

The design of continuous-flow photochemical systems may seem overwhelming for one who is considering flow chemistry for the first time. However, this should not be the case, as a typical setup consists only of a few simple components that can be assembled – even by nonexperts – in less than an hour. There are three major parts: 1) the reagent delivery and mixing zone, 2) the reactor part, and 3) the quenching and analysis zone, with the final collection of the reaction mixture (Figure 1.6).

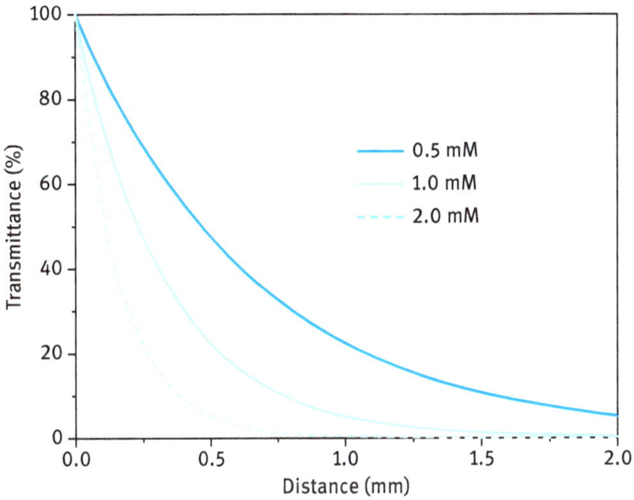

Figure 1.5: Attenuation of light as a function of distance in a photocatalytic reaction using Ru $(bpy)_3Cl_2$ (c = 0.5, 1, and 2 mM, ε = 13,000 cm^{-1} M^{-1}) utilizing the Bouguer–Lambert–Beer correlation (Equation (12.3)).

The first part of the system is where the reactants are introduced and mixed prior to the injection into the photomicroreactor. If the reaction mixture is homogeneous and no background reactions are occurring when mixing all the reagents, a single syringe pump can be used to introduce the stock solution [14]. The only precaution that needs to be taken is to keep the reservoir in the dark to avoid any photochemical reaction occurring prematurely. When large volumes of solution need to be introduced, HPLC or peristaltic pumps may be considered. Immiscible reagents or solutions (e.g., organic-aqueous mixtures) can be introduced separately and mixed using cheap T-, Y-, or cross-shaped mixers. The ratio between the different liquids can be modulated simply by tuning the flow rates set at the individual pumps. When gaseous reagents are employed, they are generally delivered using a mass flow controller, a device which allows precisely dosing that gas into the reactor. Consequently, the exact stoichiometry of the gas can be controlled, which is rather challenging in conventional batch reactors [15]. In addition, the solubility of the gas in the liquid phase can be enhanced by increasing the pressure of the system by using back-pressure regulators at the exit of the reactor. At elevated pressures, HPLC pumps are required to infuse the liquids into the reactor.

The second part of a photochemical system is the reactor, where the actual photochemical reaction takes place. A typical photochemical microreactor can be assembled by using transparent capillary tubing (Figure 1.7A and B) made from inert materials (e.g., fluorinated polymers) and a suitable light source (see Section 1.5). Such photochemical reactors can be easily made by the operator according to

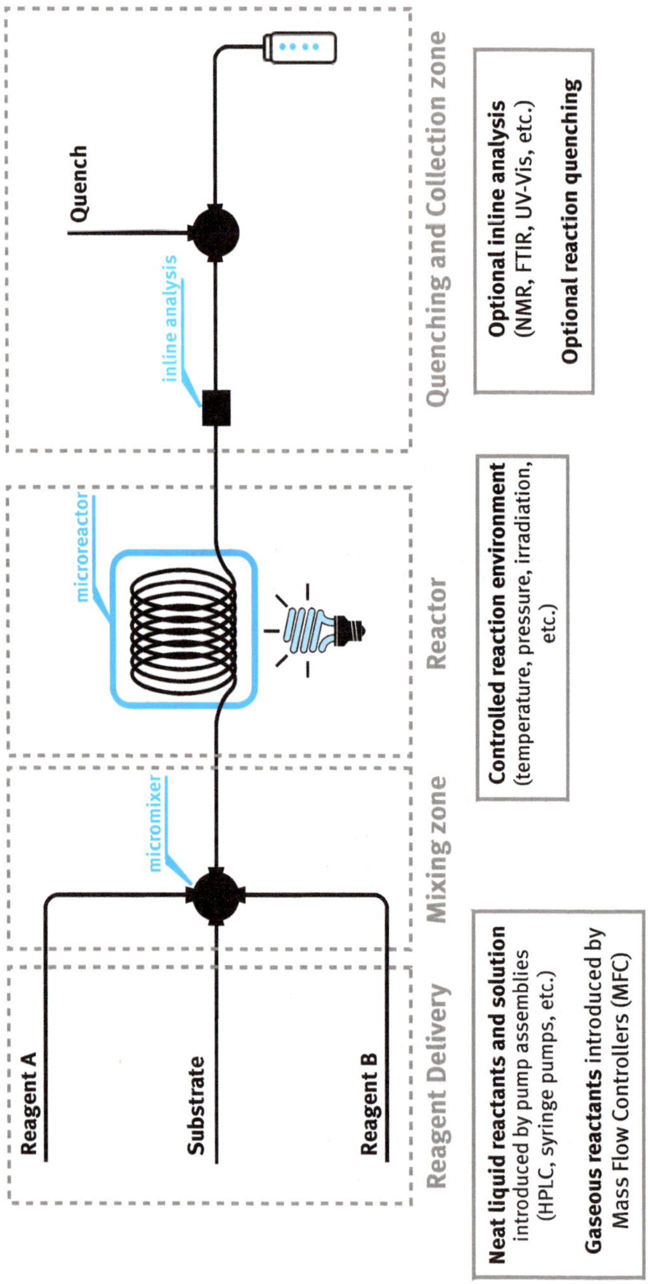

Figure 1.6: Schematic representation of a typical continuous-flow photochemical setup.

established literature procedures [16]. The advantage of Do-It-Yourself assembled reactors is that they are cheap and provide a great insight in the technology. Moreover, this further allows the practitioner to repair setups quickly and to customize the design to the specific needs of the photochemical transformation. Alternatively, commercially available reactors can be used (Figure 1.7C); they are more expensive but enable standardization of the photochemical protocols. Flow reactors can also be made of glass (Figure 1.7D); these reactors can be equipped with static mixers to increase mass transfer.

Once the reaction mixture exits the irradiated zone, it is often immediately quenched, which constitutes the third part of the setup (Figure 1.6). The sample can be collected and further analyzed off-line. Alternatively, one can integrate suitable sensors, such as in- or on-line UV-Vis, FT-IR, MS or NMR, to analyze the reaction mixture in-line, which is of high interest in studying, for example, reaction kinetics [17]. When reactive intermediates or by-products are still present, an additional quench can be merged with the reaction stream. This provides a safe strategy to handle unstable or hazardous reagents and avoids the build-up of those compounds in the collection vessel [18].

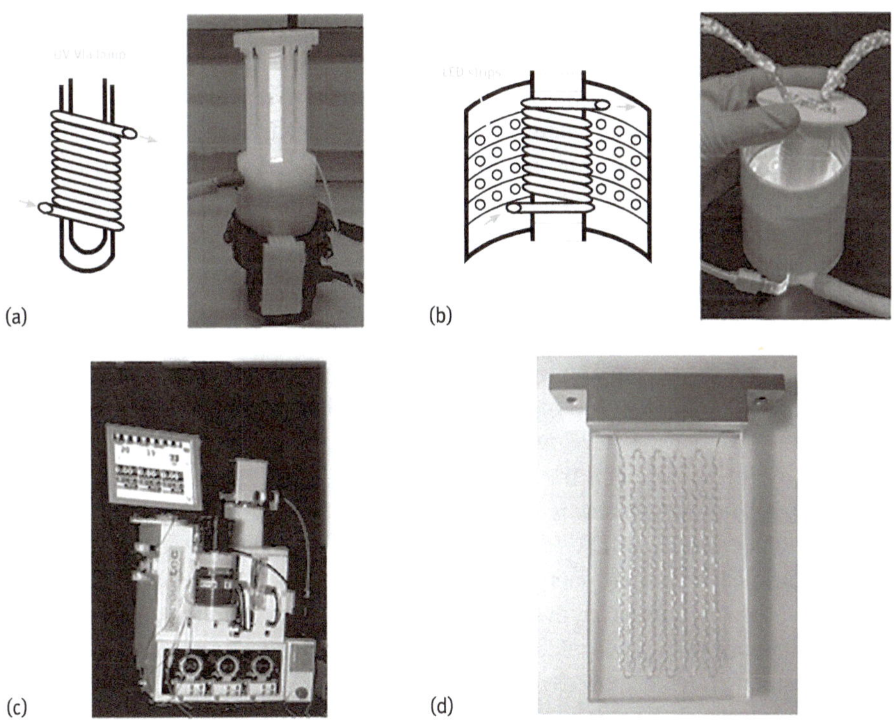

Figure 1.7: Examples of photomicroreactors: (A) PFA tubing coiled around the lamp, (B) LED strips surrounding a coiled PFA tubing, (C) commercially available photomicroreactor (Vapourtec UV 150), and (D) typical flat microreactor made from silicon with embedded static mixers (Little Things Factory).

i (*For further details on this issue, please see Volume 1, Chapter 3, Title: Technology overview/Overview of the devices*)

1.5 The selection of the right light source

The selection of the proper light source is one of the key design considerations for the construction of a photochemical setup [19]. As mentioned earlier, it is crucial to match the emission of the light source with the absorption characteristics of the photoactive substances in the reaction mixture to avoid by-product formation. Other important considerations in the light source selection are cost price, energy efficiency, light intensity, lifetime, and dimensions of the light source.

Without a doubt, the use of light emitting diodes (LEDs) has revolutionized the light industry [20], and their emergence has been one of the key reasons why photocatalysis witnessed a remarkable renaissance in the past decade. LEDs are made of layered crystalline semiconductor material and emit light using the concept of electroluminescence. A single LED chip has a narrow emission spectrum (10–30 nm) and is available nearly everywhere in the visible spectrum as also in the near-UV region. These LEDs are energy efficient and have a long lifetime, making them a lower cost choice for photochemical processes. In addition, due to their small dimensions, LEDs can be aligned with microchannels, which increases the energy efficiency of the reactor [11, 21]. For UVB (280–315 nm) and UVC (230–280 nm) applications, LED solutions remain immature due to a low level of performance (low ratio between power emitted and input power, ≈ 1–3%) and poor thermal and lifetime stability [22]. For applications in these UV regions, classical medium- and low-pressure mercury light sources are still frequently used.

From the vantage point of sustainability and low cost, use of solar irradiation has great appeal to drive photochemical reactions [23]. However, it is very challenging to use this light source due to its diffuse nature (caused by light scattering and reflections), its variability (day/night cycles and passing clouds), and broad spectral distribution. Most reactors developed for continuous-flow solar photochemistry utilize parabolic refractors and solar trackers to increase the irradiation intensity in reactor channels [23a]. Recently, a new reactor concept called luminescent solar concentrator-based photomicroreactors (LSC-PM) was developed which allowed addressing these issues associated with solar energy (Figure 1.8) [24]. LSCs are transparent materials that are doped with fluorescent dyes [25]. After absorption of solar light, the re-emitted fluorescent light can be waveguided within the high refractive index-material towards certain positions (Figure 1.8A). By embedding reactor channels in LSC materials, the luminescent light can be waveguided to these channels leading to a higher photon flux [26]. It should be noted that fluorescent light can be tuned to a narrow wavelength range with the absorption maximum of the photocatalyst. Such LSC-PM reactors have been developed in different colors to match the needs of different photocatalysts

(Figure 1.8B and C) [27]. To cope with light intensity fluctuations, an Arduino-based self-optimizing system was developed, which adjusted in real time, the residence time based on the observed photon flux; hence, a constant yield and conversion could be guaranteed even when clouds were passing by [28].

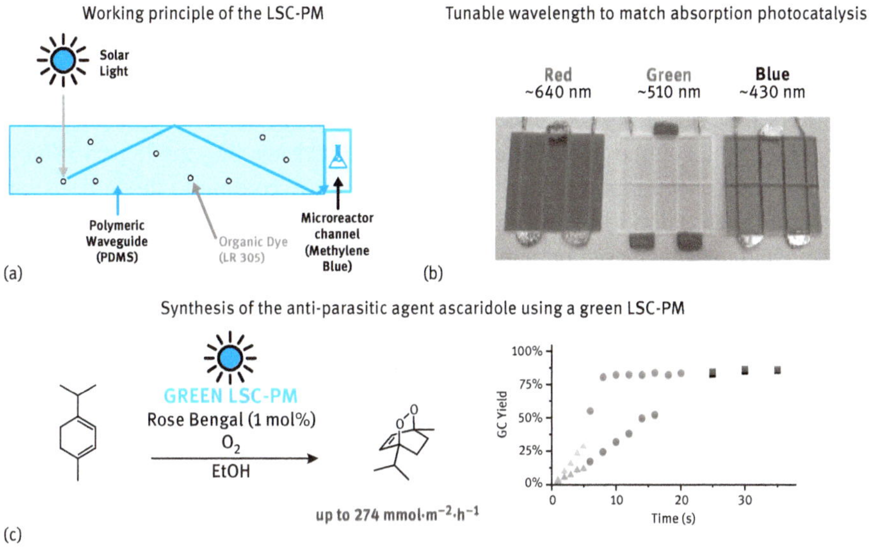

Figure 1.8: Luminescent solar concentrator-based photomicroreactors (LSC-PM) to harvest solar energy for photochemical applications. (A) Working principle of the LSC-PM. (B) Picture of LSC-PMs with different dye dopants. (C) The solar synthesis of ascaridole in a green LSC-PM; the yield in the LSC-PM is shown in red, while the blue curve is carried out in a non-doped microreactor for comparison.

1.6 How flow can make an impact on synthetic organic photochemistry – concrete examples

In the previous sections, we discussed the specific reasons why microreactor technology can make an impact on photochemistry and photocatalysis. In this chapter, we have selected some notable examples and have organized them in different sections, depending on the state of matter: homogeneous or multiphase reaction conditions. However, it is not our aim to be exhaustive, and for more examples, we refer to comprehensive reviews on this subject [1,2].

1.6.1 Homogeneous reaction conditions

An interesting example of homogeneous continuous-flow photochemistry is the decagram scale synthesis of dimethyl 1,4-cubanedicarboxylate (**1**), reported by Linclau et al. [29]. The compound is a valuable building block and serves as a nonaromatic spacer in polymer chemistry and as a phenyl bioisostere in medicinal chemistry [30]. The crucial, yet difficult to scale, step in the synthesis of this molecule is the photo-induced intramolecular [2 + 2] cycloaddition of intermediate (**2**) (Figure 1.9). The authors first carried out an extensive optimization in a small capillary microreactor (FEP, 2 mL volume, ID = 0.7 mm), while also investigating the impact of residence time and concentration of the starting materials. According to the authors, an increase in concentration did not lead to a more productive transformation. The residence time had to be increased proportionally with increasing starting material concentration, leading to an optimal residence time of 45 and 90 min for starting material concentrations of 0.1 and 0.2 M, respectively. From this observation, it can be rationalized that the light source supplies just enough photons to carry out the transformation within the given time frame. Higher productivities should be feasible with more intense light sources (see discussion in Section 1.3, Bunsen–Roscoe law). Using the conditions optimized in the small-scale reactor, the scale-up (108 mmol, 34 g of 2) of this reaction was successfully carried out in a 54 mL reactor, where three reactors of each 18 mL were connected in series (FEP, ID = 0.7 mm). In this method, the complete synthesis of the dimethyl 1,4-cubanedicarboxylate was realized on a 12.8 g scale (58 mmol).

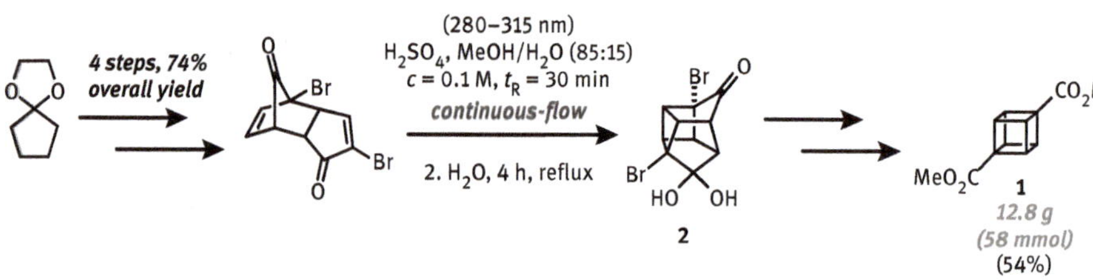

Figure 1.9: Total synthesis of dimethyl 1,4-cubanedicarboxylate with a photo-induced intramolecular [2 + 2] cycloaddition, which was scaled in flow.

1.6.2 Multiphase reaction conditions

Multiphase photochemical reactions are extremely difficult to carry out in batch reactors due to the segregation of the two phases. Typically, these reactions are vigorously stirred to minimize interfacial mass transfer limitations. However, at larger scale, it is challenging to ensure a reproducible interfacial area, making such transformations difficult to scale. In contrast, in flow reactors, these multiphase conditions result in a so-called segmented flow (Taylor flow or slug flow), where the two segments alternate

[31]. The lengths of these segments can be easily tuned by varying the flow rate between the two immiscible phases. In addition, within these segments, secondary fluid patterns that increase the mixing efficiency within the segment and the diffusion of reactive species between two segments [32] are established.

1.6.2.1 Liquid–liquid segmented flow

Beeler et al. reported a beneficial effect of flow on the aqueous-organic [2 + 2] photocycloaddition of cinnamates and cinnamides (Figure 1.10) [33]. This reaction is important in the synthesis of bioactive truxinic and truxillic derivatives. Remarkable rate acceleration was observed in flow, which was ascribed to the higher photon density in the thin film region. The results showed that the segmented flow regime leads to a higher conversion and a faster reaction compared to a homogeneous reaction medium. Interestingly, the higher the ratio of aqueous to organic phase, the longer the thin organic film became, and the higher the conversion was, within a given timeframe.

1.6.2.2 Gas–liquid segmented flow

The use of gases as reactants in organic synthesis is often challenging, due to the slow diffusion of these gases in the liquid phase. Especially when highly reactive species that need to be quenched by a gas are generated, it is crucial that gas–liquid mass transfer limitations are minimized to avoid by-product formation. High mass transfer rates can be achieved in a segmented flow regime, where active mixing is observed due to the so-called Taylor recirculation patterns [15, 34]. Noël et al. exploited this concept to enable the aerobic sp^3 C–H oxidation via hydrogen atom transfer (HAT) photocatalysis using tetrabutyl ammonium decatungstate as a photocatalyst (Figure 1.11) [35]. The reactivity of decatungstate originates from the formation of highly electrophilic oxygen centers that can abstract hydrogen atoms yielding carbon-centered radicals. The radicals can be subsequently trapped by oxygen to enable the installment of carbonyl moieties. The flow protocol proved to be robust and tolerated different functional groups such as esters, ethers, and amines. Also, both activated and inactivated positions can be selectively functionalized, due to the electronic and steric bias of the decatungstate photocatalyst [36]. This also allowed for the selective C–H oxidation of interesting natural compounds such as ambroxide, pregnelonone acetate, eucalyptol, camphor, sclareolide, and artemisinin.

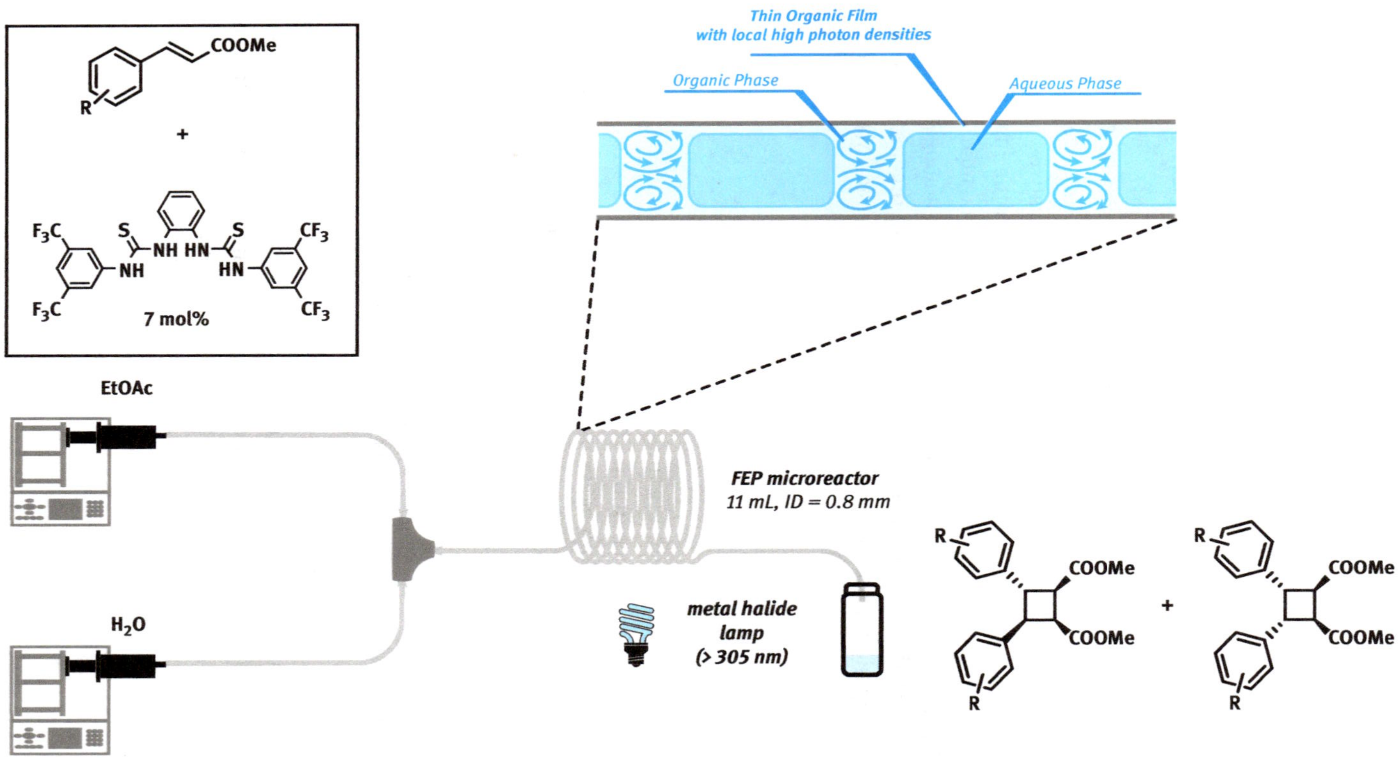

Figure 1.10: Rate acceleration was observed [2 + 2] photo-cycloaddition of cinnamates in a segmented flow regime due to local high photon densities.

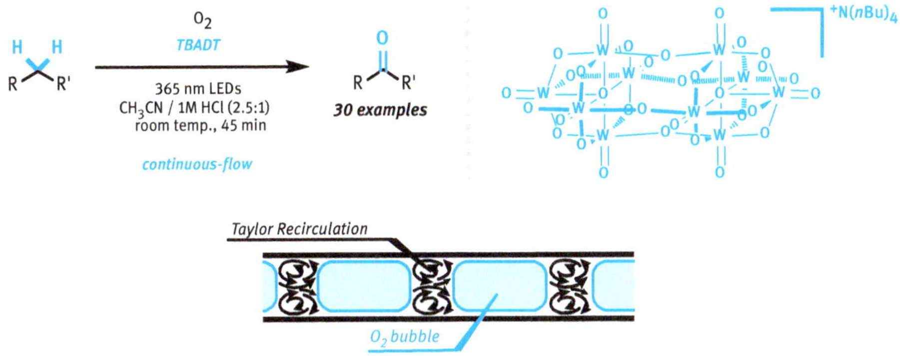

Figure 1.11: Selective sp^3 C–H aerobic oxidation enabled by decatungstate photocatalysis.

(For further details on this issue, please see Volume 1, Chapter 1, Title: Fundamentals of Flow Chemistry)

1.6.2.3 Pressurized gas–liquid flow

In flow reactors, gas–liquid reaction mixtures can be readily pressurized through the use of back-pressure regulators. The increase in pressure can force the gaseous reactants into the liquid phase, improving its availability for the chemical transformation. As an example of this strategy, Noël et al. used decatungstate-enabled HAT photocatalysis to activate volatile alkanes [37]. The activation of gaseous alkanes is very challenging at low temperatures due to the intrinsic inertness of these C–H bonds. Hence, gaseous alkanes have mainly been used as fuels in vehicles or are burned for heat or electricity generation. However, due to their low cost and high availability, these compounds hold great promise in their conversion into value-added commodity chemicals. The use of continuous-flow photomicroreactors has proven to be key within the alkane activation strategy, as it allows simultaneous improvement in the irradiation efficiency (needed to excite decatungstate) and increase in the reaction pressure until complete liquefaction of the gases to establish a suitable contact with the decatungstate photocatalyst (through use of back-pressure regulators) [38] (Fig 1.12). After HAT, the nucleophilic carbon-centered radicals so generated, derived from isobutane, propane, ethane and methane could be scavenged with a diverse set of Michael acceptors, leading to the corresponding hydroalkylated compounds with good to excellent isolated yields.

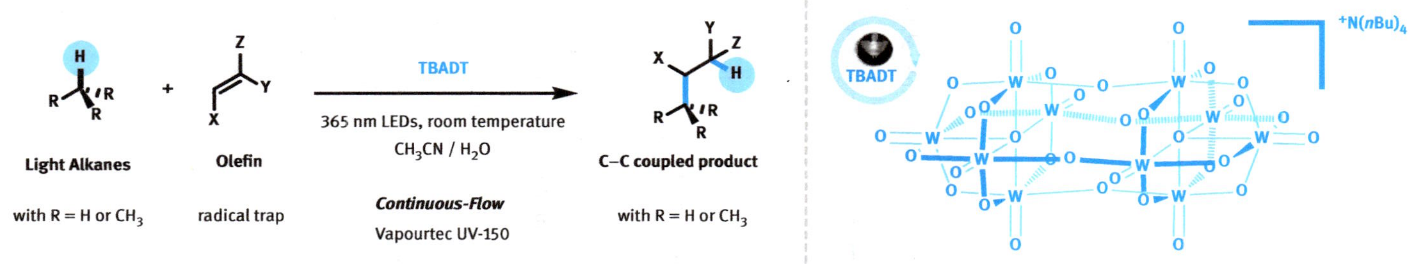

Figure 1.12: Decatungstate-enabled C(sp^3)–H functionalizations of light hydrocarbons in flow.

1.6.2.4 Handling heterogeneous photocatalysts in flow

Semiconductors are an important class of solid photocatalysts that should allow for straightforward recuperation and reuse of catalytic material [39]. However, solid–liquid reactions are a constant challenge in flow chemistry due to the high risk of channel blockage [40]. This is especially problematic when using capillaries with narrow cross sections.

One strategy in handling such solid photocatalysts is to pack them in a cartridge that can be irradiated [41]. George, Poliakoff, Rossen et al. immobilized a homogeneous porphyrin-based photosensitizer on Amberlyst-15 via electrostatic interactions [42] (Fig. 1.13). The supported catalyst was subsequently charged in a sapphire tube (10 mm OD, 1 mm wall thickness, 120 or 240 mm length) and was irradiated with white LEDs. This packed column was used in singlet oxygen-promoted cyclization for the synthesis of the antimalarial artemisinin (**3**) starting from dihydroartemisinic acid (**4**). Interestingly, the Amberlyst support also provided the required acidity to enhance the selectivity of the photochemical process. Only a small amount of the supported porphyrin leached from the reactor during the reactions (2.3 mg leaching for 1 g of product).

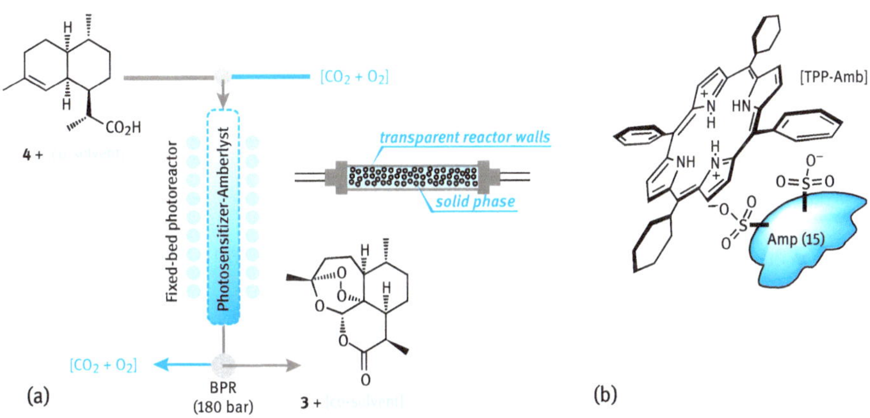

Figure 1.13: (A) Schematic representation of the continuous-flow reactor, which consists of a transparent sapphire tube containing the immobilized photocatalyst, surrounded by a concentric transparent cooling jacket, three banks of high-power white LEDs, and a back-pressure regulator (BPR). (B) Immobilized porphyrin-based photosensitizer onto Amberlyst-15.

One issue with packed-bed photocatalytic reactors is that the limited light penetration through the bed results in suboptimal use of the photocatalyst. Hence, the suspension of the photocatalyst in the liquid reaction stream could solve some of these issues. However, settling of the catalyst could lead to clogging of the microchannels. Gilmore, Seeberger et al. exploited the toroidal fluid recirculation patterns observed in a Taylor flow regime, to sustain the suspension of a modified graphitic nitride photocatalyst (CMB-C_3N_4) [43] (Fig. 1.14). The catalyst could be effectively recovered by simple filtration or centrifugation. This strategy was used for the photocatalytic decarboxylative fluorination.

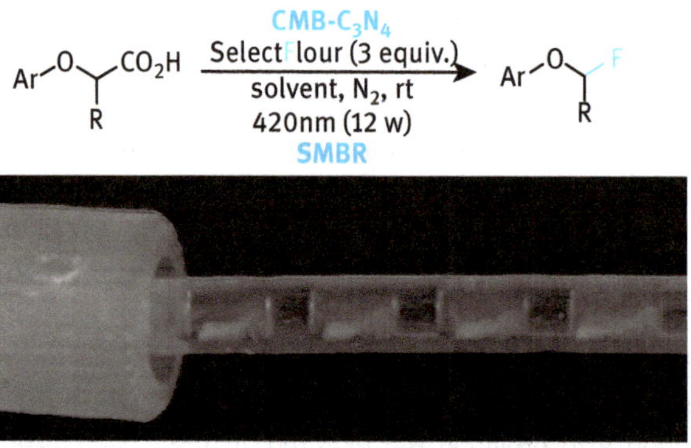

Figure 1.14: Photocatalytic decarboxylative fluorination using graphitic nitride, suspended in the liquid reaction mixture using toroidal vortices observed in a segmented flow regime.

1.7 Scale-up of photochemical processes

Numerous examples have demonstrated the efficacy of continuous-flow micro- and milli-reactors in increasing reaction yield and selectivity, both in academic [44] and industrial [45] settings. However, only a limited set of examples have been successfully scaled up and implemented in the industry [46].

The volume of a microreactor can be expressed using the following equation:

$$V = \frac{\pi}{4} \cdot N \cdot L \cdot D^2 \tag{1.5}$$

where N is the number of the channels connected in parallel, L is the length of the channel, and D is the hydrodynamic channel diameter. Despite its simplicity, this equation can reveal the main strategies in scaling microreactor technology and, thus, also photochemical processes (Figure 1.15) [47].

The first option is to increase the total number of channels (N) that are placed in parallel. This is called numbering-up and has been frequently used in the past [48]. Numbering-up was deemed to be one of the easiest strategies in scaling microreactor technology, as transport phenomena remain constant. However, one of the major engineering challenges in this strategy is ensuring equal reaction conditions in all individual reactors (i.e., identical mass, heat, and photon transfer). This can only be achieved when a good flow distributor, which partitions the reaction stream over the different reactors, is present. Different flow distributor designs have been developed depending on the nature of reaction mixture (homogeneous versus multiphase). A full

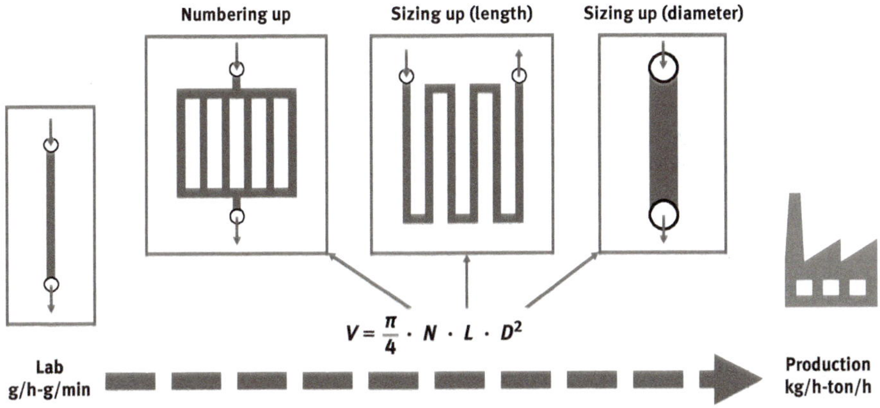

Figure 1.15: Different strategies in scale microreactor technology from a single-channel laboratory scale to a production scale.

discussion on this matter is beyond the scope of this chapter, and we refer to a suitable review for further reading [47].

The second option is called sizing-up and is about increasing the dimensions of the reactor. This can be done either by increasing the length of the reactor (L) or its diameter (D). The easiest strategy is to increase the length of the reactor to increase its volume. Although the channel diameter remains unchanged in such a scenario and, thus, optimal irradiation of the reaction medium can be ensured, the fluid velocity and Reynolds number scale together with the channel length, resulting in improvement of the mass and heat transfer performances. However, there is always an upper limit to the pressure that a reactor and the pumping system can handle. The longer and the narrower the capillary, the higher the pressure drop will be (Hagen–Poiseuille law). Consequently, more expensive pumps and more energy are required to operate the flow process. In the pharmaceutical industry, a pressure drop of 70 psi (=4.8 bar) is usually considered as a reasonable upper limit. Finally, it should also be noted that smaller channels are also more prone to blockage by solid materials in the reaction mixture, which is a significant cost risk for these larger reactors [49].

While many photochemical advantages are associated with the small size of the microchannels, it can be immediately understood that increasing the diameter (D) can become an issue due to the observed attenuation effect of photon transport (Bouguer–Lambert–Beer law). However, by combining high photon flux light sources [50] and increased mass transfer, these hurdles can be partially overcome, and the diameter can be increased (up to one-to-several mm). Increased mass transfer can be exploited to bring the reactants to the reactor walls where the highest photon fluxes are observed. Examples of such reactors are spinning disk reactors (Figure 1.16A) [51], vortex reactors [52], oscillatory reactors (Figure 1.16B) [53], or reactors with embedded static mixers (Figure 1.16C) [54].

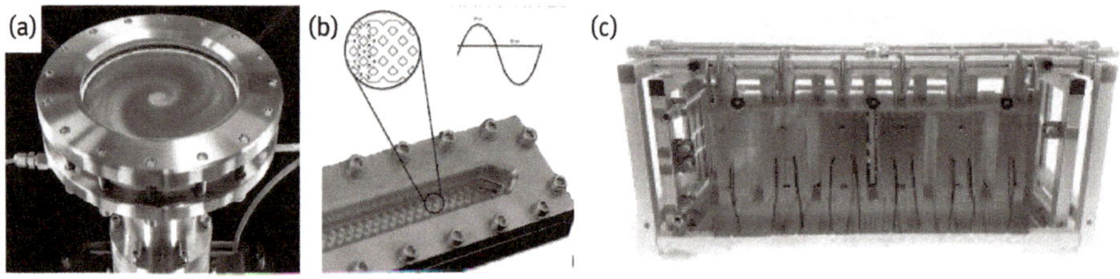

Figure 1.16: (A) Photospinning disk reactor, (B) oscillatory reactors, for example, HANU® reactor is depicted, and (C) Corning® Advanced-Flow™ G1 photoreactor as an example of combining numbering-up and sizing-up strategies.

As each scale-up strategy has its merits, there is not a single strategy that can reach the scale-up factors required for industrial implementation. Numbering-up can maintain excellent irradiation profiles and mass transfer characteristics, but the total number of channels in parallel is limited, due to challenges in uniform flow distribution. Alternatively, sizing-up of the reactor diameter can ensure a higher throughput, but due to the attenuation effect, its increase is restricted, as well. Furthermore, sizing-up of the reactor length suffers from prohibitively high-pressure drops over the reactor. Hence, it is clear that different approaches need to be combined to reach a satisfying production capacity [47]. The total scale-up factor of the reactor volume can be expressed as the product of individual scale-up factors of the channel number (S_N), channel length (S_L), and the channel diameter (S_D):

$$S = S_N \cdot S_L \cdot S_D^2 \tag{1.6}$$

For example, for a photochemical process, scaling the diameter with a factor 3, increasing the length by 5, and placing 10 of those reactors in parallel will result in a scale-up factor of 450. Such a scenario can be realized, for example, with Corning's Advanced Flow reactors, which have modules of different sizes labelled LF, G1, G2, G3, and G4, resulting in a gradual increase of the reaction scale with a factor of 400 on a single plate level (Figure 1.16 C).

1.8 Use of automation protocols in combination of photochemical flow reactors

The synthesis of complex organic molecules is often a daunting and time-consuming challenge, which requires the orchestration of numerous reactions and purification stages. In order to reduce repetitive laboratory work, the automation of chemistry has been at the forefront of many research activities, in the past decade [55]. Automation should allow gathering more data with greater accuracy, in a shorter amount of time

[56]. If such tedious repetitive tasks can be eliminated, more time becomes available for the researchers to carry out more creative work [57]. In addition, it should be possible to carry out some hazardous processes in a more secure environment using automation protocols and robotics, without exposing the researcher to potentially hazardous situations. Recently, some examples have also shown the power of automation in the field of photochemistry and photocatalysis.

Stern–Volmer analysis is an important kinetic experiment in the field of photocatalysis, which allows determination of the rate at which the excited state of the photocatalyst is quenched (Fig 1.17 A). In this experiment, the fluorescence of the photocatalyst is measured with increasing amounts of quencher. After plotting the fluorescence intensity versus the quencher concentration, a linear plot can be fitted from which the quenching rate constant can be derived, using its slope. The task is repetitive, yet it is often plagued with inconsistency due to oxygen interference and inevitable human errors. Noël et al. have developed a continuous-flow platform which allows carrying out these experiments in a fully automated fashion [58]. In only 15 min, data for the desired Stern-Volmer plot could be gathered and analyzed. Due to the confined spaces of the microfluidic platform, oxygen could be completely excluded. Hence, the results were consistent over different experiments, showing a high level of accuracy (R^2) and reproducibility for the derived quenching rate constant. Interestingly, using an autosampler, different potential quenchers could be automatically screened in a time- and labor-efficient fashion. This procedure is of value in elucidating the operative photocatalytic mechanism [59] or even in inventing novel photocatalytic reactions [60].

Another repetitive and time-consuming task for synthetic organic chemists is the optimization of reaction conditions. Jensen, Jamison et al. developed a reconfigurable system that can optimize diverse chemical reactions in automated fashion [61] (Fig 1.17 B). The system is a plug-and-play device, which can be rapidly reconfigured to meet the needs of a specific reaction. It harbors different bays, where different standardized reactions and purification modules can be clicked. Amongst the different reactions that have been showcased, a photocatalytic cyanation of tetrahydroquinoline has been optimized using a SNOBFIT (stable noisy optimization by branch and fit) algorithm and an online HPLC to analyze the reaction yield. Optimized reaction conditions were found after 33 individual experiments, requiring a total optimization time of 7 h.

While accelerated reaction conditions are observed often for photochemical transformations in flow, some reactions do require a longer residence time, and thus, continuous-flow operation becomes challenging. For such transformations, an oscillatory flow strategy, where a liquid segment is moved back and forth in the microreactor can be advantageous, as it allows a more flexible residence time, while preserving the excellent mass- and heat-transport advantages of the flow operation [62]. Kennedy, Stephenson et al. have used such a reactor operation in combination with droplet microfluidics to carry out many different photocatalytic reactions simultaneously [63] (Fig 1.17 C). Reaction droplets of 5 nL were formulated by a liquid handler and separated from each other with perfluorodecalin. Hence, more than hundred reaction droplets, each with unique reaction conditions, could be loaded in the photochemical capillary microreactor (100 μm inner diameter), after which they were exposed to visible light irradiation. Oscillatory flow was imposed to ensure excellent mixing inside every droplet. After the set time was reached, the individual samples were withdrawn from the reactor and analyzed sequentially, using electrospray ionization–mass spectrometry (Figure 1.17).

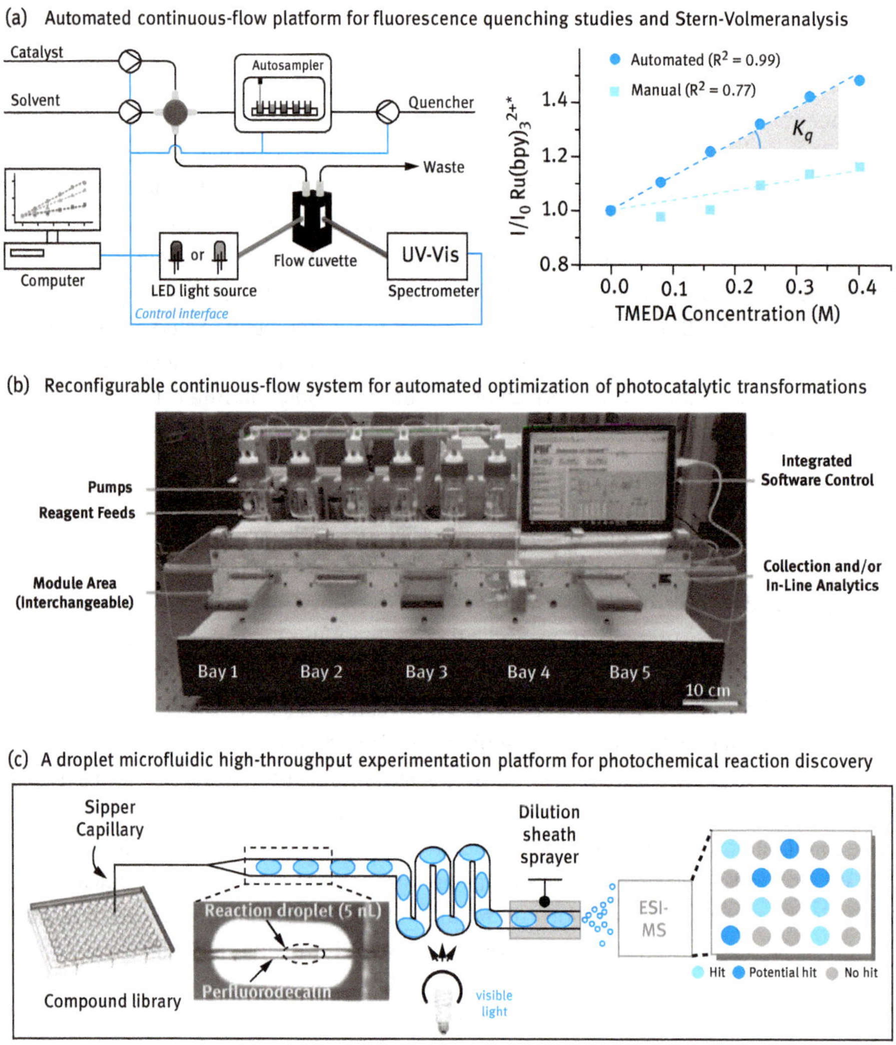

Figure 1.17: Automated flow chemistry allows (A) carrying out fluorescence quenching studies, (B) optimizing photocatalytic reaction conditions, and (C) discovering new photocatalytic transformations without human intervention.

1.9 Summary

Flow chemistry is rapidly gaining popularity amongst researchers in academia and industry. Flow chemistry has established itself as the prime go-to technology for some applications, making it popular across disciplines. In our opinion, photochemistry is one of these applications where flow technology has made an undeniable impact. In this chapter, we have provided an in-depth analysis into the reasons why flow chemistry became so popular in photochemistry and photocatalysis. A good understanding of these fundamental principles is crucial to get the maximum out of the technology.

Microreactor technology has shown to enable acceleration of the photochemical reaction kinetics, effectively reducing the reaction times, and to facilitate scaling of photochemical transformations, even beyond the gram scale. However, in combination with other advantages of the microscale environment, synergetic effects have been observed. For example, a combination of photochemical activation and the use of gaseous reagents allow effective expansion of the synthetic toolbox and development of new reactions, which are very challenging to carry out in conventional batch vessels. Finally, flow photochemistry is, now, a mature discipline, and it is ready for further integration by academic institutions into their chemistry programs [64]. This will further ensure the build-up of crucial know-how and will solidify uptake of the technology in academic and industrial settings [65].

Further readings

- Noel T. Photochemical Processes in Continuous-Flow Reactors: From Engineering Principles to Chemical Applications, World Scientific Publishing Company, 2017.
- Stephenson C., Yoon T., MacMillan, D. Visible Light Photocatalysis in Organic Chemistry, Wiley-VCH, 2018.
- Koenig B. Chemical photocatalysis, De Gruyter, 2020.
- Fagnoni M., Protti S., Ravelli D. Photoorganocatalysis in Organic Synthesis, World Scientific Publishing Company, 2019.
- Cambié D.; Bottecchia C.; Straathof N. J. W.; Hessel V.; Noël T. Applications of Continuous-Flow Photochemistry in Organic Synthesis, Material Science, and Water Treatment, Chem. Rev. 2016, 116, 10276–10341. DOI: 10.1021/acs.chemrev.5b00707.
- Sambiagio C.; Noël T. Flow Photochemistry: Shine Some Light on Those Tubes!, Trends in Chemistry 2020, 2(2), 92–106. DOI: 10.1016/j.trechm.2019.09.003
- Williams J. D.; Kappe C. O. Recent Advances towards Sustainable Flow Photochemistry, Curr. Opin. Green Sust. Chem. 2020, 10, 100351. DOI: 10.1016/j.cogsc.2020.05.001
- Buzzetti L.; Crisenza G. E. M.; Melchiorre P. Mechanistic studies in photocatalysis, Angew. Chem. Int. Ed. 2018, 58, 3730–3747. DOI: 10.1002/anie.201809984
- McAtee, R. C.; McClain E. J.; Stephenson C. R. J. Illuminating photoredox catalysis, Trends in Chemistry 2019, 1, 111–125. DOI: 10.1016/j.trechm.2019.01.008
- Riente Paiva P.; Noel T. Application of Metal Oxide Semiconductors in Light-Driven Organic Transformations. Catal. Sci. Technol. 2019, 9, 5186–5232 DOI: 10.1039/c9cy01170f

Study questions

1.1 Describe the rate constant for photochemical reactions and the three apparent regimes that can be distinguished.

1.2 What are the general parameters to consider when designing a photochemical flow reactor, and why are these important?

1.3 Jamison and coworkers reported this transformation to prepare amino acids via photoredox catalysis (*Nature Chemistry* **2017**, *9*, 453–456):

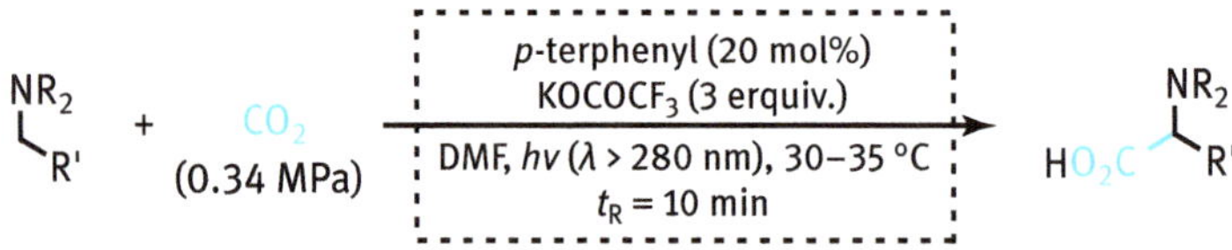

Explain in detail why this reaction is ideally suited to be performed using microreactor technology.

1.4 Explain how solid-based photocatalysts can be used in flow without clogging of the microchannels.

1.5 Describe the different strategies to scale photochemical transformations using continuous-flow reactors. Also, explain how industrial volumes can be met.

1.6 What are the advantages of automating photochemical transformations in continuous-flow reactors?

References

[1] (a) Bottecchia C, Noël T, Photocatalytic modification of amino acids, peptides, and proteins, Chem – A Eur J, 2019, 25, 26–42; (b) Shaw MH, Twilton J, MacMillan DWC, Photoredox catalysis in organic chemistry, J Org Chem, 2016, 81, 6898–6926; (c) Yoon TP, Ischay MA, Du J, Visible light photocatalysis as a greener approach to photochemical synthesis, Nat Chem, 2010, 2, 527–532.

[2] (a) Li P, Terrett JA, Zbieg JR, Visible-light photocatalysis as an enabling technology for drug discovery: a paradigm shift for chemical reactivity, ACS Med Chem Lett, 2020, 11, 2120–2130; (b) McAtee RC, McClain EJ, Stephenson CRJ, Illuminating photoredox catalysis, Trends Chem, 2019, 1, 111–125; (c) Bogdos MK, Pinard E, Murphy JA, Applications of organocatalysed visible-light photoredox reactions for medicinal chemistry, Beilstein J Org Chem, 2018, 14, 2035–2064; (d) Romero NA, Nicewicz DA, Organic photoredox catalysis, Chem Rev, 2016, 116, 10075–10166.

[3] (a) Sambiagio C, Noël T, Flow photochemistry: shine some light on those tubes!, Trends Chem, 2020, 2, 92–106; (b) Cambié D, Bottecchia C, Straathof NJW, Hessel V, Noël T, Applications of continuous-flow photochemistry in organic synthesis, material science, and water treatment, Chem Rev, 2016, 116, 10276–10341; (c) Su Y, Straathof NJW, Hessel V, Noël T, Photochemical transformations accelerated in continuous-flow reactors: basic concepts and applications, Chem Eur J, 2014, 20, 10562–10589.

[4] (a) Williams JD, Kappe CO, Recent advances toward sustainable flow photochemistry, Curr Opin Green Sust Chem, 2020, 25, 100351; (b) Rehm TH, Reactor technology concepts for flow

photochemistry, ChemPhotoChem, 2020, 4, 235–254; (c) Di Filippo M, Bracken C, Baumann M, Continuous flow photochemistry for the preparation of bioactive molecules, Molecules, 2020, 25, 356; (d) Van Gerven T, Mul G, Moulijn J, Stankiewicz A, A review of intensification of photocatalytic processes, Chem Eng Process, 2007, 46, 781–789.

[5] (a) Rehm TH, Flow photochemistry as a tool in organic synthesis, Chem Eur J, 2020. 10.1002/chem.202000381; (b) Noël T, A personal perspective on the future of flow photochemistry, J Flow Chem, 2017, 7, 87–93.

[6] Poplata S, Tröster A, Zou Y-Q, Bach T, Recent advances in the synthesis of cyclobutanes by olefin [2 + 2] photocycloaddition reactions, Chem Rev, 2016, 116, 9748–9815.

[7] (a) Kärkäs MD, Porco JA, Stephenson CRJ, Photochemical approaches to complex chemotypes: applications in natural product synthesis, Chem Rev, 2016, 116, 9683–9747; (b) Bach T, Hehn JP, Photochemical reactions as key steps in natural product synthesis, Angew Chem Int Ed, 2011, 50, 1000–1045.

[8] Bloh JZ, Holistic A, approach to model the kinetics of photocatalytic reactions, Front Chem, 2019, 7, 128.

[9] Bonfield HE, Knauber T, Lévesque F, Moschetta EG, Susanne F, Edwards LJ, Photons as a twenty-first century reagent, Nat Commun, 2020, 11, 804.

[10] (a) Taştan Ü, Seeber P, Kupfer S, Ziegenbalg D, Photochlorination of toluene – the thin line between intensification and selectivity. Part 2: selectivity, React Chem Eng, 2021. 10.1039/D0RE00366B; (b) Haas CP, Roider T, Hoffmann RW, Tallarek U, Light as a reaction parameter – systematic wavelength screening in photochemical synthesis, React Chem Eng, 2019, 4, 1912–1916.

[11] Roibu A, Morthala RB, Leblebici ME, Koziej D, Van Gerven T, Kuhn S, Design and characterization of visible-light LED sources for microstructured photoreactors, React Chem Eng, 2018, 3, 849–865.

[12] Su Y, Kuijpers KPL, König N, Shang M, Hessel V, Noël T, Mechanistic A, investigation of the visible-light photocatalytic trifluoromethylation of heterocycles using CF3I in flow, Chem Eur J, 2016, 22, 12295–12300.

[13] Corcoran EB, McMullen JP, Lévesque F, Wismer MK, Naber JR, Photon equivalents as a parameter for scaling photoredox reactions in flow: translation of photocatalytic C–N cross-coupling from lab scale to multikilogram scale, Angew Chem Int Ed, 2020, 59, 11964–11968.

[14] While syringe pumps are expensive, they can be made from cheap parts using a combination of 3D printing technology and microcontrollers, see:; (a) Samokhin AS, Syringe pump created using 3D printing technology and arduino platform, J Anal Chem, 2020, 75, 416–421; (b) Neumaier JM, Madani A, Klein T, Ziegler T, Low-budget 3D-printed equipment for continuous flow reactions, Beilstein J Org Chem, 2019, 15, 558–566.

[15] Mallia CJ, Baxendale IR, The use of gases in flow synthesis, Org Process Res Dev, 2016, 20, 327–360.

[16] (a) Britton J, Jamison TF, The assembly and use of continuous flow systems for chemical synthesis, Nat Protoc, 2017, 12, 2423–2446; (b) Straathof NJW, Su Y, Hessel V, Noël T, Accelerated gas-liquid visible light photoredox catalysis with continuous-flow photochemical microreactors, Nat Protoc, 2016, 11, 10–21; (c) Most researchers describe in detail their reactor in the Supporting Information, which is accessible free-of-charge.

[17] Baumann M, Integrating continuous flow synthesis with in-line analysis and data generation, Org Biomol Chem, 2018, 16, 5946–5954.

[18] (a) Kockmann N, Thenée P, Fleischer-Trebes C, Laudadio G, Noël T, Safety assessment in development and operation of modular continuous-flow processes, React Chem Eng, 2017, 2, 258–280; (b) Movsisyan M, Delbeke EIP, Berton JKET, Battilocchio C, Ley SV, Stevens CV,

Taming hazardous chemistry by continuous flow technology, Chem Soc Rev, 2016, 45, 4892–4928; (c) Gutmann B, Cantillo D, Kappe CO, Continuous-flow technology – a tool for the safe manufacturing of active pharmaceutical ingredients, Angew Chem Int Ed, 2015, 54, 6688–6728.

[19] Sender M, Ziegenbalg D, Light sources for photochemical processes – estimation of technological potentials, Chem Ing Tech, 2017, 89, 1159–1173.

[20] Tsao JY, Han J, Haitz RH, Pattison PM, The Blue LED Nobel Prize: historical context, current scientific understanding, human benefit, Ann Phys, 2015, 527, A53–A61.

[21] Su Y, Talla A, Hessel V, Noël T, Controlled photocatalytic aerobic oxidation of thiols to disulfides in an energy-efficient photomicroreactor, Chem Eng Technol, 2015, 38, 1733–1742.

[22] Tokode O, Prabhu R, Lawton LA, Robertson PKJ, Led UV, sources for heterogeneous photocatalysis, In: Environmental Photochemistry Part III. The Handbook of Environmental Chemistry, Bahnemann D, Robertson P, Eds, Springer, Berlin, Heidelberg: Berlin, 2014, Vol. 35.

[23] (a) Cambié D, Noël T, Solar photochemistry in flow, Top Curr Chem, 2018, 376, 45; (b) Oelgemöller M, Solar photochemical synthesis: from the beginnings of organic photochemistry to the solar manufacturing of commodity chemicals, Chem Rev, 2016, 116, 9664–9682; (c) Protti S, Fagnoni M, The sunny side of chemistry: green synthesis by solar light, Photochem Photobiol Sci, 2009, 8, 1499–1516.

[24] Cambié D, Zhao F, Hessel V, Debije MG, Noël T, A leaf-inspired luminescent solar concentrator for energy-efficient continuous-flow photochemistry, Angew Chem Int Ed, 2017, 56, 1050–1054.

[25] (a) Papakonstantinou I, Portnoi M, Debije MG, The hidden potential of luminescent solar concentrators, Adv Energy Mater, 2020. 10.1002/aenm.202002883; (b) Meinardi F, Bruni F, Brovelli S, Luminescent solar concentrators for building-integrated photovoltaics, Nat Rev Mater, 2017, 2, 17072; (c) Debije MG, Verbunt PPC, Thirty years of luminescent solar concentrator research: solar energy for the built environment, Adv Energy Mater, 2012, 2, 12–35.

[26] (a) De Oliveira GX, Lira JOB, Cambié D, Noël T, Riella HG, Padoin N, Soares C, CFD analysis of a luminescent solar concentrator-based photomicroreactor (LSC-PM) with feedforward control applied to the synthesis of chemicals under fluctuating light intensity, Chem Eng Res Des, 2020, 153, 626–634; (b) Cambié D, Zhao F, Hessel V, Debije MG, Noël T, Every photon counts: understanding and optimizing photon paths in luminescent solar concentrator-based photomicroreactors (LSC-PMs), React Chem Eng, 2017, 2, 561–566.

[27] Cambié D, Dobbelaar J, Riente P, Vanderspikken J, Shen C, Seeberger PH, Gilmore K, Debije MG, Noël T, Energy-efficient solar photochemistry with luminescent solar concentrator based photomicroreactors, Angew Chem Int Ed, 2019, 58, 14374–14378.

[28] Zhao F, Cambié D, Hessel V, Debije MG, Noël T, Real-time reaction control for solar production of chemicals under fluctuating irradiance, Green Chem, 2018, 20, 2459–2464.

[29] Collin DE, Jackman EH, Jouandon N, Sun W, Light ME, Harrowven DC, Linclau B, Decagram synthesis of dimethyl 1,4-cubanedicarboxylate using continuous-flow photochemistry, Synthesis 2021, 53, 1307–1314.

[30] (a) Locke GM, Bernhard SSR, Senge MO, Nonconjugated hydrocarbons as rigid-linear motifs: isosteres for material sciences and bioorganic and medicinal chemistry, Chem Eur J, 2019, 25, 4590–4647; (b) Eaton PE, Cubanes: starting materials for the chemistry of the 1990s and the new century, Angew Chem Int Ed, 1992, 31, 1421–1436.

[31] (a) Liu Y, Chen G, Yue J, Manipulation of gas-liquid-liquid systems in continuous flow microreactors for efficient reaction processes, J Flow Chem, 2020, 10, 103–121;

(b) Kashid MN, Kiwi-Minsker L, Microstructured reactors for multiphase reactions: state of the art, Ind Eng Chem Res, 2009, 48, 6465–6485.
[32] Noël T, Su Y, Hessel V, Beyond organometallic flow chemistry: the principles behind the use of continuous-flow reactors for synthesis, In: Organometallic Flow Chemistry, Noël T, Ed, Springer International Publishing, Cham, 2016, 1–41.
[33] Telmesani R, White JAH, Beeler AB, Liquid-liquid slug-flow-accelerated [2+2] photocycloaddition of cinnamates, ChemPhotoChem, 2018, 2, 865–869.
[34] (a) Han S, Kashfipour MA, Ramezani M, Abolhasani M, Accelerating gas–liquid chemical reactions in flow, Chem Commun, 2020, 56, 10593–10606; (b) Ramezani M, Kashfipour MA, Abolhasani M, Minireview: flow chemistry studies of high-pressure gas-liquid reactions with carbon monoxide and hydrogen, J Flow Chem, 2020, 10, 93–101.
[35] Laudadio G, Govaerts S, Wang Y, Ravelli D, Koolman HF, Fagnoni M, Djuric SW, Noël T, Selective C(sp3)–H aerobic oxidation enabled by decatungstate photocatalysis in flow, Angew Chem Int Ed, 2018, 57, 4078–4082.
[36] Ravelli D, Fagnoni M, Fukuyama T, Nishikawa T, Ryu I, Site-selective C–H functionalization by decatungstate anion photocatalysis: synergistic control by polar and steric effects expands the reaction scope, ACS Catal, 2018, 8, 701–713.
[37] Laudadio G, Deng Y, Van Der Wal K, Ravelli D, Nuño M, Fagnoni M, Guthrie D, Sun Y, Noël T, C (sp3)–H functionalizations of light hydrocarbons using decatungstate photocatalysis in flow, Science, 2020, 369, 92.
[38] Govaerts S, Nyuchev A, Noel T, Pushing the boundaries of C–H bond functionalization chemistry using flow technology, J Flow Chem, 2020, 10, 13–71.
[39] Riente P, Noël T, Application of metal oxide semiconductors in light-driven organic transformations, Catal Sci Technol, 2019, 9, 5186–5232.
[40] (a) Chen Y, Sabio JC, Hartman RL, When solids stop flow chemistry in commercial tubing, J Flow Chem, 2015, 5, 166–171; (b) Hartman RL, Managing solids in microreactors for the upstream continuous processing of fine chemicals, Org Process Res Dev, 2012, 16, 870–887.
[41] Bottecchia C, Erdmann N, Tijssen PMA, Milroy L-G, Brunsveld L, Hessel V, Noël T, Batch and flow synthesis of disulfides by visible-light-induced TiO2 photocatalysis, ChemSusChem, 2016, 9, 1781–1785.
[42] Amara Z, Bellamy JFB, Horvath R, Miller SJ, Beeby A, Burgard A, Rossen K, Poliakoff M, George MW, Applying green chemistry to the photochemical route to artemisinin, Nat Chem, 2015, 7, 489–495.
[43] Pieber B, Shalom M, Antonietti M, Seeberger PH, Gilmore K, Continuous heterogeneous photocatalysis in serial micro-batch reactors, Angew Chem Int Ed, 2018, 57, 9976–9979.
[44] Plutschack MB, Pieber B, Gilmore K, Seeberger PH, The Hitchhiker's guide to flow chemistry, Chem Rev, 2017, 117, 11796–11893.
[45] (a) Baumann M, Moody TS, Smyth M, Wharry S, A perspective on continuous flow chemistry in the pharmaceutical industry, Org Process Res Dev, 2020, 24, 1802–1813; (b) Hughes DL, Applications of flow chemistry in the pharmaceutical industry – highlights of the recent patent literature, Org Process Res Dev, 2020, 24, 1850–1860.
[46] Berton M, De Souza JM, Abdiaj I, McQuade DT, Snead DR, Scaling continuous API synthesis from milligram to kilogram: extending the enabling benefits of micro to the plant, J Flow Chem, 2020, 10, 73–92.
[47] Dong Z, Wen Z, Zhao F, Kuhn S, Noël T, Scale-up of micro- and milli-reactors: an overview of strategies, design principles and applications. Chem Eng Sci : X 2021, 100097.
[48] (a) Zhao F, Cambié D, Janse J, Wieland EW, Kuijpers KPL, Hessel V, Debije MG, Noël T, Scale-up of a luminescent solar concentrator-based photomicroreactor via numbering-up, ACS Sustain Chem Eng, 2018, 6, 422–429; (b) Kuijpers KPL, van Dijk MAH, Rumeur QG,

Hessel V, Su Y, Noël T, A sensitivity analysis of a numbered-up photomicroreactor system, React Chem Eng, 2017, 2, 109–115; (c) Su Y, Kuijpers K, Hessel V, Noël T, A convenient numbering-up strategy for the scale-up of gas–liquid photoredox catalysis in flow, React Chem Eng, 2016, 1, 73–81.
[49] Yizheng C, Jasmine CS, Ryan LH, When solids stop flow chemistry in commercial tubing, J Flow Chem, 2015, 5, 166–171.
[50] (a) Lévesque F, Di Maso MJ, Narsimhan K, Wismer MK, Naber JR, Design of a kilogram scale, plug flow photoreactor enabled by high power LEDs, Org Process Res Dev, 2020. 10.1021/acs.oprd.0c00373; (b) Elliott LD, Berry M, Harji B, Klauber D, Leonard J, Booker-Milburn KI, A small-footprint, high-capacity flow reactor for UV photochemical synthesis on the kilogram scale, Org Process Res Dev, 2016, 20, 1806–1811.
[51] Chaudhuri A, Kuijpers KPL, Hendrix RBJ, Shivaprasad P, Hacking JA, Emanuelsson EAC, Noël T, Van Der Schaaf J, Process intensification of a photochemical oxidation reaction using a Rotor-Stator Spinning Disk Reactor: a strategy for scale up, Chem Eng J, 2020, 400, 125875.
[52] Lee DS, Amara Z, Clark CA, Xu Z, Kakimpa B, Morvan HP, Pickering SJ, Poliakoff M, George MW, Continuous photo-oxidation in a vortex reactor: efficient operations using air drawn from the laboratory, Org Process Res Dev, 2017, 21, 1042–1050.
[53] (a) Wen Z, Maheshwari A, Sambiagio C, Deng Y, Laudadio G, Van Aken K, Sun Y, Gemoets HPL, Noël T, Optimization of a decatungstate-catalyzed C(sp3)–H alkylation using a continuous oscillatory millistructured photoreactor, Org Process Res Dev, 2020, 24, 2356–2361; (b) Rosso C, Gisbertz S, Williams JD, Gemoets HPL, Debrouwer W, Pieber B, Kappe CO, An oscillatory plug flow photoreactor facilitates semi-heterogeneous dual nickel/carbon nitride photocatalytic C–N couplings, React Chem Eng, 2020, 5, 597–604.
[54] Bianchi P, Petit G, Monbaliu J-CM, Scalable and robust photochemical flow process towards small spherical gold nanoparticles, React Chem Eng, 2020, 5, 1224–1236.
[55] Sanderson K, Automation: chemistry shoots for the Moon, Nature, 2019, 568, 577–579.
[56] (a) Coley CW, Eyke NS, Jensen KF, Autonomous discovery in the chemical sciences part I: progress, Angew Chem Int Ed, 2020, 59, 22858–22893; (b) Coley CW, Eyke NS, Jensen KF, Autonomous discovery in the chemical sciences part II: outlook, Angew Chem Int Ed, 2020, 59, 23414–23436.
[57] Mennen SM, Alhambra C, Allen CL, Barberis M, Berritt S, Brandt TA, Campbell AD, Castañón J, Cherney AH, Christensen M, Damon DB, Eugenio De Diego J, García-Cerrada S, García-Losada P, Haro R, Janey J, Leitch DC, Li L, Liu F, Lobben PC, MacMillan DWC, Magano J, McInturff E, Monfette S, Post RJ, Schultz D, Sitter BJ, Stevens JM, Strambeanu II, Twilton J, Wang K, Zajac MA, The evolution of high-throughput experimentation in pharmaceutical development and perspectives on the future, Org Process Res Dev, 2019, 23, 1213–1242.
[58] Kuijpers KPL, Bottecchia C, Cambié D, Drummen K, König NJ, Noël T, A fully automated continuous-flow platform for fluorescence quenching studies and Stern–Volmer analysis, Angew Chem Int Ed, 2018, 57, 11278–11282.
[59] Capaldo L, Ravelli D, The dark side of photocatalysis: one thousand ways to close the cycle, Eur J Org Chem, 2020, 2020, 2783–2806.
[60] Hopkinson MN, Gómez-Suárez A, Teders M, Sahoo B, Glorius F, Accelerated discovery in photocatalysis using a mechanism-based screening method, Angew Chem Int Ed, 2016, 55, 4361–4366.
[61] Bédard A-C, Adamo A, Aroh KC, Russell MG, Bedermann AA, Torosian J, Yue B, Jensen KF, Jamison TF, Reconfigurable system for automated optimization of diverse chemical reactions, Science, 2018, 361, 1220.

[62] Coley CW, Abolhasani M, Lin H, Jensen KF, Material-efficient microfluidic platform for exploratory studies of visible-light photoredox catalysis, Angew Chem Int Ed, 2017, 56, 9847–9850.
[63] Sun AC, Steyer DJ, Allen AR, Payne EM, Kennedy RT, Stephenson CRJ, A droplet microfluidic platform for high-throughput photochemical reaction discovery, Nat Commun, 2020, 11, 6202.
[64] Blanco-Ania D, Rutjes FPJT, Continuous-flow chemistry in chemical education, J Flow Chem, 2017, 7, 157–158.
[65] Kuijpers KPL, Weggemans WMA, Verwijlen CJA, Noël T, Flow chemistry experiments in the undergraduate teaching laboratory: synthesis of diazo dyes and disulfides, J Flow Chem, J Flow Chem 2021, 11, 7–12.

Martin Linden, Maximilian M. Hielscher, Balázs Endrődi,
Csaba Janáky and Siegfried R. Waldvogel

2 Electrochemical processes in flow

2.1 Electrochemical aspects in flow

As it would go far beyond the scope of this chapter to give a comprehensive survey on electrosynthetic chemistry, we will only give a brief overview of the key principles. A more detailed insight can be gained within the recommended literature [1–8].

2.1.1 General electrochemical aspects

Increasing ecological and economic concerns in modern society challenge chemists to develop more efficient and environmentally benign synthesis techniques. Due to the use of inexpensive electricity as terminal redox agent, electrosynthetic processes come along with a significant cut in reagent waste and, thus, offer an appealing and green alternative to classical synthetic approaches. Since the reagent is applied close to the electrode, runaway reactions can be avoided and inherent safety is given [1–5, 9, 10]. In a routine laboratory electrochemical cell, a working electrode and a counter electrode (WE and CE, respectively) are placed in a finite distance from each other. For batch-type cells, a reference electrode is optionally inserted in the cell in the close vicinity of the WE.

Importantly, the electron transfer itself, even though it is a key step, is not the only factor affecting the electrochemical conversion of a substrate (Fig. 2.1, top). The characteristics of the electrode surface, for example, surface morphology or any catalytic behavior of the material, as well as the accessible potential window and inhibiting effects like overpotentials, may promote a certain reaction path. In addition, consecutive reactions of the generated intermediates are influenced by the conditions prevalent in the electrolyte. As most electroconversions contain a cascade of electron transfers and chemical steps, the (de-)stabilization of the intermediates by the electrolyte and mixing effects (mass transfer) have high influence on the selectivity. This (de-)stabilization may happen, for example, by solvation effects (Tab. 2.1), as it can be observed in the anodic dehydrogenative phenol–phenol and phenol–arene C–C cross-coupling. Here, the solvent system is successfully used to manipulate the oxidation potential and nucleophilicity of the employed arenes [4, 9, 11]. Furthermore, the electrodegradation of the solvent may happen by evolution of hydrogen (protic solvents) or oxygen (water only). However, depending on the potential applied, the backbone of the molecule can be decomposed, too.

https://doi.org/10.1515/9783110693690-002

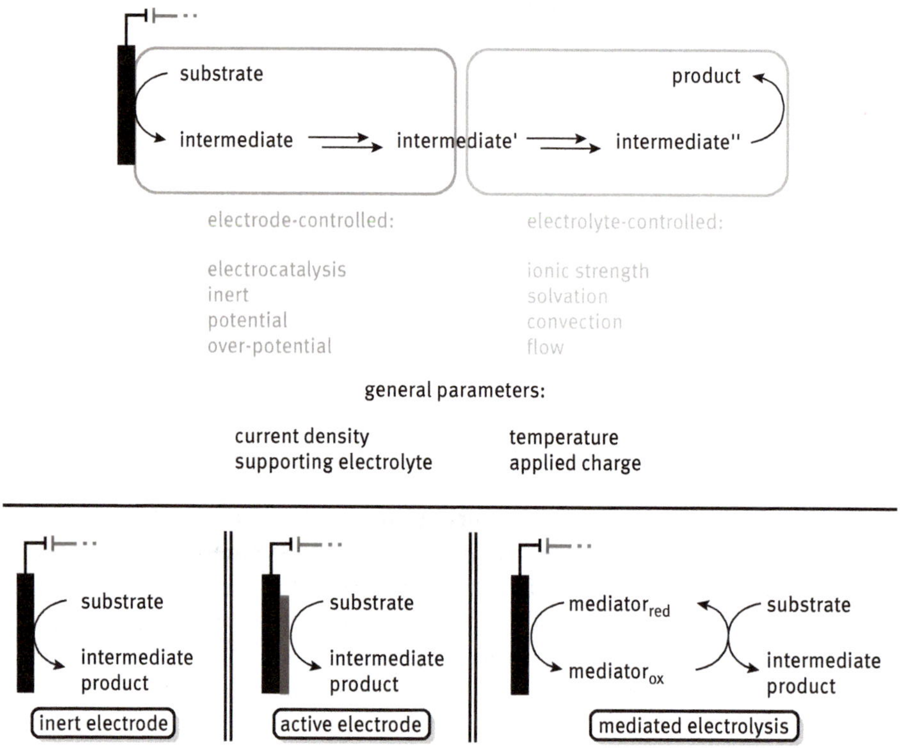

Fig. 2.1: Top: Exemplary reaction cascade in electrosynthetic processes and important parameters that influence the reaction. Bottom: Different modes of electron transfer. Adapted with permission from [9]. Copyright 2020 American Chemical Society.

Therefore, the stability of the solvent under the given conditions (electrode material, supporting electrolyte, etc.), also known as the electrochemical window (Fig. 2.2), has to be respected. Still, in some cases the degradation of an (inexpensive) solvent may be a counter reaction worth considering or can even be helpful, for example, in the case of electrogenerated bases [12]. On the other hand, the supporting electrolyte is not only an agent in ensuring sufficient conductivity of the system. By formation of a layer on the electrode surface, its accessibility can be tuned toward the wanted species and, thus, concurring reactions can be suppressed [13]. Coordinating effects of the anions may contribute to the abovementioned stabilization of intermediates and the acid–base properties of the solution can be tailored. Furthermore, the supporting electrolyte has to be inert to reduction and oxidation as well as to reactions with the electrogenerated species [14].

Since the electrode surface is the place where the (initial) electron transfer from or into the electrolyte comes to pass, the material is of high interest for controlling the electrochemical steps of the reaction. Generally speaking, any conductive material can serve as an electrode. The stability, costs, malleability, and, especially, the

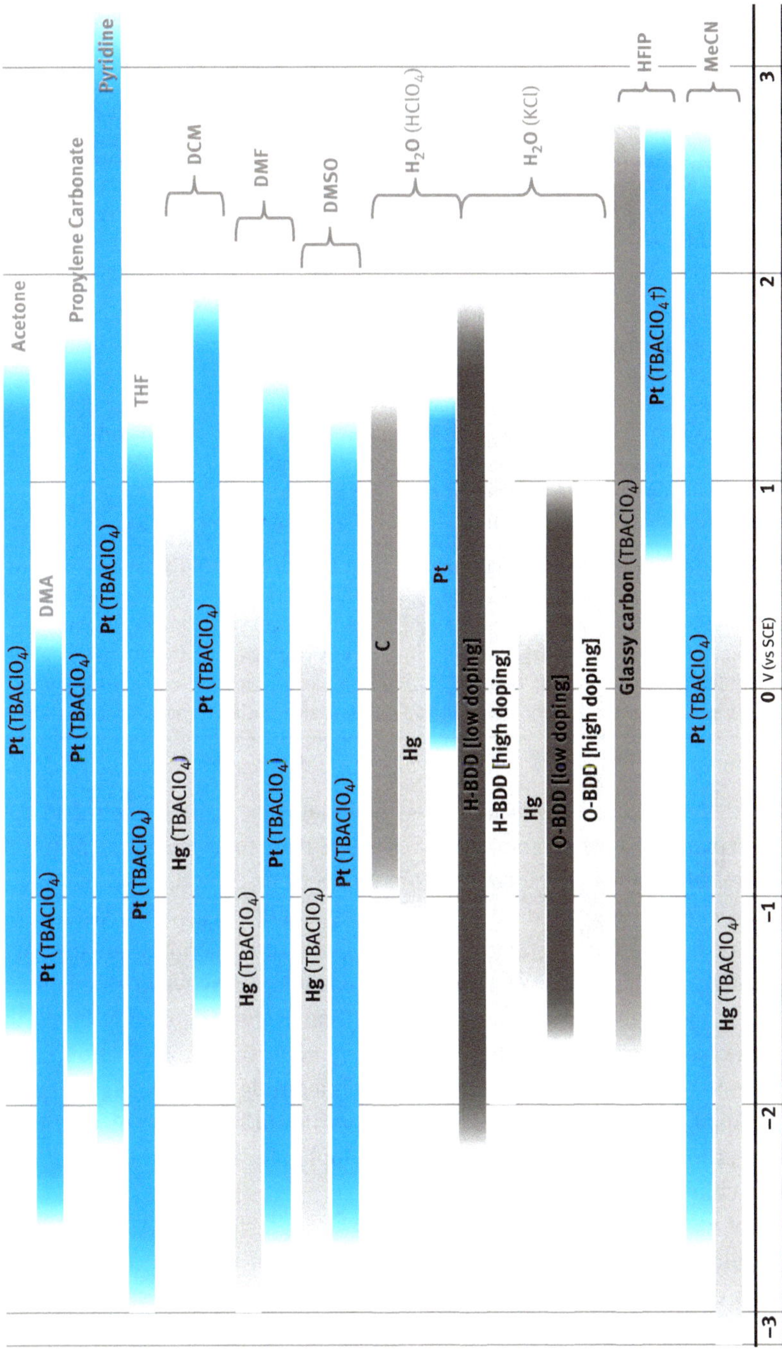

Fig. 2.2: Electrochemical window of different solvents at various electrode materials. (†current density cut-off at j=±0.1 mA/cm^2) Reprinted with permission of the authors, D.M. Heard and A.J.J. Lennox, from [12].

Tab. 2.1: Influence of the solvent's acid/base properties on ion solvation, proton transfer, and redox stability.

	Solvents with strong acidity	Solvents with weak acidity	Solvents with weak basicity	Solvents with strong basicity
Ion solvation properties	Solvation of small anions is easy (small anions are not reactive)	Solvation of small anions is difficult (small anions are reactive)	Solvation of small cations is difficult (small cations are reactive)	Solvation of small cations is easy (small cations are not reactive)
Proton donating ability	Proton donation from solvent is easy (narrow pH region on the basic side, strong bases are leveled, very weak acids cannot be titrated)	Proton donation from solvent is difficult (wide pH region on the basic side, strong bases are differentiated, very weak acids can be titrated)	Proton acceptance from solvent is difficult (wide pH region on the acidic side, strong acids are differentiated, very weak bases can be titrated)	Proton acceptance from solvent is easy (narrow pH region on the acidic side, strong acids are leveled, very weak bases cannot be titrated)
Redox properties	Reduction of solvent is easy (narrow potential region on the negative side, strong reducing agents are unstable)	Reduction of solvent is difficult (wide potential region on the negative side, strong reducing agents are stable)	Oxidation of solvent is difficult (wide potential region on the positive side, strong oxidizing agents are stable)	Oxidation of solvent is easy (narrow potential region on the positive side, strong oxidizing agents are unstable)

often unique reactivity led to a cutback to a few materials that made it into broad application. While platinum is often employed in academia due to its appealing physical and electrochemical features, significantly more inexpensive carbon-based materials (graphite, glassy carbon (GC)) are used when it comes to larger scales [12]. Metal anodes with low oxidation potentials (Mg, Al, Zn, etc.), so-called sacrificial anodes, are stoichiometrically degraded during electrolysis and come into play when an auxiliary effect of the metal ions generated is needed. Even though they are readily found in literature on batch electrochemistry, they would imply high maintenance effort (regular exchange of the electrode, variation in interelectrode gap, leakage issues, etc.) in flow cells. Additionally, the interelectrode gap would alter in a nonuniform way. Therefore, they are of no importance in flow applications and will not be considered in this chapter.

The easiest way to obtain an electrochemical conversion is the direct electrolysis (Fig. 2.1, bottom left). The substrate is oxidized or reduced directly by a heterogeneous electron transfer at the electrode surface, and the selectivity of the electroconversion is dominated by the applied potential [9, 15]. Due to the simple setup, this method is often used [3–5, 12]. As mentioned above, overpotentials may still have high impact on the reaction outcome. This includes overpotential of the substrate as well as overpotentials of concurring reactions like hydrogen or oxygen evolution reaction (HER/OER). However, the complex nature of the processes on the electrode surface often obliges to rely on empirical experience. While there are standardized, tabulated values for HER/OER overpotentials at diverse electrode materials (Fig. 2.3) and in different media as a basis of estimation, a targeted trial-and-error approach is usually the most promising procedure for most other conversions [16]. An introduction to statistically driven optimization approaches is given in the *reaction optimization* section of this chapter (Section 2.4). Which reaction has to be promoted or suppressed at the respective electrode is a matter of consideration. For example, discharging of protons is a common counter reaction for anodic processes in protic media. Thus, cathodes with low HER overpotential are usually employed in those systems. Vice versa, a cathodic reduction in a protic medium requires a WE with high overpotential for HER to avoid charge dissipation [12]. In this context, boron-doped diamond (BDD) has drawn increasing attention as a commercially available and sustainable high-performance material over the last years, due to its extraordinary high overpotential for both HER and OER, thus enabling a broad electrochemical window for various solvents [16–19].

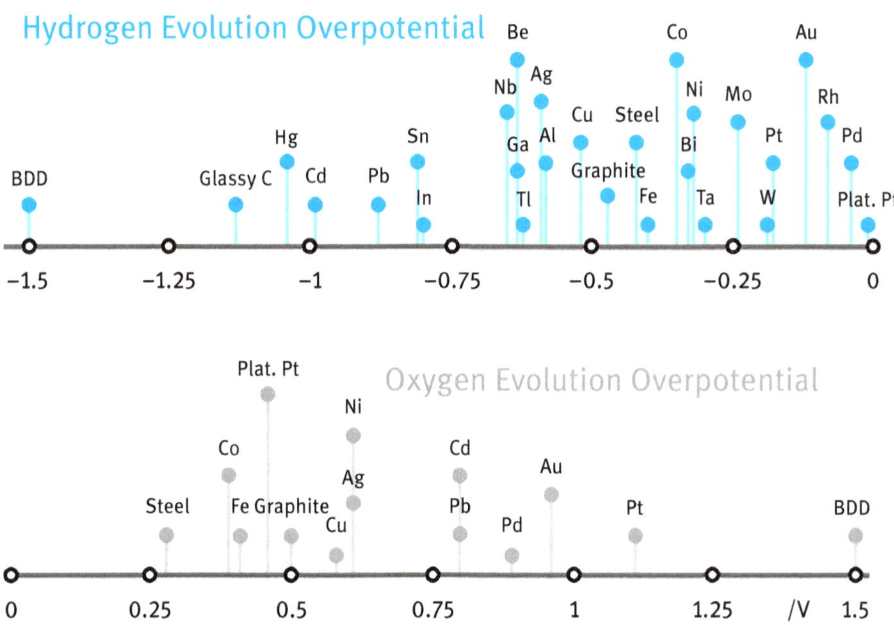

Fig. 2.3: HER and OER overpotential at various electrodes. HER overpotential recorded at 1 mA cm^{-2}, 25 °C, in 1 M aqueous HCl. OER overpotential recorded at 1 mA cm^{-2}, 25 °C, in 1 M aqueous KOH. Reprinted with permission of the authors, D.M. Heard and A.J.J. Lennox, from [12].

For larger molecules, the ad-/desorption behavior of intermediates on the electrode surface largely influences the selectivity of the reaction. This was observed for Kolbe and Hofer-Most electrolysis on platinum or graphite electrodes, where the stronger adsorption of the radicals formed in the first step is assumed to facilitate a second electron abstraction, thus leading to cationic species instead of radicals. Furthermore, the electrode may even be catalytically involved in the reaction, whereby certain intermediates are preferably stabilized in relation to others, or promote a concerted or step-wise process of electron abstraction/addition and stabilization of the radical ion by fragmentation [12].

Nevertheless, the reactivity of some substrates may demand more sophisticated systems to achieve the desired conversion or fine-tune the selectivity of the reaction. Speaking of the electrode, this can be realized by active electrode reactions (Fig. 2.1, bottom center). The selectivity of active electrodes originates in the redox potential and the reactivity of an active layer on the electrode surface. This layer can be understood to be a redox filter. It is usually prepared beforehand to provide a uniform electrode surface and is regenerated in situ during the electrolysis. The big advantage of an active electrode is the introduction of high selectivity without contamination of the electrolyte. Thus, no additional workup effort or reagents are needed. Often, stoichiometric reagents can be replaced by similar immobilized species electro(re)generated in this technique. The most prominent example is the nickel oxide hydroxide anode, which is widely employed, for example, in the conversion of primary and secondary alcohols to carboxylic acids and ketones [20–23]. Similarly, molybdenum(V) reagents can be mimicked by employment of molybdenum electrodes with extraordinary selectivity in arene coupling reactions [24, 25].

A frequent challenge of electrosynthetic chemistry is the requirement of low current densities to achieve efficient conversions and, therefore, prolonged reaction times. Using three-dimensional structured electrodes can mitigate this effect. While maintaining the same geometrical dimensions, these porous materials (foams, meshes, felts, fabrics, etc.) feature a significantly higher electrochemically active surface area. Therefore, a higher geometrical current density may be applied with the same or better results. Especially in flow electrochemistry, where the internal mass transfer is strongly connected to the flow rate, this setup enables new possibilities. Additionally, structured electrodes here may serve as a retention mixing unit, hence, boosting mass transfer within the cell [11, 20].

If neither direct electrolysis nor active electrodes show satisfactory performance, the use of a dissolved electron transfer catalyst (mediator) is indicated. By using a mediator, the conversion of the substrate itself is decoupled from the electrode (Fig. 2.1, bottom right). The mediator is activated by heterogeneous electron transfer at the electrode; however, the substrate transformation usually occurs as a homogenous reaction in the bulk electrolyte. Mediated electrolysis is mostly employed in case of substrates with high overpotential, or, when the reactivity of a mediator offers a unique reaction pathway or selectivity [4, 5, 7, 15]. These properties were successfully used, for example, in various electrochemical fluorination techniques, including *gem*-difluorinations and the synthesis of fluoromethyl substituted heterocycles [4, 7, 26–29] or selective oxidation of a primary alcohol to an aldehyde in the presence of a secondary alcohol [7, 23, 30].

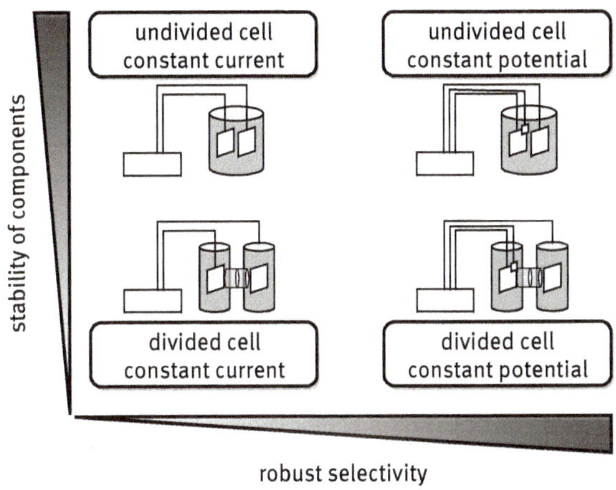

Fig. 2.4: Different cell designs and modes of operation versus their impact on the reaction. Reprinted with permission from [31]. Copyright 2018 American Chemical Society.

Besides manipulating the mechanisms at the electrode and in the electrolyte, the cell design and the mode of operation may be changed to influence the outcome of the reaction (Fig. 2.4). For ease in use, nowadays, electrosynthetic chemistry commonly prefers galvanostatic conditions. Here, a simple two electrode setup in combination with a current source can be used. The electrolysis time is determined by the application of a sufficient amount of charge, which is enforced by adjusting the cell voltage to maintain the desired current. Contrary to the galvanostatic setup, the potential of the WE is maintained as a constant in potentiostatic electrolysis. This allows precise control over which species is converted. However, this mode suffers from a significantly more complex and expensive setup with WE, CE, and a reference electrode. Additionally, a current drop in proceeding conversion due to a decrease in available substrate results in extensive reaction times. Therefore, with regard to scalability and technical application, galvanostatic electrosynthesis is often the technique of choice [2, 4, 5, 10, 15, 31].

Many conversions can be done in undivided cells – for example, a simple beaker or a two-plate electrode arrangement in an electrochemical flow cell, minimizing the apparatus effort. Consecutive or concurring transformations of the substrate and the product at the CE may, however, interfere with the selectivity and efficiency of the reaction. This can be circumvented by switching to divided cells. The cell is split into an anode and a cathode compartment by a separator. Suitable separator materials range from simple glass, ceramic, or polymer (PE/PP) frits to functionalized materials like Nafion® or Thomapor®. While the former only slow down the mixing of the two compartments, the latter bear ionic groups and, thus, can discriminate between differently charged species. It has to be noted, that the employment of a separator results in a severe increase in the cell voltage due to the additional internal resistance. Inexpensive glass and ceramic frits show lower voltage drops than unfunctionalized polymers; however, their brittleness and inflexibility might raise challenges during scale-up. Nafion®, Thomapor®, and similar materials represent a suitable intermediate option, but come with the drawback of comparatively high cost. If a separator is needed, a careful balance must be reached between increased cell voltage, selectivity, scalability, material costs, and recyclability [15, 31].

2.1.2 Electrolysis in flow

The main advantage of flow electrolysis compared to batch processes is, as known from other applications, the high surface-to-volume ratio. Besides the improved heat exchange, which is beneficial for the management of the heat produced by the internal resistance of a cell, the contact area between reactant solution and electrode features the possibility of fast conversion and, thus, high productivity of the cell. Moreover, the short residence time within the actual cell facilitates precise control over the charge applied to each small fraction of the substrate solution. In addition, a prolonged exposure to an electrode material can be avoided and slow degradation reactions efficiently suppressed. This way, reactions of substrates prone to over-oxidation/-reduction can be realized and controlled much more easily, since the product is removed from the cell [10, 11, 32–36].

In the context of continuous manufacturing and multistep reactions in flow, a single pass of the electrolyte through the cell is of high interest [10, 11]. However, in contrast to classical chemistry, where an added reagent can be consumed later on, in electrochemical processes, the charge provided has to be consumed right away while the substrate (or the mediator) is at the electrode, hence, establishing a flow setup, within the cell. Therefore, a short residence time in the interelectrode gap may be challenging if low current densities have to be applied [11, 37]. If substrate and product are easily separated from the mixture and the overall conversion is of minor importance, one might consider using a flow cell with a high throughput to synthesize the desired amount of product and reuse the unconverted substrate. If this approach cannot be pursued, the residence time and, thus, the applied charge may be increased by decreasing the flow rate or using a longer channel for electrolysis. However, a decreased flow rate results in an inferior mass transfer regime, which may have a negative influence on conversion and selectivity. Electrolysis cells with long channels (meandering, serpentine, ammonite cell) can address this problem, but to some extent, at the cost of a more complex setup [11]. They will be discussed in more detail in Section 2.1.4 of this chapter. Instead of extending the channel length within one cell, a cascade of cells can be employed, or the electrolyte can be pumped through a single cell, multiple times. While passing the electrolyte through multiple electrolysis cycles, a proportional fraction of charge is applied in each passage [10, 35, 37]. In a single cell with one long channel, the local current density fluctuates along the channel with the changing concentrations of the electrochemically active species. Hence, the electrolysis conditions change significantly [11, 34]. In the cascading mode, the current density can be controlled in each cell or pass-through independently, and more controlled reaction conditions can be ensured. Additionally, higher flow rates and better heat control can be realized [37]. Just like pumping from one vessel to another and reusing the electrolyte in cascading mode, the electrolyte can be stored in one reservoir and cycled through one cell. Instead of having a defined composition before/after every pass-through, the electrolyte constitution changes gradually during the process. However, this gradual change leads to very mild conditions and high energy efficiency.

Both cycling and cascading modes can be employed for the handling of suspensions and emulsions. The high flow rates possible in these systems, combined with stirring and/or sonication of the reservoir, provide, in most cases, sufficient mixing of the phases and prevent separation. In addition, electrolysis in segmented flow offers the possibility of enhanced mass transfer from and to the electrode surface and in between the phases by toroidal vortices at the phase boundaries between the segments (Fig. 2.5). Such systems have also been successfully employed in single pass mode [32].

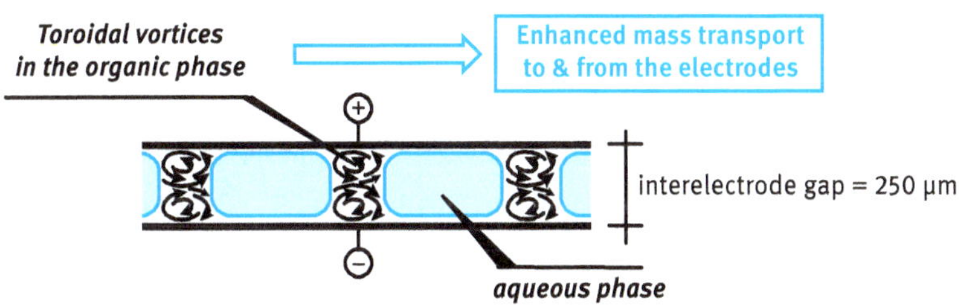

Fig. 2.5: Enhanced mass transfer in segmented flow by vortices at the phase boundary. Reprinted with permission from [32]. Copyright 2019 American Chemical Society.

As mentioned in Section 2.1.1, the evolution of hydrogen is a widespread and convenient counter reaction for anodic processes. This and other gas-evolving reactions are a major issue in flow electrochemistry. While the evolution of gas bubbles may enhance the mass transfer, the gas can block a significant part of the electrode, if it is not removed from the surface properly. The insulating properties of the gas, therefore, lead to a nonuniform and irreproducible distribution of the local current densities [10, 11, 32]. In cascading or cycling mode, this problem is readily solved. The high applicable flow rates ensure efficient removal of the gas from the electrode, which is, in turn, removed from the system via the headspace of the reservoir [20, 37]. In single pass flow electrolysis – if the system is not robust enough to tolerate the effects of the gas – the bubbles may be (partially) removed by pressurizing the whole system. The higher pressure causes smaller gas segments and, thus, a more uniform current distribution and decreased internal resistance. On the other hand, more gas is dissolved in the electrolyte at elevated pressure. Therefore, conversion, yield, and current efficiency decrease due to reoxidation of the dissolved hydrogen at catalytically active anodes [10].

Other issues in electrochemistry involve fouling and corrosion of the electrodes. Corrosion can usually only be prevented by using an appropriate electrode material. Some applications have been published with electrodes unstable within the electrolyte itself; however, they are stabilized by the applied potential [38, 39]. Other examples are known, where electrode corrosion is prevented by the supporting electrolyte [40–42]. Fouling is a common process in electrochemistry. Side reactions like polymerization form an insoluble layer on the electrode surface. This passivation of the electrode may compromise the performance of the reactor (nonuniform current distribution, increased resistance, etc.), but may be needed if the layer provides an enhanced selectivity. Unwanted fouling can be prevented by use of self-cleaning materials like BDD [17, 19, 43, 44]. However, the highly reactive species formed in this internal cleaning process (e.g., hydroxyl radicals for aqueous media at BDD anodes) limits the application of these systems to conversions with robust substrates and products [45].

2.1.3 Follow-up conversions

Just like other transformations, an electrosynthetic process, often, is just one out of many steps in a synthetic route. Therefore, downstream conversions have to be done. Directly consecutive reactions may happen without workup in between. In electrochemistry, these may be performed either directly during the electrolysis (in-cell) or after the application of charge is finished (ex-cell) [10, 11, 32, 36, 46].

For in-cell synthesis, the reaction partner for the follow-up conversion needs to be stable under the electrolytic conditions. On the other hand, in-cell reactions can be used to generate and directly consume unstable reagents. Many electrosynthetic processes can be understood as in-cell conversions. As mentioned above (Section 2.1.1), the electrogenerated species usually undergoes chemical steps like trapping of a generated cation by a nucleophile (e.g., solvent) or reprotonation of an electrogenerated base. Example reactions are the methoxylation of furanes, *N*-protected pyrrolidines or toluenes, the synthesis of diaryliodonium salts, and non-Kolbe reactions [4, 11, 32, 33, 47]. Further, oxidations of cyanhydrines and the addition products of aldehydes and *N*-heterocyclic carbenes (*Breslow* intermediates) [46, 48] and carboxylations of electrogenerated carbanions have been published [33]. If the reaction partner is not stable for electrolysis, ex-cell processing can be employed. Especially in flow, the ex-cell approach offers appealing possibilities due to the opportunity for processing unstable electrogenerated species right away. Consequently, a lot of work has been done on this field [11, 32, 33]. The most prominent examples of these processes are electrogenerated high performance oxidizers like peroxodicarbonate [19, 49], periodate [50], and hypervalent organic iodine species [19, 32, 36] with corresponding follow-up conversions. Another prominent example of an ex-cell approach is the cation flow method. Metastable carbocations are produced electrochemically at low temperatures and may be stored under inert conditions for further reactions or processed right away in flow. Examples range from allylation and arylation reactions to cationic polymerization. Additionally, more recent developments enabled similar reactions as an in-cell version (Fig. 2.6). In a highly laminar flow, a solution of the cation precursor is fed from the anode side, while the nucleophile is contained in the cathode feed. Due to the absence of turbulences, mixing of the two streams solely relies on diffusion. Thus, the cation can be generated in the presence of the nucleophile without the latter being oxidized [11, 32, 33, 36].

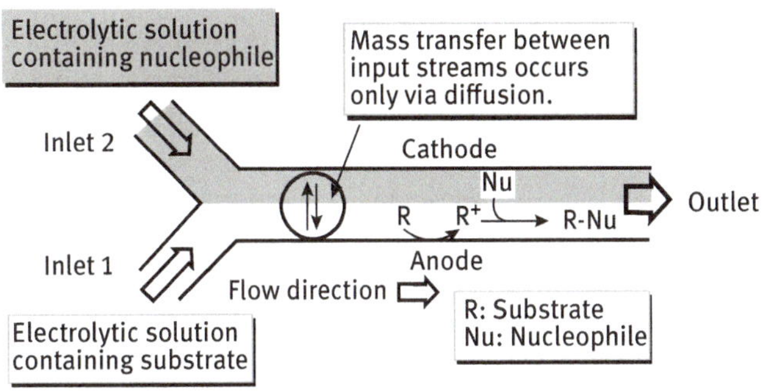

Fig. 2.6: Cation flow method as an in-cell approach. Reprinted with permission from [51]. Copyright 2007 American Chemical Society.

2.1.4 Availability of lab-scale flow electrolyzers

The majority of narrow gap flow cells in academia are self-made, mostly in cooperation with the departments' machine shop. The widespread emergence of possibilities such as CNC machining and 3D-supported planning allows the precise machining of flow cells [32, 34, 35, 52]. In the plate-frame approach, especially (see Section 2.2.2), the cell is easily built by screwing together two half cells with a spacer in between. For a divided setup, a membrane and a second spacer may be employed. If the spacer is structured, even serpentine channels can be realized [53]. In some cases, the flow setup was made commercially available [34, 54]. Common suppliers are IKA (IKA ElectraSyn Flow, flow cell and related equipment) [55], Vapourtec (Vapourtec Ion Electrochemical Reactor, flow cell and related equipment) [56], Cambridge Reactor Design (ammonite cells) [57], and CONDIAS (diamond-based electrodes) [58]. The IKA flow cells come along with a modular design, enabling easy and fast exchange of different electrode materials, since several half cells of different materials are included in the packages.

They are among the less expensive devices, allowing an easy entry in this field. The Ammonite Electrolysis cell offers a very long pathway for the reaction to take place. This ensures high single-pass conversion and/or driving multistep processes. For gas diffusion electrode (GDE) containing electrolyzer cells, among others, Electrocell GmbH [59], Dioxide Materials [60], and ThalesNanoEnergy [61] have their products in the market. These cells allow standardized comparison of components (catalysts, membranes, and gas diffusion layers (GDLs)) in different industrially relevant processes (e.g., water splitting, CO_2 reduction).

2.2 Design of flow electrolyzers

Whether we consider technology scale-up or reliable rapid screening of materials in the laboratory, continuous-flow reactors have multiple benefits compared to their batch counterparts. Among others, these include increased mass transfer and improved mixing of different phases, better temperature and heat transfer control, and more precise influence on reaction mixture residence time in the reactor. When moving from batch-type experiments to continuous-flow cells, the architecture and design of the reactor (electrochemical cell) must first be clearly defined. This includes features such as the type of electrolyte used (liquid/gas), reactor material-reaction mixture compatibility, whether it can be pressurized or not, the applicable temperature and flow rate, and the possibility of using a reference electrode. The design of flow electrolyzers has developed continuously, over the past decades. As a major flag carrier, the chlor-alkali process went through several evolutionary steps, starting from diaphragm-separated cells, through using asbestos separator and mercury cells, to the currently dominantly applied membrane-separated cells [62]. The most important driver in this case was the energy efficiency of the process, hence the production cost of chlorine and NaOH. As for the energy efficiency of an electrochemical process, the most straightforward definition is the voltage efficiency of the cell. This is the ratio of the thermodynamic voltage and the total voltage applied to the electrolysis cell (V_{cell}) to drive the electrochemical reactions at a given rate. Note that the thermoneutral voltage is not only sufficient for driving the reactions, but also to produce enough heat to maintain constant temperature. Beyond the thermodynamic requirements, V_{cell} also includes losses related to the anodic and cathodic electrochemical processes (activation losses, overpotentials), to the cell components (ohmic losses, IR drop), and to mass transport. In aiming for high energy efficiency, each of these effects should be minimized (see Fig. 2.7 for details) [63].

It is important to distinguish between the large-quantity production of commodity chemicals, where large current densities are of prime importance, and the continuous-flow electrochemical production of fine chemicals (e.g., pharmaceutical compounds). In the latter case, the purity of the product stream, and hence, the process selectivity

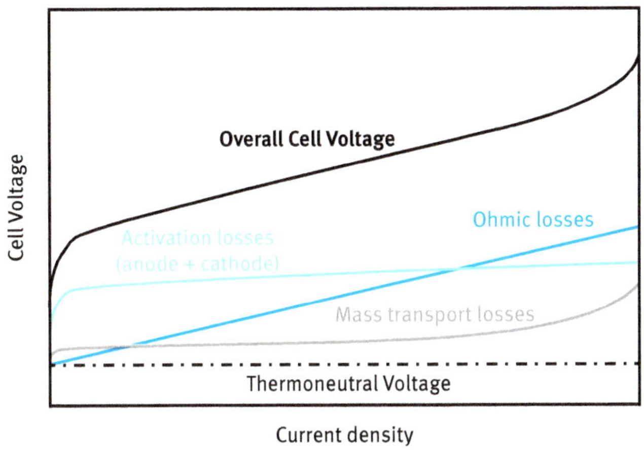

Fig. 2.7: Voltage loss mechanisms in electrochemical cells. Reproduced with permission by Elsevier [63].

might be an even more important parameter than the reaction rate. In what follows, some typical cell arrangements will be discussed.

2.2.1 Industrial narrow gap cells

To minimize the energy input and to avoid the use of large solution volumes, the electrodes are generally placed very close to each other (narrow gap). A typical example of this is the production of sodium chlorate, where the electrodes are positioned at 2–3 mm distance from each other, in an interdigitated manner. This way, the IR-drop is minimized, and the reactions are driven on both sides of the electrodes, hence increasing the surface area. Narrow gap cells offer a inexpensive and easy design. On the other hand, such cells have a few clear drawbacks. One important issue is the cross-talk of the electrode processes. Products formed on one electrode can be re-converted on the other, hence decreasing the Faraday efficiency (FE) of the product formation [64]. Such narrow gap cells have been found to allow the use of less conductive media (lower supporting electrolyte concentration), but this, eventually, still limits the maximum current density (typically in the range of a few hundred mA cm^{-2}). At higher current densities, the insufficient removal of the products or byproducts, for example, hydrogen, from the electrodes becomes a major obstacle. Using concentrated electrolytes complicates the product separation process and invokes extra costs related to the handling of high density/viscosity solutions.

While some robust reactions are performed in flow cells with large electrode gap, the output of reactions with oxidation-sensitive intermediates or products can be strongly dependent on the electrode gap. The interaction between the liquid phase and the solid electrode results in the formation of a rather immobile stagnant

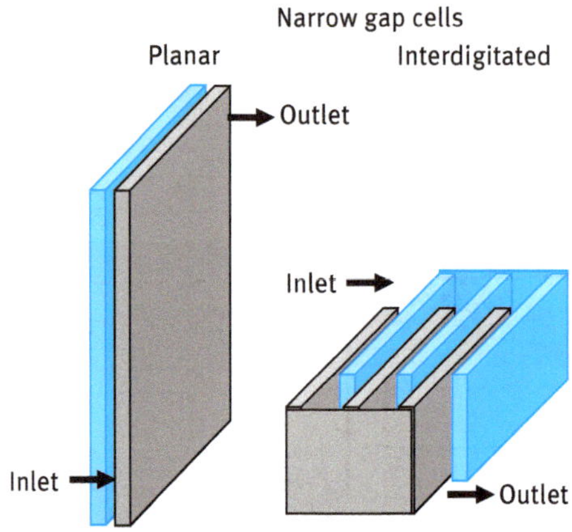

Fig. 2.8: Two typical arrangements of industrial narrow gap electrolyzer cells.

layer with a thickness in the µm range [65]. Within this layer, mass transport takes place only by diffusion, in contrast to convection within the bulk. The size of the stagnant layer, besides other properties, depends on the speed of motion for the bulk, whereby faster movement reduces the thickness of the stagnant layers. This leads to smaller diffusion paths. The corresponding diffusion coefficient is first-order dependent on the temperature [66]. With constant current density and transferred amount of charge, a smaller electrode gap leads to higher flow rates in the cell and thus also to a smaller stagnant layer. In addition, the required amount of supporting electrolyte can be reduced or completely avoided by narrow gaps. This was demonstrated well by *Waldvogel* and coworkers [34]. They investigated the domino-oxidation-reduction sequence for synthesis of nitriles from aldoximes. The batch process suffered from dehalogenation of the starting material; flow electrolysis with an electrode gap of 1 mm showed similar behavior or bad conversion. Since dehalogenation, in this case, is a slow chemical follow-up reaction, it is not directly evident from CV experiments. A smaller gap size of 0.12 mm allows for supporting electrolyte-free electrolysis and allows a short residence time, which suppresses dehalogenation. By optimizing the ratios of the solvent mixture, the reaction achieved a yield of 80%. Dehalogenated species could only be detected in traces.

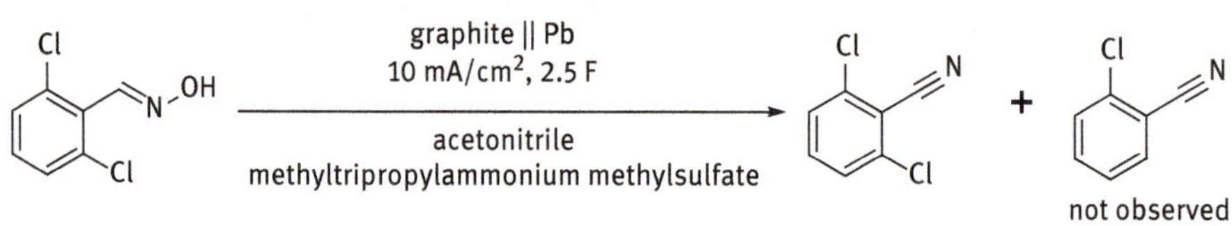

2.2.2 Membrane/diaphragm-separated electrolyzer cells and plate-frame approach

Physically separating the electrodes circumvents product mixing, hence the back-transformation of anodic/cathodic products on the other electrode. A typical historical example for such separator is a glass or asbestos diaphragm, which was successfully applied in the early years of the chlor-alkali process [62]. A more current and widely applied technology that uses a diaphragm separator is alkaline water electrolysis [67]. Including such a separator between the electrodes increases the product formation FE by avoiding product mixing. On the other hand, the extra cell element induces extra losses due the increased cell resistance (*R*). To minimize the IR drop in the cell, the concept of zero-gap cells was developed. In such electrolyzer cells, the electrodes are pressed directly to the separator with the catalyst layers facing each other (Fig. 2.9). This way, the resistance of the electrolyzer is minimized. In addition, the electrolysis is conducted at elevated temperatures to achieve maximum conductive performance of the separator, for example, in chlor-alkali electrolysis at 85–90 °C.

Instead of further increasing the lateral dimension of the electrodes, the electrolysis area can be increased by building electrolyzer stacks. In this method, several cells are built on top of each other, applying bipolar plates (acting as anode on one side and as cathode on the other) between the cells. A serial electrical connection is established between the cells, while the reactant is typically fed in parallel. In industrial applications, such electrolyzer stacks are built on large frames (Fig. 2.9). In such application, the electrodes (and other cell components, e.g., gaskets, membranes) are first secured on the frame and are subsequently pressed together with the rest of the cell [68].

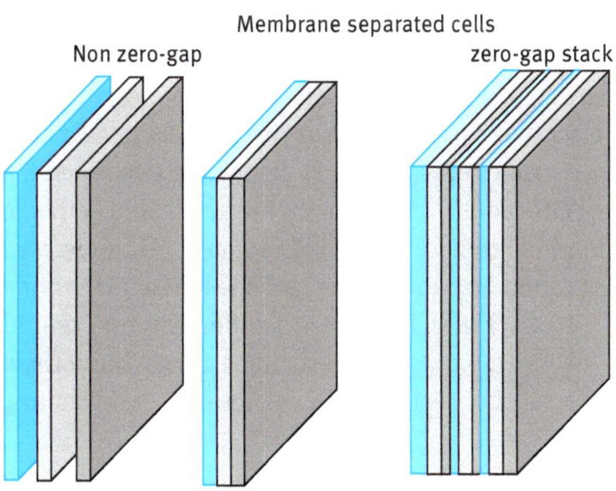

Fig. 2.9: Possible embodiments of membrane-separated electrolyzers and electrolyzer stacks.

To further decrease the cell resistance, polymer-based ion exchange membranes are typically used in zero-gap electrolyzers. As there is no liquid electrolyte between the electrodes, the ion conduction is determined by the chemical nature of the used membrane. Hence, these electrolyzer cells can be categorized by the ion-exchange properties of the membrane as follows: cation exchange membrane (CEM), anion exchange membrane (AEM), or bipolar membrane (BPM)-based electrolyzers (Fig. 2.10) [69].

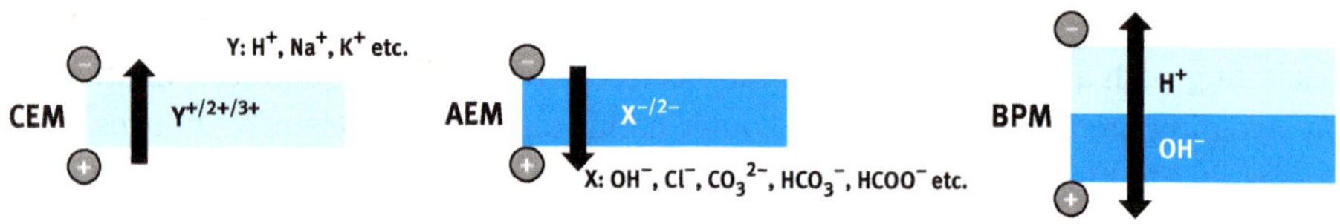

Fig. 2.10: Working principle of different ion exchange membrane types in electrolyzer cells. CEM – cation exchange membrane, AEM – anion exchange membrane, BPM – bipolar membrane.

When applying a CEM to separate the cell, the ion conduction is maintained by the migration of cations from the anode to the cathode. These are either cations from the applied electrolyte (e.g., Na^+ ions in case of Na_2SO_4 electrolysis or the chlor-alkali process in a CEM cell), or protons, either from the electrolyte solution, or formed as anodic by-products. A typical example for the latter is CEM water electrolysis, where H^+ is the by-product during oxygen evolution ($2H_2O \rightarrow O_2 + 4H^+ + 4e^-$) [70]. H^+ ions form on the anode, and as they are the charge carriers crossing the membrane, they are also present on the cathode side of the membrane in high local concentration. This results in a highly acidic pH on both electrodes. Using AEMs, on the other hand, leads to reversed direction ion conduction – in this case, anions cross the membrane from the cathode to the anode [71, 72]. In typical electrochemical reduction processes, such as water, CO_2 or O_2 reduction, OH^- ions are formed. This –contrary to the CEM membrane case – results in an alkaline local pH on both electrodes. In recent years, BPMs are being applied in electrolysis processes to avoid product crossing. The typical operation of BPMs is based on water dissociation at the interface between the AEM and the CEM, and consequent OH^- and H^+ migration to the anode and cathode, respectively [73–75]. This results in acidic cathode and alkaline anode pH. A reversed combination of an AEM and a CEM would lead to alkaline cathode and acidic anode pH. In this case, OH^- (or CO_3^{2-}/HCO_3^-) and H^+ ions (or other cations) would move toward the interface of the two membranes and recombine there. The formed product must be removed from the space between the two membranes, necessitating the inclusion of a buffer layer.

Notably, ion conduction is always associated with the transport of water through the membrane. As an example, when applying a Nafion® membrane (CEM), two to three water molecules are carried from the anode to the cathode with each H^+ ion transported. This changes the electrolyte composition (if any) on both electrodes, which must be considered during continuous operation. Also note that a free space is required in both the anode and the cathode compartments to deliver the reactants to the catalyst surface and to allow the products to leave the cell. Accordingly, porous electrodes are generally used as catalyst supports. This includes frits, meshes, foams, or other porous structures made of conducting materials, typically (electro)chemically resistant metals or carbon.

2.2.3 Gas diffusion electrodes

Supplying reactants to the electrodes in gas phase boosts the reaction rate by circumventing the rate limitations related to transport processes in liquids [76, 77]. In such cases, the reaction occurs on a triple phase boundary, where the solid catalyst particle, the gas reactant, and a liquid phase (or solid electrolyte) are mutually present [78]. The most recent understanding of the reaction conditions within the triple phase boundary is that a thin liquid layer forms on the catalyst surface. The reactant is dissolved in this liquid layer, leading to a very thin diffusion layer that ensures high mass transport rate to the catalyst particles.

When designing GDEs, several requirements should be simultaneously fulfilled [79]. The most important of these are the high electrical conductivity of the electrode, the gas permeation properties of the structure, the catalyst layer composition, and the hydrophilicity of the different components and the GDE, as a whole. The first of these is somewhat trivial, as high electrical conductivity of the electrodes is a prerequisite in any electrochemical process. To achieve this, metal foams, meshes, or other porous structures might be used. Carbon-based structures (carbon felts, carbon papers, etc.), however, offer higher flexibility, lower price, and easier processability, hence these are the most typically used GDLs in GDEs.

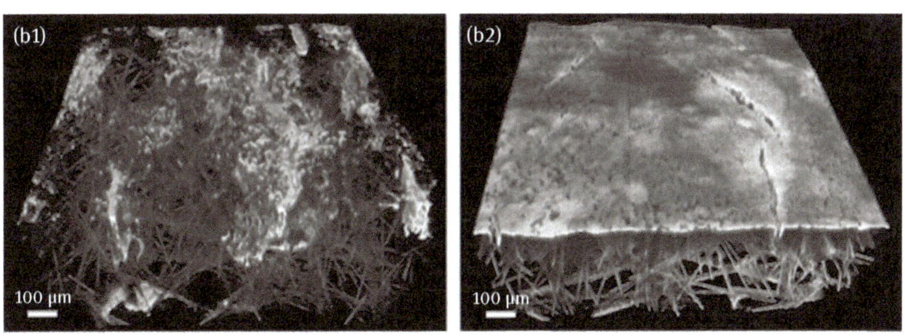

Fig. 2.11: Three-dimensional microtomographic models of a GDE, showing the structure of the microporous layer (b1) and the microporous layer with the catalyst. Copyright 2013 Wiley-VCH [80].

Carbon-based GDEs are usually formed of multiple layers with specific functionality (Fig. 2.11). A macroporous layer ensures that the reactant gas can enter the structure easily, and the products can exit the cell. A microporous layer on the other side serves as support for the catalyst and is also responsible for good electrical connection between the GDL and the catalyst particles. To avoid water accumulation in the GDE (flooding of the structure, increased diffusion layer thickness, and the consequent reaction rate decrease), the carbon GDL is most usually infused by a hydrophobic material such as PTFE.

The structure of the catalyst layer is of prime importance in achieving high reaction rate. Large electrochemically active surface area (ECSA), proper electrical conductivity, high ion conductivity, porosity, and homogeneity are all crucial properties of GDEs that define the electrochemical performance of an electrolyzer cell. Without going into too much detail regarding the specific function of each constituents, the chemical nature, physical dimensions, and morphological features of the catalyst particles, the chemical quality, the amount of binder used, and the total layer thickness must be tailored for the specific application. As for the catalyst layer formation on a GDL, multiple methods are widely applied. These include physical methods, such as brush-, spray-, or drop-coating or physical vapor deposition, and chemical methods such as electrodeposition, chemical deposition, or chemical vapor deposition. Note that even the catalyst layer preparation method can have a significant effect on the electrochemical properties of the GDE [80].

2.2.4 Electrochemical system design

While most studies focus on the electrochemical cell itself, there are many other components which together enable electrochemical processes. Beyond the obvious potentiostat/galvanostat (or power supply in simple cases), depending on the cell configuration, up to two liquid and gas inlets, product stream processing parts (including liquid/gas separators and product collection units), pressure and temperature sensors and controllers, a gas- and liquid flow meter, and often complex in-line analytical systems, are utilized. To have a controlled electrolysis process (yielding reproducible results), all these units need to function together smoothly. Here, we show one specific example of CO_2 electroreduction as the cathodic process, which is coupled with a liquid phase oxidation process [81]. In this particular process, one or two (peristaltic) pumps are used to deliver the liquid electrolytes (depending on the cell configuration). Optionally, gas phase reactants can also be fed to the anode of the cell, just like in the cathodic CO_2 feed. These fluids are tempered before entering the cell to maintain constant operational conditions. Subsequently, the anode product stream leaves the cell through a pressure regulating valve to be processed further. First, the gas products are removed, and then, the dissolved/liquid phase products are separated from the anolyte. The composition of the anode product stream shall be analyzed continuously to ensure proper operation. This should include online and offline methods both for the liquid and gas phases. These generally encompass gas/liquid chromatography (coupled with mass spectrometry), NMR, FTIR, and Raman spectroscopies, but other product specific spectroscopic methods can also be applied. Another engineering challenge will be to isolate the valuable products from the anolyte, which can be subsequently recirculated in the process.

2.3 Electrochemical processes in flow

2.3.1 Industrial electrochemical processes in flow

Performing electrolysis in flow cells is an easy way to continuously refresh the reactant solution/gas mixture around the electrodes, thus providing constant and precisely controlled reaction conditions. This is one of the most important differences when compared to batch processes, where the continuous reactant depletion causes dynamically varying conditions. Another important difference is the continuous removal of products to avoid their accumulation. This way, unwanted consecutive chemical/electrochemical reactions are avoided. When discussing industrial flow-electrochemical processes, one must distinguish between the mass production of commodity chemicals and the production of specialty chemicals, which is performed on a significantly smaller scale. From the first class, the chlor-alkali, the chlorate, water-, and CO_2 electrolysis processes are discussed here briefly.

The chlor-alkali process is one of the largest electrochemical industrial processes in terms of both annual production and energy consumption [62]. During the process, a concentrated NaCl solution (brine) is continuously fed to the anode, and diluted caustic soda (30%) is supplied to the cathode of a divided electrolyzer cell. The separator is typically a combination of two cation exchange membranes (perfluorosulfonic acid/perfluorocarboxylic acids) in the currently used cells. At the anode, which is typically a dimensionally stable anode (DSA), chlorine is formed and is extracted from the cell in gaseous state. The exhausted brine solution exits the cell and is recirculated in the inlet line. At the cathode, water is reduced to form hydrogen (being responsible for about 4% of the global H_2 production) and hydroxide ions. These ions, together with the Na^+ ions crossing the cation exchange membrane to maintain the ion conduction between the electrodes, form the other important product of the process, a concentrated caustic soda. An alternative that is gaining traction is the use of oxygen depolarized cathodes (ODC) [82]. In the ODC technology, the cathode is a porous GDE that is fed by pure oxygen gas. The reduction of oxygen occurs at a significantly less negative electrode potential compared to water reduction; hence the cell voltage is decreased.

Industrial water electrolysis cells are of similar design as the chlor-alkali cell, whether we consider alkaline water electrolysis or PEM water electrolysis processes [83]. In both cases, hydrogen is produced on the cathode, while oxygen forms on the anode. The two compartments are separated by means of a diaphragm (or

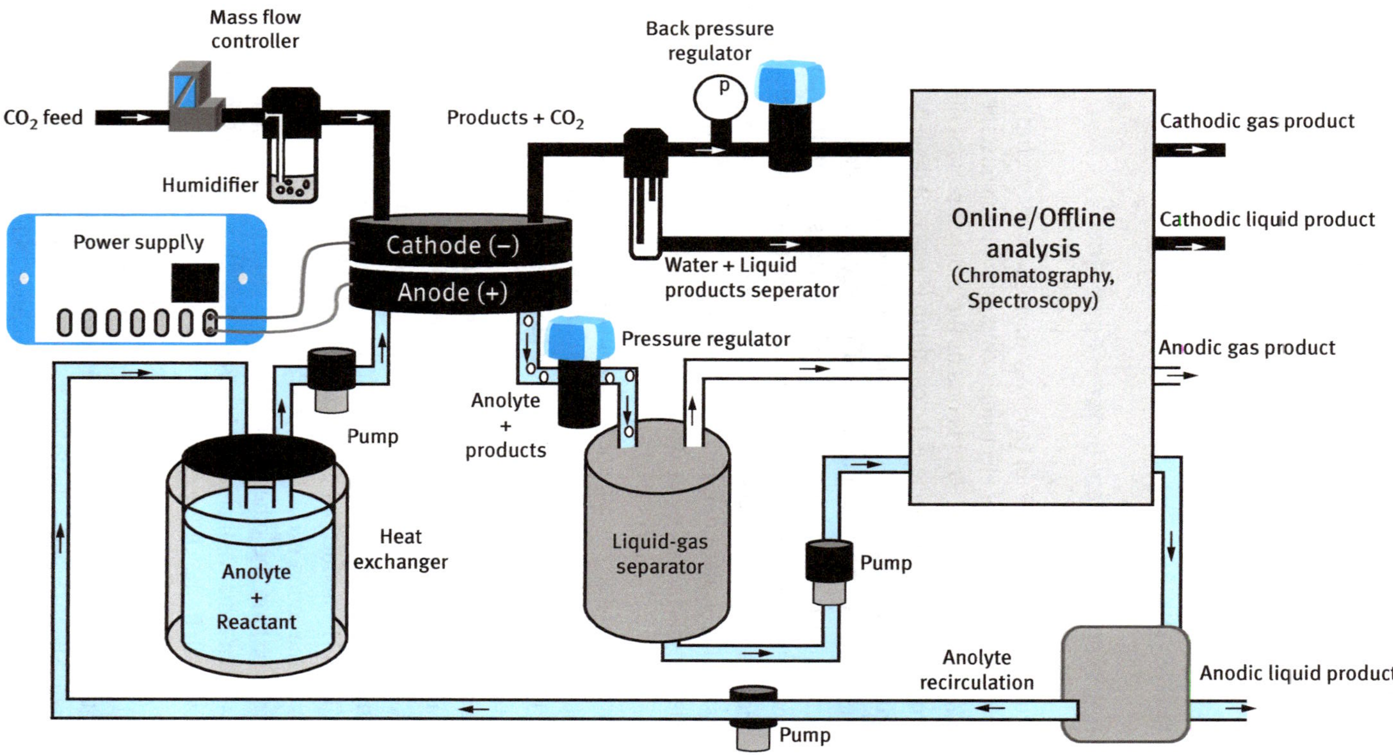

Fig. 2.12: Exemplary operational framework for CO_2R paired with a value-added anode process in a zero-gap design. It can be extended to cells using liquid catholyte, by adding an extra pump to the cathode circuit [77]. Copyright 2020 Elsevier.

alkaline membrane) in case of alkaline cells; while a PEM membrane is inserted between the two electrodes in PEM electrolyzer cell. Alkaline water electrolysis is a mature technology with large plants installed worldwide already. Using alkaline electrolyte allows the use of non-noble metal electrodes that significantly reduce capital costs of this technology. However, with the rapid membrane development, PEM water electrolyzers allow operation at higher current densities and lower cell voltages, hence result in higher energy efficiency and production rate. This technology is in the industrialization phase, with several large plants (MW range) installed already.

Of late, significant R&D resources are being dedicated to finding active catalysts and suitable cell designs for the industrial-scale electroreduction of carbon dioxide, thus turning a harmful greenhouse gas into useful chemicals. Solid oxide electrolysis cells (SOEC) offer very high process efficiencies in the direct electrochemical conversion of CO_2 to CO or water to H_2 [84]. Due to the high operating temperatures, the kinetics and thermodynamics of these processes are both very favorable. The SOEC technology is approaching maturity, with larger pilot plants already installed worldwide. This technology necessitates the application of heat tolerant structural materials and high reaction temperatures, which results in high operational and capital investment cost. This is expected to decrease with the mass production of these cells; hence, SOECs are envisaged to compete with low temperature cells in the future – a battle, which will eventually be decided by the cost and the nature of the given application.

Among the low temperature cells, two types have emerged over the years: (i) microfluidic cells, in which thin liquid layer(s) flow between the electrodes, and (ii) zero-gap designs, in which only the anode is fed with liquid electrolyte, but all the cell constituents are pressed together [63]. In both cases, CO_2 is fed in gas phase through a GDE cathode. This technology is still in the R&D phase, but larger and/or multilayer cells have already been operating at sufficiently high current densities, although the lifetime of such devices has to be improved.

The cells used for sodium chlorate production represent another class with no separator placed between the electrodes. Instead, the cathodes and anodes of these cells are arranged in an interdigitated manner with a few mm thick solution layer between them. The chemistry of the cell is similar to that of the chlor-alkali process: H_2 evolves on the cathode, and chlorine forms on the anode. The solution pH is however significantly different, which leads to the dissolution of chlorine and its subsequent chemical transformation to sodium chlorate in an external reaction vessel [85].

Synthesis of different organic molecules (e.g., adiponitrile, *p*-aminophenol, or anthraquinone) constitutes another important group of industrial flow-electrochemical processes [53, 86]. No general cell structure, process conditions, or electrode type can be defined for this group; these parameters must always be tailored to the given reaction.

2.3.2 Electroconversion of small molecules

Using the earlier described modular flow cell, *Waldvogel* and coworkers demonstrated the synthesis of nitriles from the corresponding aldoximes in a direct domino oxidation–reduction sequence [34]. Conventional approaches depend on harsh reagents like thionyl chloride or acetic anhydride. While the batch electrolysis faced the challenge of additional dehalogenation reaction, the optimization of flow electrolysis allows a supporting electrolyte-free electrolysis with excellent selectivity. The aldoxime is oxidized at a graphite anode, giving nitrile oxide, which is subsequently reductively deoxygenated at a lead cathode. During optimization, an acetonitrile water mixture of 11:1 (v/v) sufficiently suppressed dehalogenation. The solvent was recovered by evaporation, and the product was obtained by simple crystallization with 63% yield.

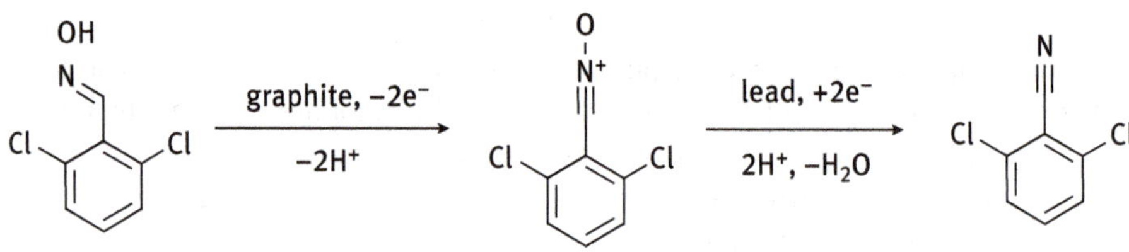

Fig. 2.13: Domino-oxidation–reduction sequence for the synthesis of nitriles from aldoximes in a narrow-gap cell.

Using a separator, they were able to trap the nitrile oxide with methyl acetoacetate in a 1,3-dipolar cycloaddition, yielding isoxazole with 60% as building block, for example, dicloxacillin synthesis. *Waldvogel* and coworkers used the same type of cell to investigate phenol-phenol cross-coupling at BDD anodes [35]. In combination with fluorinated alcohols like 1,1,1,3,3,3-hexafluoroisopropanol (HFIP), especially, a wide electrochemical window and good stabilization of the generated phenoxyl radicals is achieved. Established approaches use transition metal catalysts in a reductive fashion for the synthesis of such compounds. Additionally, leaving functionalities are required to be initially installed at the coupling partners. The direct electrochemical dehydrodimerization reaction overcomes these economical drawbacks and makes them sustainable. They showed efficient cross-coupling of different phenols and naphthol derivatives in yields up to 86%. Additionally, scale up to a flow cell with an anodic surface of 4 cm × 12 cm resulted in stable yields, showing the high robustness and scalability of this reaction.

Noel and coworkers reported a series of transformations starting from thioethers and thiols, and starting with the oxidation of sulfides to sulfoxides at iron anodes [87]. Using polarography, they identified the necessary cell voltages for selective oxidation to the corresponding sulfoxides or sulfones. They then expanded the protocol with the

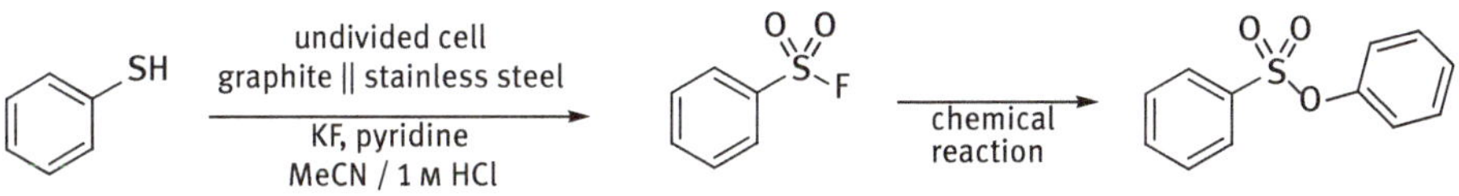

Fig. 2.14: Electrochemical oxidation of thiols in the presence of potassium fluoride toward sulfonyl fluorides. Chemical follow-up reaction toward phenyl sulfonate derivatives.

preparation of sulfonamides from amines and thiols at a graphite anode [88]. The thiol is oxidized to the disulfide, which is then attacked by the generated aminium radical, under formation of a sulfenamide. The sulfenamide is then oxidized via the sulfinamide to the sulfonamide. They reported the electrochemical oxidation of thiols in the presence of potassium fluoride to obtain sulfonyl fluorides usable in the popular sulfur(VI)-fluoride exchange (SuFEx) click chemistry reactions [89]. The reaction is carried out in a biphasic system with 1 M HCl and acetonitrile, using an undivided, self-made flow cell with a graphite anode and stainless steel cathode with yields up to 81%. Since the corresponding sulfonyl fluorides are, sometimes, volatile compounds, they circumvent the isolation by an uninterrupted flow protocol for the SuFEx click reaction toward phenyl sulfonate derivatives, with up to 73% isolated yield.

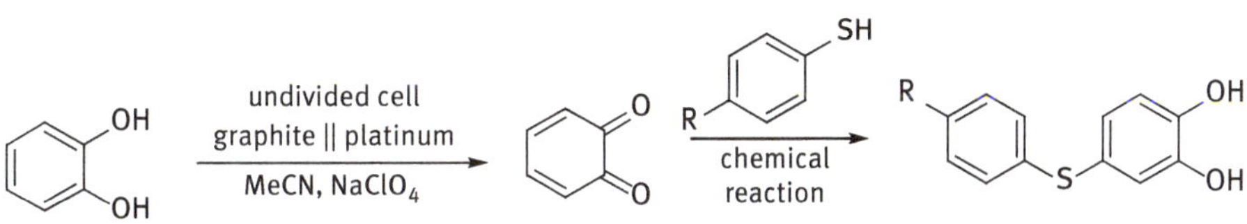

Fig. 2.15: Anodic oxidation of *o*-quinones toward unstable *o*-benzoquinone. Follow-up Michael addition.

In a similar fashion, *Atobe* and coworkers successfully synthesized diphenyl sulfides via unstable *o*-benzoquinone, which they generated electrochemically and later trapped in a Michael addition reaction with benzenethiols [90]. *o*-Quinones are versatile building blocks but need to be generated in situ, since they tend to decompose, isomerize, or polymerize upon storage. Due to this high reactivity and the competing oxidation due to similar oxidation potentials, batch-type electrolysis (in- and ex-cell) ended up with low yields. A direct oxidation of the dihydroquinones in flow is carried out at a graphite anode, and the stream is immediately joined with the thiol one, resulting in the corresponding Michael addition product, with yields up to 88%.

2.3.3 Electrosynthesis with high value addition

Besides new pathways for classical synthesis with an improved atom economy and an inherent safety, electrosynthesis can be utilized in a variety of redox-based transformations. The over-stoichiometric use of terminal oxidizers or reduction agents can be circumvented, with carbon-based electrode materials allowing metal-free synthesis that is useful for later drug synthesis.

Waldvogel and coworkers have lately optimized the electrochemical synthesis of periodate at a BDD anode [50]. Periodate is used in a broad scope of the synthesis of active pharmaceutical ingredients. Previous approaches suffer from toxic anode materials like lead dioxide or expensive starting materials. By using a Nafion® membrane, they avoided the otherwise-necessary toxic anti-reducing agents. The synthesis was optimized by design of experiments (DoE), resulting in a highly efficient conversion with yields up to 98%. The use of BDD electrodes makes this synthesis heavy metal free, which drastically lowers the costs for the purification of the product.

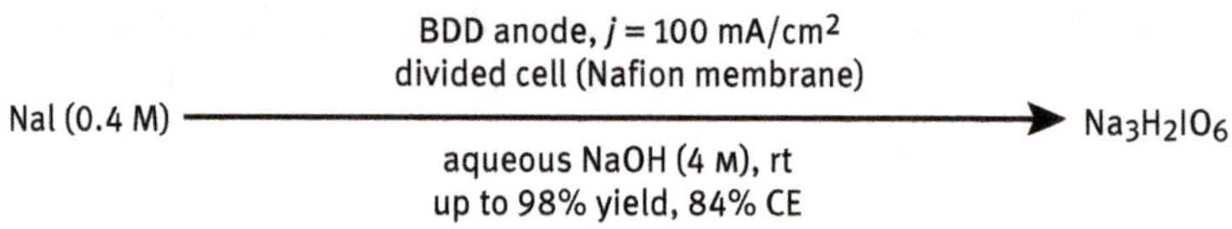

Fig. 2.16: Oxidation of sodium iodide to periodate at a BDD anode.

The processing of natural feedstock from abundant sources in the sense of sustainable production of base chemicals will become an increasingly important field of research, in the near future. The US Department of Energy identified 12 high interest sugar-based chemicals to produce value-added chemicals from biomass that is derived from, for example, hemicellulose, cellulose, and lignin. Latest research shows that flow electrolysis is a precious platform for a sustainable transformation of biomass feedstock.

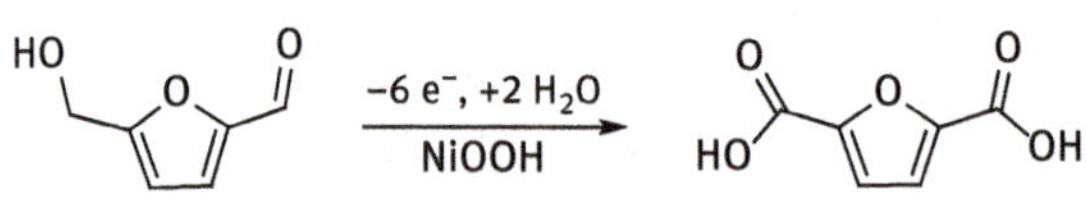

Fig. 2.17: Two-step oxidation sequence of (HMF) toward (FDCA) at activated nickel hydroxide electrode.

Giling and coworkers developed a two-cell reactor rig for the two-step oxidation sequence of 5-(hydroxymethyl)furfural (HMF) to obtain 2,5-furandicarboxylic acid (FDCA) [91]. During the optimization, different noble anode materials (Au, Au_3Pd_2, and Pt) were tested; finally, nickel with deposited active NiOOH layer gave outstanding yields of up to 90% with high faradaic efficiencies of 80%. They were able to scale up the reaction using nickel foam electrodes in a continuous setup with integrated product separation and a production of about 30 g FDCA per hour. The input stream with 15 wt% HMF is oxidized under basic conditions, resulting in a 5 wt% FDCA stream that is acidified, resulting in direct product crystallization.

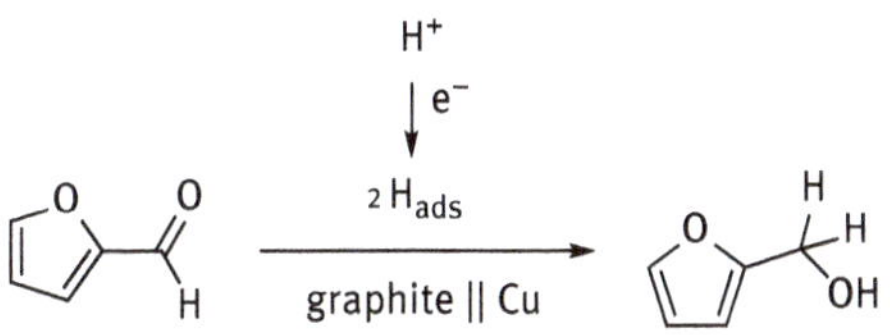

Fig. 2.18: Electrocatalytic hydrogenation of furfural to furfuryl alcohol. *Noel* and coworkers exploited the Volmer reaction to generate furfuryl alcohol from furfural in an electrocatalytic hydrogenation fashion [92].

Furfuryl alcohol is used as a monomer in furan resin synthesis [93]. During the reduction, the dimerization of the neutral radical needs to be suppressed. The optimization showed that under basic conditions, using a copper-graphite electrode setup yields furfuryl alcohol of 90%. The use of small amounts of potassium ethanolate (0.05 M) as base and supporting electrolyte was beneficial for the reaction, resulting in high faradaic efficiency of 90%.

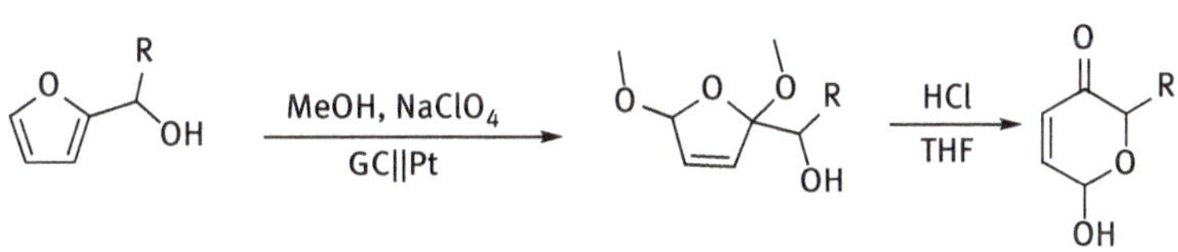

Fig. 2.19: Anodic conversion of furfuryl alcohol in presence of MeOH toward 2,5-dimethoxy-dihydrofurfuryl alcohols. Subsequent Achmatowicz reaction yields hydroxypyrones.

Starting from furfuryl alcohols, *Robertson* and coworkers developed an flow oxidation protocol for the 2,5-dimethoxy-dihydrofurfuryl alcohols [94]. A subsequent acidic hydrolysis in Achmatowicz reaction fashion yields the corresponding hydroxypyrones like maltol. Maltol and derivatives are widely used as flavor enhancers. For the electrolysis, they used a self-designed cell made from PEEK as mounting material. A GC anode and a platinum cathode were used and fixed with Araldite® epoxy resin glue. With optimized electrolysis conditions, they yielded dimethoxy derivatives of up to 71% isolated yield.

Hammond and coworkers investigated the electrochemical depolymerization of lignosulfonate in a technical-scale reactor [95]. Lignosulfonate is an abundant by-product of the pulping industry. In oxidative conditions, a variety of phenolic monomers like vanillin can be obtained. They used a FM01-LC reactor commercially available from ICI, which can be equipped with different electrode materials. The optimization of reaction allowed electrolysis of 150 g at up to 12 A, 145 °C/500 kPa/3 M NaOH. The yield of vanillin was similar to that of terminal oxidizers in batch processes of about 5–7% w/w.

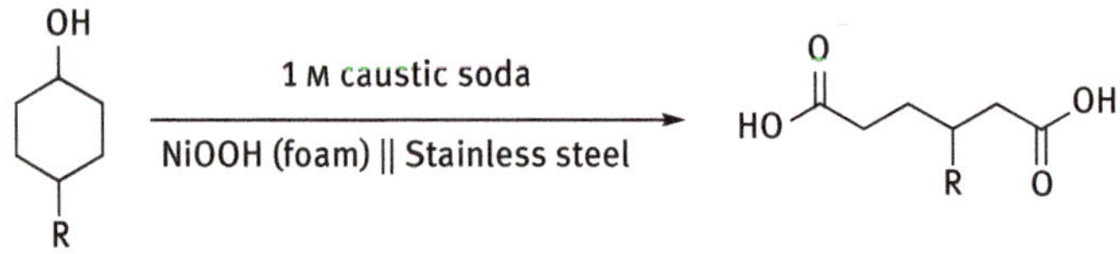

Fig. 2.20: Electrochemical oxidation of alkylated cyclohexanols to alkylated adipic acid derivatives at NiOOH foam electrodes.

Catalytic reductive fractionation of lignin gives rise to a broad scope of alkylated cyclohexanols. Lately, *Waldvogel* and coworkers have developed a protocol for the electrochemical oxidation of alkylated cyclohexanols to alkylated adipic acid derivatives at NiOOH foam electrodes in batch and flow electrolyzers [20]. The 3-methyladipic acid was obtained in yields up to 64% in a cycling electrolysis, providing efficient removal of the developing hydrogen. Sodium hydroxide was employed as a base as well as supporting electrolyte, making this an elegant and cost efficient pathway to obtain dicarboxylic acids, which can be later employed as monomer in polyester and polyamide synthesis.

2.3.4 Electrochemical synthesis of drug metabolites

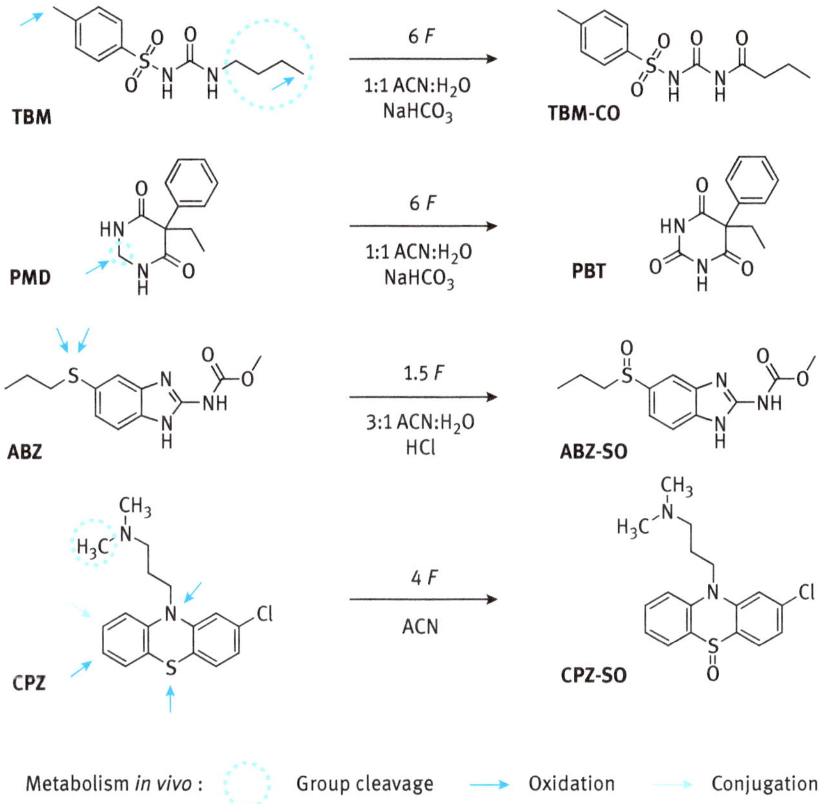

Fig. 2.21: Oxidation products of the flow electrolysis of tolbutamide (TBM), primidone (PMD), albendazole (ABZ), and chlorpromazine (CPZ). Reprinted with permission from [96]. Copyright 2013 American Chemical Society.

Electrochemical synthesis offers a low-cost approach for synthesizing and studying drug metabolites [97]. Importantly, this way, the use of expensive enzymes is avoided. Furthermore, electrochemical flow cells offer an easy way of continuously producing such metabolites, which can be analyzed by a connected in-line analytical system. Also, such a system allows the detection and characterization of unstable (or reactive) metabolites that, in a biological matrix, would otherwise rapidly bind covalently to proteins [98]. Not all metabolic pathways can be mimicked by electrochemical methods. Therefore, these studies cannot fully circumvent biological studies. Furthermore, some electrochemically induced metabolic pathways lack their counterpart in biological systems, which could, therefore, be somewhat misleading. In addition, most of such electroconversions only provide traces for analysis but not enough material for extensive studies. A valid approach is the installation of stabilizing groups [99].

2.3.5 Paired and consecutive electrolysis

To increase process efficiency and decrease production costs in any electrochemical process, it is desired to form valuable products on both electrodes, while applying as the lowest possible cell voltage. When it comes to industrially relevant cathodic reactions such as H_2 production or CO_2 electroreduction, water oxidation (i.e., oxygen evolution) reaction is the most often used green anodic pair. This requires high energy input (i.e., large cell voltage), and generates a product of little commercial value (i.e., oxygen). Pairing appropriate alternative anode processes with these reductions is of major interest, as it might result in a significantly increased total product value. On the other hand, this significantly increases the system complexity, necessitating the continuous composition monitoring of both the anodic and cathodic product streams, and the multistep separation of these products. This implies scientific and engineering challenges, but a properly designed and optimized paired electrolysis process offers the possibility of direct industrial implementation. As a promising example, either CO_2 reduction or H_2 production might be coupled with anodic glycerol oxidation to glyceraldehyde, a product that is more valuable than O_2 [100, 101] by orders of magnitude. Furthermore, the standard redox potential of this process is almost 1 V lower than that in water oxidation, offering operation at significantly lower cell voltages when compared to the case of anodic water oxidation.

Finding suitable catalysts for multistep electrochemical processes involving the transfer of multiple electrons (and multiple protons) could be challenging. In certain cases, forming an intermediate in one electrolysis cell and further reacting it in a subsequent setup applying different catalyst material is favorable in terms of energy requirements (including process selectivity, hence the costs related to product separation). Reducing CO_2 to an intermediate and further reacting it in a subsequent chemical/electrochemical step offers a way to synthesize a large variety of functional chemicals that serve as building blocks for organic synthetic chemistry. An example is the production of ethylene oxide in a two-step electrochemical method – first reducing CO_2 to C_2H_4, and subsequently, selectively oxidizing the product in a separate cell [102]. This is a green alternative for ethylene oxide production, which is an important bulk chemical that is produced in large quantities – ~20 million metric tons per year.

2.4 Strategies for screening and optimization

The optimization of reactions in electrochemical flow cells can present a challenge due to the high number of parameters involved. Classic optimization strategies, like one variable at a time (OVAT), often fail to reach the global maximum (or minimum) of a system response, since they cannot resolve interactions between the parameters. An interaction is the influence of one parameter setting on the system response when changing the value of a second parameter. This can result in a high number of experiments needed for optimization, as well as unnecessary allocation of resources and time. Statistically driven approaches like DoE provide a remedy. We do not want to describe the theory in detail in this chapter, since it is covered in literature [103, 104], but we want to provide a guide and a jump start with a few real-life examples.

The basic idea is to treat a reaction as a black box process, if no deeper understanding of the mechanisms is available prior to the optimization. The process is influenced by reaction parameters – both controllable like the current density as well as probably not controllable like the humidity in the lab. If possible, one should always record data on these so-called covariates. The setting of these parameters results in system responses like yield or selectivity, once the process is carried out. In contrast to OVAT, the influence of all parameters on the system response is analyzed at once; therefore each parameter is investigated at two levels (high (+) and low (–) setting), resulting in a balanced experimental design. These so-called full factorial designs require a total number of 2^n experiments, where n is the number of parameters. When the experimental plan is carried out, one can start with the calculation of the main effects. A main effect of a parameter represents the mean change of a system response when the setting is changed from low to high, calculated by the difference of the two subsets, one containing all experimental results with the parameter on high setting, and the second containing experimental results with the parameter on low setting. An interaction causes a different system response when changing one parameter from low to high depending on the setting of a second parameter and is calculated as the difference of the two effects at both levels of the second parameter. It needs to be noted that main effects are usually dominating in contrast to the effect sizes of the interactions; they decrease with increasing order of interaction. This acted as motivation for fractional factorial designs, which include only a subset of the full design, resulting in lower numbers of experiments needed, but for the price of the confounding of main effects and interactions [105]. Based on the amount of reduction, the models are classified into three ascending resolution levels: III–V.

Besides the calculation of the main effects, the balanced fashion of the design allows a linear regression to yield a predictive model within the investigated space. Furthermore, the calculated effect sizes can be probed by using analysis of variance (ANOVA). The working assumption is the normal distribution of the model error over the whole design, justified by the central limit theorem. In the case of fractional designs, nonsignificant parameters can be excluded from the model, resulting in a stronger design (projection property).

Within the range of a maximum (or minimum) of a system response, a linear regression is no longer sufficient to fit the resulting curvature of the responses hyperplane in the parameter space. This can be checked by a center point experiment. The result is compared with the predicted value from the linear regression. In case of a large residuum, the design needs to be extended by additional experiments to fulfill the needs of a second order regression. Typical designs are the central composite design or Box–Behnken design [106, 107].

DoE needs to be understood as a recursive procedure. In the early stage of an optimization, a fractional design or screening design should be carried out to measure the main effect sizes and exclude all parameters that show no influence on the system response. With a reduced number of parameters, designs of higher resolution can be used. Within the range of a maximum (or minimum), a second order model can be used to fine-tune the parameter settings [108]. If more than one system response is of interest, the desirability functions offer multitarget optimization. For each response, its desirability function scales between 0 (if the predicted response is outside the acceptable range) and 1 (if it matches to desired value). The individual desirability functions are then multiplied to get an overall desirability that needs to be maximized. Most statistical software suites provide an algorithm to solve this optimization problem, resulting in optimized parameter settings.

Wirth and coworkers recently showed an elegant combination of their electrochemical flow cell with online 2D HPLC analysis [54]. They used the memory effect of chirality in *N*-aryl carbonylated amino acid derivatives under Kolbe electrolysis to generate enantiomerically enriched alkoxylated amides, via an acyliminium ion intermediate. Due to the supporting electrolyte (SE)-free conditions, they were able to measure yield and enantioselectivity as system responses directly, via (chiral) HPLC. Initial experiments in batch type cells resulted in poor yields and selectivities due to the unstable acyliminium intermediate. They started by pre-screening of electrode materials, substrate concentrations, and flow rates ending up with graphite and GC as promising anode materials. Afterward, using a 2^{5-1} fractional design, they investigated the influence of the concentration of starting material, flow rate, applied charge, temperature, and – as a category parameter – the anode material (graphite and GC). In the case of yield, the amount of charge and the anode material were dominating parameters, whereas, in the case of enantioselectivity, the anode material was mainly influential together with a second order interaction involving the temperature. They finally investigated the influence of the flow rate, amount of charge, and temperature in a full factorial 2^3 design, using GC as anode material ending up with optimized conditions with yield up to 100% and enantioselectivities up to 70 % *ee*.

Waldvogel and coworkers investigated the anodic C–C cross-coupling of 2-methoxy-4-methylphenol and 2,4-dimethylphenol as a model system [109]. This cross-coupling reaction was a subject of research for a longer period; linear optimization strategies yielded 44% of the corresponding biphenol. Based on this primary research, they used a 2^{7-3} design to investigate the influence of the electrode gap, temperature, current density, amount of charge, concentration of SE, and concentration as well as ratio of the substrates. In 35 experiments, they found the electrode gap as the dominating parameter, where a small gap was beneficial for the yield. Due to DoE-based investigation, an isolated yield of 85% was easily achieved. Furthermore, the main effect analysis revealed no significant influence of the substrate concentration or current density; therefore, a near-threefold increase in space–time yield was reached.

2.5 Options for industrialization and scale-up

Industrialization of electrochemical reactions give rise to different challenges in the various processes. Many conversions discussed in this chapter are industrially relevant, even at a relatively small scale. Synthesis of drug molecules or selective redox transformation of key intermediates can be readily performed in microfluidic cells, in up to kilogram scale. In these cases, the consecutive reactions and the cross-talk between the two electrodes are the most important challenges during scale-up. Interdigitated electrodes, as well as special long-path designs can help increasing the conversion. In strong contrast, large units and systems are needed in the chlor-alkali electrolysis, water splitting, and CO_2 conversion processes, to have industrially relevant size. Moreover, while product selectivity is the key driver in the first category, energy efficiency, conversion rate, and materials cost become dominant for large-scale technologies. This is something that has to be kept in mind even at the research stage, because electrolysis cells shall be developed, which operate (i) at high current density (conversion rate), (ii) at low cell voltage (i.e., high energy efficiency), (iii) with high Faradaic efficiency (selectivity), and (iv) with high conversion efficiency. Notably, even though these four parameters together describe the overall performance of an electrolysis cell, very seldom are all of these reported in the scientific literature [71].

Another important concept to highlight is the scale-out approach. This means that the result of the scale-up process is not a giant cell, but, rather, multiple cell stacks. One stack is made of several (up to over 100) electrochemical cells, where the key component is a bipolar plate (acting as cathode in one cell and anode in the subsequent cell). This approach overcomes the need for manufacturing very large electrodes and cell components, as also difficulties related to the management of the electrochemical process itself. Stack cells enabled the industrialization of both water splitting and the chlor-alkali process, and the concept has been recently extended to new processes. For example, *Endrődi* and *Janáky* published on the implementation of this strategy to CO_2 electrolyzers, which can accelerate technology development to scale up electrochemical CO_2 reduction to an industrially relevant level. A zero-gap CO_2-stacked cell which consists of multiple layers and can operate with a pressurized CO_2 gas feed without the need for any liquid catholyte was demonstrated. The flexibility of the presented design allows different connections between the layers of the electrolyzer based on the distribution of the reactant CO_2 gas. Connecting the cells in parallel (Fig. 2.22A), the gas is equally distributed among them; hence, pure CO_2 is fed to each cathode. On the other hand, when connecting the gas channels in series (Fig. 2.22B), the total gas flux enters the first layer, and

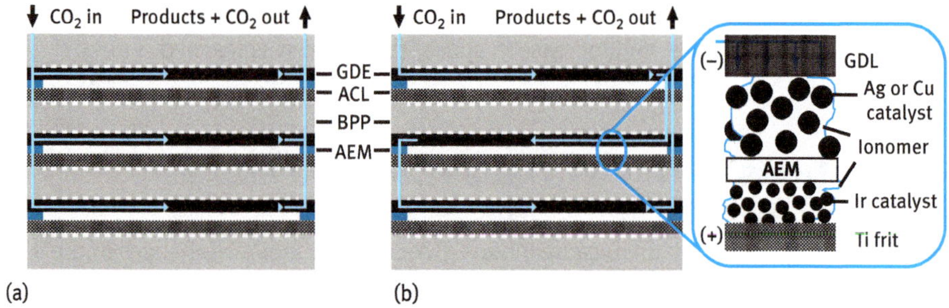

Fig. 2.22: CO_2 gas channel structure in an electrolyzer stack consisting of three layers in the (A) parallel and (B) serial connection configurations. BPP, bipolar plate; ACL, anode catalyst layer; GDE, gas diffusion electrode; GDL, gas diffusion layer; AEM, anion exchange membrane. Copyright 2019 American Chemical Society [110].

the off-gas (remnant CO_2 + products) continues to the subsequent layer(s), thus allowing very high conversion efficiencies.

Further readings

- Hamann CH, Hamnett A, Vielstich W. Electrochemistry. 2nd ed. Weinheim: Wiley-VCH, 2007.
- Schmidt VM. Elektrochemische Verfahrenstechnik. Wiley, 2003.
- Yoshida J-I, Suga S. Basic Concepts of “Cation Pool” and “Cation Flow” Methods and Their Applications in Conventional and Combinatorial Organic Synthesis. Chem. Eur. J. 2002,8,2650.
- Zoski CG. Handbook of electrochemistry. Amsterdam: Elsevier, 2007.
- Hammerich O, Speiser B, eds. Organic electrochemistry. 5th ed. Boca Raton, London, New York: CRC Press, 2016.
- Wiebe A, Gieshoff T, Möhle S, Rodrigo E, Zirbes M, Waldvogel SR. Electrifying Organic Synthesis. Angew. Chem., Int. Ed. 2018,57,5594–619. (open access)
- Möhle S, Zirbes M, Rodrigo E, Gieshoff T, Wiebe A, Waldvogel SR. Modern Electrochemical Aspects for the Synthesis of Value-Added Organic Products. Angew. Chem., Int. Ed. 2018,57,6018–41. (open access)
- Noël T, Cao Y, Laudadio G. The Fundamentals Behind the Use of Flow Reactors in Electrochemistry. Acc Chem Res 2019,52,2858–69. (open access)
- Waldvogel SR, Lips S, Selt M, Riehl B, Kampf CJ. Electrochemical Arylation Reaction. Chem Rev 2018,118,6706–65. (open access)
- Röckl JL, Pollok D, Franke R, Waldvogel SR. A Decade of Electrochemical Dehydrogenative C,C-Coupling of Aryls. Acc Chem Res 2020,53,45–61.
- Pollok D, Waldvogel SR. Electro-organic synthesis – a 21st century technique. Chem. Sci. 2020,519,379. (open access)
- Heard DM, Lennox AJJ. Electrode Materials in Modern Organic Electrochemistry. Angew. Chem., Int. Ed. 2020. (open access)
- Yan M, Kawamata Y, Baran PS. Synthetic Organic Electrochemical Methods Since 2000: On the Verge of a Renaissance. Chem Rev 2017,117,13230–319.

- Pletcher D, Green RA, Brown RCD. Flow Electrolysis Cells for the Synthetic Organic Chemistry Laboratory. Chem Rev 2018,118,4573–91. (open access)
- Endrődi B, Bencsik G, Darvas F, Jones R, Rajeshwar K, Janáky C. Continuous-flow electroreduction of carbon dioxide. Prog. Energy Combust. Sci. 2017,62,133–54. (open access)

Study questions

?

2.1 Electrochemical processes are widely used. Please provide some examples.

2.2 Please describe the principle of an electrochemical cell, indicate electrodes, and explain physical and chemical processes at the individual electrode surfaces. Use water as electroactive species in acidic and alkaline solution to evolve hydrogen and oxygen.

2.3 Why is it necessary to add a so-called supporting electrolyte into an electrolyte to perform a conventional electrolysis? Is it possible to reduce the amount of supporting electrolyte when a narrow-gap cell is used?

2.4 Please make a suggestion for avoiding or suppressing undesired electrode reactions.

2.5 Please describe strategies for handling gas evolution within the flow electrolyzers.

2.6 What are suitable electrodes for anodic conversions of organic compounds?

2.7 Carbon electrodes can be considered as essentially metal free. Which materials do you know and what are the specific features, thereof?

2.8 Platinum is often used in lab electrolyzers. Why should chloride-containing electrolytes be avoided?

2.9 What is a paired electrolysis? Please provide examples.

2.10 Please calculate the necessary applied charge for a homo-coupling reaction in dependence of the molar quantity of the substrate A. How does this change when considering a cross-coupling of components A and B? Which criterion is used to decide which of the substance quantities is used in the calculation? How is the flow rate calculated for a given charge quantity and current density, in relation to the volume of the reactor chamber (for a single pass)?

References

[1] Frontana-Uribe BA, Little RD, Ibanez JG, Palma A, Vasquez-Medrano R, Organic electrosynthesis: a promising green methodology in organic chemistry, Green Chem, 2010, 12, 2099.

[2] Pollok D, Waldvogel SR, Electro-organic synthesis – a 21st century technique, Chem Sci, 2020, 519, 379.

[3] Yan M, Kawamata Y, Baran PS, Synthetic organic electrochemical methods since 2000: on the verge of a renaissance, Chem Rev, 2017, 117, 13230–13319.

[4] Wiebe A, Gieshoff T, Möhle S, Rodrigo E, Zirbes M, Waldvogel SR, Electrifying Organic Synthesis, Angew Chem, Int Ed, 2018, 57, 5594–5619.

[5] Möhle S, Zirbes M, Rodrigo E, Gieshoff T, Wiebe A, Waldvogel SR, Modern electrochemical aspects for the synthesis of value-added organic products, Angew Chem, Int Ed, 2018, 57, 6018–6041.

[6] Hammerich O, Speiser B, eds, Organic Electrochemistry, 5th ed., Boca Raton, London, New York, CRC Press, 2016.

[7] Francke R, Little RD, Redox catalysis in organic electrosynthesis: basic principles and recent developments, Chem Soc Rev, 2014, 43, 2492–2521.

[8] Marken F, Atobe M, eds, Modern Electrosynthetic Methods in Organic Chemistry, Boca Raton, London, New York, CRC Press Taylor & Francis Group, 2019.

[9] Röckl JL, Pollok D, Franke R, Waldvogel SR, A decade of electrochemical dehydrogenative C,C-coupling of aryls, Acc Chem Res, 2020, 53, 45–61.

[10] Maljuric S, Jud W, Kappe CO, Cantillo D, Translating batch electrochemistry to single-pass continuous flow conditions: an organic chemist's guide, J Flow Chem, 2020, 10, 181–190.

[11] Pletcher D, Green RA, Brown RCD, Flow electrolysis cells for the synthetic organic chemistry laboratory, Chem Rev, 2018, 118, 4573–4591.

[12] Heard DM, Lennox AJJ, Electrode materials in modern organic electrochemistry, Angew Chem, Int Ed, 2020.

[13] Edinger C, Kulisch J, Waldvogel SR, Stereoselective cathodic synthesis of 8-substituted (1R,3R,4S)-menthylamines, Beilstein J Org Chem, 2015, 11, 294–301.

[14] Izutsu K, Electrochemistry in Nonaqueous Solutions, Weinheim, Wiley-VCH, 2002.

[15] Hilt G, Basic strategies and types of applications in organic electrochemistry, ChemElectroChem, 2020, 7, 395–405.

[16] Martínez-Huitle CA, Waldvogel SR, Trends of Organic Electrosynthesis by Using Boron-Doped Diamond Electrodes, In: Yang N, ed, Novel Aspects of Diamond, Cham, Springer International Publishing, 2019, 173–197.

[17] Waldvogel SR, Mentizi S, Kirste A, Boron-doped diamond electrodes for electroorganic chemistry, Top Curr Chem, 2012, 320, 1–31.

[18] Yang N, Yu S, Macpherson JV et al., Conductive diamond: synthesis, properties, and electrochemical applications, Chem Soc Rev, 2019, 48, 157–204.

[19] Lips S, Waldvogel SR, Use of boron-doped diamond electrodes in electro-organic synthesis, ChemElectroChem, 2019, 6, 1649–1660.

[20] Waldvogel SR, Rauen A, Weinelt F, Sustainable electroorganic synthesis of lignin-derived dicarboxylic acids, Green Chem, 2020.

[21] Schäfer H-J, Oxidation of organic compounds at the nickel hydroxide electrode, In: Dewar MJS, Dunitz JD, Hafner K, et al., eds, Electrochemistry I, Berlin, Heidelberg, Springer Berlin Heidelberg, 1987, 101–129.

[22] Kaulen J, Schäfer H-J, Oxidation of alcohols by electrochemically regenerated nickel oxide hydroxide. Selective oxidation of hydroxysteroids, Tetrahedron, 1982, 38, 3299–3308.

[23] Francke R, Quell T, Wiebe A, Waldvogel SR, Oxygen-Containing Compounds, Alcohols, Ethers, and Phenols, In: Hammerich O, Speiser B, eds, Organic Electrochemistry, 5th edition, revised and expanded, Boca Raton, London, New York, CRC Press, 2016, 981–1033.

[24] Beil SB, Müller T, Sillart SB et al., Active molybdenum-based anode for dehydrogenative coupling reactions, Angew Chem, Int Ed, 2018, 57, 2450–2454.

[25] Beil SB, Breiner M, Schulz L et al., About the selectivity and reactivity of active nickel electrodes in C–C coupling reactions, RSC Adv, 2020, 10, 14249–14253.

[26] Fuchigami T, Fujita T, Electrolytic partial fluorination of organic compounds. 14. The first electrosynthesis of hypervalent iodobenzene difluoride derivatives and its application to indirect anodic gem-difluorination, J Org Chem, 1994, 59, 7190–7192.

[27] Berger M, Herszman JD, Kurimoto Y et al., Metal-free electrochemical fluorodecarboxylation of aryloxyacetic acids to fluoromethyl aryl ethers, Chem Sci, 2020, 11, 6053–6057.

[28] Haupt JD, Berger M, Waldvogel SR, Electrochemical fluorocyclization of N-allylcarboxamides to 2-oxazolines by hypervalent iodine mediator, Org Lett, 2019, 21, 242–245.

[29] Herszman JD, Berger M, Waldvogel SR, Fluorocyclization of N-propargylamides to oxazoles by electrochemically generated ArIF2, Org Lett, 2019, 21, 7893–7896.

[30] Nutting JE, Rafiee M, Stahl SS, Tetramethylpiperidine N-Oxyl (TEMPO), Phthalimide N-Oxyl (PINO), and related N-Oxyl species: electrochemical properties and their use in electrocatalytic reactions, Chem Rev, 2018, 118, 4834–4885.
[31] Waldvogel SR, Lips S, Selt M, Riehl B, Kampf CJ, Electrochemical arylation reaction, Chem Rev, 2018, 118, 6706–6765.
[32] Noël T, Cao Y, Laudadio G, The fundamentals behind the use of flow reactors in electrochemistry, Acc Chem Res, 2019,52,2858–69.
[33] Atobe M, Tateno H, Matsumura Y, Applications of flow microreactors in electrosynthetic processes, Chem Rev, 2018, 118, 4541–4572.
[34] Gütz C, Stenglein A, Waldvogel SR, Highly modular flow cell for electroorganic synthesis, Org Process Res Dev, 2017, 21, 771–778.
[35] Gleede B, Selt M, Gütz C, Stenglein A, Waldvogel SR, Large, highly modular narrow-gap electrolytic flow cell and application in dehydrogenative cross-coupling of phenols, Org Process Res Dev, 2019.
[36] Elsherbini M, Wirth T, Electroorganic synthesis under flow conditions, Acc Chem Res, 2019, 52, 3287–3296.
[37] Selt M, Franke R, Waldvogel SR, Supporting-electrolyte-free and scalable flow process for the electrochemical synthesis of 3,3′,5,5′-tetramethyl-2,2′-biphenol, Org Process Res Dev, 2020.
[38] Wirtanen T, Rodrigo E, Waldvogel SR, Selective and scalable electrosynthesis of 2H-2-(Aryl)-benzod-1,2,3-triazoles and their N-oxides by using leaded bronze cathodes, Chemistry, 2020, 26, 5592–5597.
[39] Gütz C, Grimaudo V, Holtkamp M et al., Leaded bronze: an innovative lead substitute for cathodic electrosynthesis, ChemElectroChem, 2018, 5, 247–252.
[40] Kulisch J, Nieger M, Stecker F, Fischer A, Waldvogel SR, Efficient and stereodivergent electrochemical synthesis of optically pure menthylamines, Angew Chem, Int Ed, 2011, 50, 5564–5567.
[41] Edinger C, Waldvogel SR, Electrochemical deoxygenation of aromatic amides and sulfoxides, Eur J Org Chem, 2014, 2014, 5144–5148.
[42] Edinger C, Grimaudo V, Broekmann P, Waldvogel SR, Stabilizing lead cathodes with diammonium salt additives in the deoxygenation of aromatic amides, ChemElectroChem, 2014, 1, 1018–1022.
[43] Wesenberg LJ, Herold S, Shimizu A, Yoshida J-I, Waldvogel SR, New approach to 1,4-benzoxazin-3-ones by electrochemical C-H amination, Chemistry, 2017, 23, 12096–12099.
[44] Herold S, Möhle S, Zirbes M, Richter F, Nefzger H, Waldvogel SR, Electrochemical amination of less-activated alkylated arenes using boron-doped diamond anodes, Eur J Org Chem, 2016, 2016, 1274–1278.
[45] Fangmeyer J, Behrens A, Gleede B, Waldvogel SR, Karst U, Mass-spectrometric imaging of electrode surfaces – a view on electrochemical side reactions, Angew Chem, Int Ed, 2020, 49, 252.
[46] Folgueiras-Amador AA, Wirth T, 5 Electrosynthesis in Continuous Flow, In: Jamison K, eds, Flow Chemistry in Organic Synthesis, Stuttgart, Georg Thieme Verlag, 2018.
[47] Watts K, Gattrell W, Wirth T, A practical microreactor for electrochemistry in flow, Beilstein J Org Chem, 2011, 7, 1108–1114.
[48] Ogawa KA, Boydston AJ, Recent developments in organocatalyzed electroorganic chemistry, Chem Lett, 2015, 44, 10–16.
[49] Chardon CP, Matthée T, Neuber R, Fryda M, Comninellis C, Efficient electrochemical production of peroxodicarbonate applying DIACHEM ® diamond electrodes, Chemistry Select, 2017, 2, 1037–1040.

[50] Arndt S, Weis D, Donsbach K, Waldvogel SR, The "Green" electrochemical synthesis of periodate, Angew Chem, Int Ed, 2020, 59, 8036–8041.
[51] Horii D, Fuchigami T, Atobe M, A new approach to anodic substitution reaction using parallel laminar flow in a micro-flow reactor, J Am Chem Soc, 2007, 129, 11692–11693.
[52] Folgueiras-Amador AA, Philipps K, Guilbaud S, Poelakker J, Wirth T, An easy-to-machine electrochemical flow microreactor: efficient synthesis of isoindolinone and flow functionalization, Angew Chem, Int Ed, 2017, 56, 15446–15450.
[53] Pletcher D, Organic electrosynthesis – A road to greater application, A mini rev Electrochem Commun, 2018, 88, 1–4.
[54] Santi M, Seitz J, Cicala R, Hardwick T, Ahmed N, Wirth T, Memory of chirality in flow electrochemistry: fast optimisation with DoE and online 2D-HPLC, Chemistry, 2019, 25, 16230–16235.
[55] IKA®-Werke GmbH & CO. KG. IKA ElectraSyn flow, 2021. (https://www.ikaprocess.com/en/Products/Electro-Organic-Synthesis-Systems-cph-45/ElectraSyn-flow-csb-ES/).
[56] Vapourtec Ltd. Ion electrochemical reactor, 2021. (https://www.vapourtec.com/products/flow-reactors/ion-electrochemical-reactor-features/).
[57] Cambridge Reactor Design Ltd. Ammonite Electrolysis Cell, 2021. (https://www.cambridgereactordesign.com/ammonite/ammonite.html).
[58] CONDIAS GmbH. DIACHEM® Diamond Electrodes, 2021. (https://www.condias.de/en-gb/products/DIACHEM%C2%AE-Electrodes).
[59] ElectroCell A/S. Electrochemical Flow Cells, 2021. (https://www.electrocell.com/products/electrochemical-flow-cells).
[60] Dioxide Materials. Gas Diffusion Electrode Electrolyzers, 2021. (https://dioxidematerials.com/).
[61] ThalesNano Energy Inc. Gas Diffusion Electrode Electrolyzers, 2021. (https://thsenergy.com/).
[62] Lakshmanan S, Murugesan T, The chlor-alkali process: work in progress, Clean Technol Environ Policy, 2014, 16, 225–234.
[63] Endrődi B, Bencsik G, Darvas F, Jones R, Rajeshwar K, Janáky C, Continuous-flow electroreduction of carbon dioxide, Prog Energy Combust Sci, 2017, 62, 133–154.
[64] Endrődi B, Simic N, Wildlock M, Cornell A, A review of chromium(VI) use in chlorate electrolysis: functions, challenges and suggested alternatives, Electrochim Acta, 2017, 234, 108–122.
[65] Amatore C, Szunerits S, Thouin L, Warkocz J-S, The real meaning of Nernst's steady diffusion layer concept under non-forced hydrodynamic conditions. A simple model based on Levich's seminal view of convection, J Electroanal Chem, 2001, 500, 62–70.
[66] Newman J, Thomas-Alyea KE, Electrochemical Systems, 3rd ed, Hoboken, Wiley-Interscience, 2004.
[67] Buttler A, Spliethoff H, Current status of water electrolysis for energy storage, grid balancing and sector coupling via power-to-gas and power-to-liquids: a review, Renewable Sustain Energy Rev, 2018, 82, 2440–2454.
[68] Arenas LF, Ponce De León C, Walsh FC, Critical review – the versatile plane parallel electrode geometry: an illustrated review, J Electrochem Soc, 2020, 167, 23504.
[69] Weekes DM, Salvatore DA, Reyes A, Huang A, Berlinguette CP, Electrolytic CO2 reduction in a flow cell, Acc Chem Res, 2018, 51, 910–918.
[70] Carmo M, Fritz DL, Mergel J, Stolten D, A comprehensive review on PEM water electrolysis, Int J Hydr Energy, 2013, 38, 4901–4934.
[71] Endrődi B, Kecsenovity E, Samu A et al., High carbonate ion conductance of a robust PiperION membrane allows industrial current density and conversion in a zero-gap carbon dioxide electrolyzer cell, Energy Environ Sci, 2020.

[72] Vincent I, Bessarabov D, Low cost hydrogen production by anion exchange membrane electrolysis: a review, Renewable Sustain Energy Rev, 2018, 81, 1690–1704.

[73] Chen Y, Vise A, Klein WE et al., A robust, scalable platform for the electrochemical conversion of CO2 to formate: identifying pathways to higher energy efficiencies, ACS Energy Lett, 2020, 5, 1825–1833.

[74] Salvatore DA, Weekes DM, He J et al., Electrolysis of gaseous CO2 to CO in a flow cell with a bipolar membrane, ACS Energy Lett, 2018, 3, 149–154.

[75] Li YC, Yan Z, Hitt J, Wycisk R, Pintauro PN, Mallouk TE, Bipolar membranes inhibit product crossover in CO2 electrolysis cells, Adv Sustain Syst, 2018, 2, 1700187.

[76] Bidault F, Brett D, Middleton PH, Brandon NP, Review of gas diffusion cathodes for alkaline fuel cells, J Power Sources, 2009, 187, 39–48.

[77] Burdyny T, Smith WA, CO 2 reduction on gas-diffusion electrodes and why catalytic performance must be assessed at commercially-relevant conditions, Energy Environ Sci, 2019, 12, 1442–1453.

[78] Wu J, Sharma PP, Harris BH, Zhou X-D, Electrochemical reduction of carbon dioxide: IV dependence of the Faradaic efficiency and current density on the microstructure and thickness of tin electrode, J Power Sources, 2014, 258, 189–194.

[79] Liu K, Smith WA, Burdyny T, Introductory guide to assembling and operating gas diffusion electrodes for electrochemical CO2 reduction, ACS Energy Lett, 2019, 4, 639–643.

[80] Jhong H-R, Brushett FR, Kenis PJA, The effects of catalyst layer deposition methodology on electrode performance, Adv Energy Mater, 2013, 3, 589–599.

[81] Vass Á, Endrődi B, Janáky C, Coupling electrochemical carbon dioxide conversion with value-added anode processes: an emerging paradigm, Curr Opin Electrochem, 2021, 25, 100621.

[82] Moussallem I, Jörissen J, Kunz U, Pinnow S, Turek T, Chlor-alkali electrolysis with oxygen depolarized cathodes: history, present status and future prospects, J Appl Electrochem, 2008, 38, 1177–1194.

[83] Ursua A, Gandia LM, Sanchis P, Hydrogen production from water electrolysis: current status and future trends, Proc IEEE, 2012, 100, 410–426.

[84] Hauch A, Küngas R, Blennow P et al., Recent advances in solid oxide cell technology for electrolysis, Science, 2020, 370.

[85] Kreysa G, Ota K, Savinell RF, Encyclopedia of Applied Electrochemistry, New York, NY, Springer New York, 2014.

[86] Sequeira CAC, Santos DMF, Electrochemical routes for industrial synthesis, J Braz Chem Soc, 2009, 20, 387–406.

[87] Laudadio G, Straathof NJW, Lanting MD, Knoops B, Hessel V, Noël T, An environmentally benign and selective electrochemical oxidation of sulfides and thiols in a continuous-flow microreactor, Green Chem, 2017, 19, 4061–4066.

[88] Laudadio G, Barmpoutsis E, Schotten C et al., Sulfonamide synthesis through electrochemical oxidative coupling of amines and thiols, J Am Chem Soc, 2019, 141, 5664–5668.

[89] Cao Y, Adriaenssens B, Bartolomeu A, De A, Laudadio G, Oliveira KTD, Noël T, Accelerating sulfonyl fluoride synthesis through electrochemical oxidative coupling of thiols and potassium fluoride in flow, J Flow Chem, 2020, 10, 191–197.

[90] Kashiwagi T, Amemiya F, Fuchigami T, Atobe M, In situ electrogeneration of o-benzoquinone and high yield reaction with benzenethiols in a microflow system, Chem Commun (Camb), 2012, 48, 2806–2808.

[91] Latsuzbaia R, Bisselink R, Anastasopol A et al., Continuous electrochemical oxidation of biomass derived 5-(hydroxymethyl)furfural into 2,5-furandicarboxylic acid, J Appl Electrochem, 2018, 48, 611–626.

[92] Cao Y, Noël T, Efficient electrocatalytic reduction of furfural to furfuryl alcohol in a microchannel flow reactor, Org Process Res Dev, 2019, 23, 403–408.
[93] Gandini A, Furans in polymer chemistry, Prog Polym Sci, 1997, 22, 1203–1379.
[94] Syntrivanis L-D, Javier Del Campo F, Robertson J, An electrochemical flow cell for the convenient oxidation of Furfuryl alcohols, J Flow Chem, 2018, 8, 123–128.
[95] Smith CZ, Utley JHP, Hammond JK, Electro-organic reactions. Part 60[1]. The electro-oxidative conversion at laboratory scale of a lignosulfonate into vanillin in an FM01 filter press flow reactor: preparative and mechanistic aspects, J Appl Electrochem, 2011, 41, 363–375.
[96] Stalder R, Roth GP, Preparative microfluidic electrosynthesis of drug metabolites, ACS Med Chem Lett, 2013, 4, 1119–1123.
[97] Jurva U, Weidolf L, Electrochemical generation of drug metabolites with applications in drug discovery and development, TrAC, Trends Anal Chem, 2015, 70, 92–99.
[98] Zhou F, Van Berkel GJ, Electrochemistry combined online with electrospray mass spectrometry, Anal Chem, 1995, 67, 3643–3649.
[99] Gutmann A, Wesenberg LJ, Peez N, Waldvogel SR, Hoffmann T, Charged tags for the identification of oxidative drug metabolites based on electrochemistry and mass spectrometry, ChemistryOpen, 2020, 9, 568–572.
[100] Guima K-E, Alencar LM, Da Silva GC, Trindade MAG, Martins CA, 3D-printed electrolyzer for the conversion of glycerol into tartronate on Pd nanocubes, ACS Sustain Chem Eng, 2018, 6, 1202–1207.
[101] Verma S, Lu S, Kenis PJA, Co-electrolysis of CO2 and glycerol as a pathway to carbon chemicals with improved technoeconomics due to low electricity consumption, Nat Energy, 2019, 4, 466–474.
[102] Leow WR, Lum Y, Ozden A et al., Chloride-mediated selective electrosynthesis of ethylene and propylene oxides at high current density, Science, 2020, 368, 1228–1233.
[103] Montgomery DC, Design and Analysis of Experiments, 8th ed., Hoboken, NJ, Wiley, 2013.
[104] Lee R, Statistical design of experiments for screening and optimization, Chem Ing Tech, 2019, 91, 191–200.
[105] Box GE, Hunter JS, The 2 k – p fractional factorial designs, Technometrics, 1961, 3, 311–351.
[106] Box GEP, Behnken DW, Some new three level designs for the study of quantitative variables, Technometrics, 1960, 2, 455–475.
[107] Box GEP, Hunter JS, Multi-factor experimental designs for exploring response surfaces, Ann Math Stat, 1957, 28, 195–241.
[108] Soravia S, Quality Engineering mit statistischer Versuchsmethodik, Chem Ing Tech, 1996, 68, 71–82.
[109] Hielscher MM, Gleede B, Waldvogel SR, Get into flow: design of experiments as key technique in the optimization of anodic dehydrogenative C,C cross-coupling reaction of phenols in flow electrolyzers, Electrochim Acta, 2021, 368, 137420.
[110] Endrődi B, Kecsenovity E, Samu A et al., Multilayer electrolyzer stack converts carbon dioxide to gas products at high pressure with high efficiency, ACS Energy Lett, 2019, 4, 1770–1777.

D. V. Ravi Kumar, Suneha Patil and Amol A. Kulkarni

3 Continuous flow methods for synthesis of functional materials

This chapter aims at exploring the principles and practices of continuous flow synthesis of different materials. Continuous flow synthesis of materials is quite different from flow synthesis of organic compounds, since it involves a number of factors such as formation of a particulate phase, nucleation-growth kinetics, size and shape control, and separation of the particles (size as well as shape dependent), followed by daunting challenges of clogging of channels and incurring heavy material losses due to wall deposition. This chapter discusses the protocols for flow synthesis of various materials, namely, metal and metal-oxide nanoparticles (NPs), semiconductors, quantum dots (QDs), porous materials, catalysts, and novel functional materials such as nanohybrids, 2D materials, metal/covalent organic frameworks (MOFs), and others. This chapter also gives contemplative information on the possibility of high-throughput flow synthesis and screening of materials. Finally, we briefly cover the challenges associated with flow synthesis of materials and methods to overcome them, along with future directions where this field can be explored further.

3.1 Introduction

Functional materials have made an immense impact on the quality of human life – directly or indirectly – through various applications such as coatings on solar cells, surface polishing powders, communication devices, energy storage devices, light emitting devices, reflecting mirrors, currency notes, precious metals and their various forms and compositions, nanofluids, cosmetics, diagnostics, paints and coatings, stain-resistant fabrics, biomimetic colorants, catalysts, displays, drag-reducing lubricants, adsorbents for chromatography, lightweight materials, conducing inks, and so on. Most often, these materials are valued based on their functionality and the quality of their performances.

The performance as well as functionality of these materials depends on the dimensions of the material, namely, size and shape of the particulate matter and thickness of the film. In case of porous materials, it depends on pore size and surface area. Thus, attaining the desired dimensions consistently through controlled synthesis is the key to retain the properties of these materials. One of the best methods

Acknowledgment: AAK and SP acknowledge the funding from the Dept. of Science and Technology (GoI)'s Advanced Manufacturing Technologies (AMT) scheme.

https://doi.org/10.1515/9783110693690-003

to achieve such a consistency in their properties is through wet chemical synthesis and that too only by following a fully automated batch/semibatch protocol or through a continuous process. Although flow synthesis of materials comes with an exciting approach coupled with several advantages, it is also a relatively challenging task, primarily because of the phase change that occurs in the form of crystallization or precipitation during the synthesis. Furthermore, it involves initiation, that is, nucleation and the process of attaining equilibrium, that is, growth. Here, we provide a brief outline of how material synthesis in flow is different from the organic flow synthesis, followed by several examples of successful flow synthesis at laboratory scale.

3.1.1 Flow synthesis of materials and difference from the typical flow synthesis of organic compounds

It is important to identify some of the key differences in the implementation of flow protocols for the synthesis of organic compounds and inorganic materials. First and foremost, it is important to understand that flow synthesis of inorganic materials is not a completely homogeneous system, because it involves the formation of particulate matter. Due to this, the flow-based syntheses of inorganic materials mainly focus on uniform size distribution, consistent particle shapes/aspect ratio, minimizing the fouling of reactor, cleaning, avoiding channel clogging, and so on. These factors, in addition to effective mass and heat transfer in the reactor, narrow the residence time distribution, channel geometry, and so on, also play an important role in the continuous flow organic transformations. Unlike most of the inorganic materials syntheses, organic transformations can be a single-step process or multistep process needing a multistep flow synthesis approach, depending on the product under consideration. Moreover, minimal solvent usage, handling low reaction volumes, high conversions, and yields in the continuous flow approaches, can effectively address the challenges that are generally associated with organic transformations, including safety. Recent reports emphasize green and sustainable technologies, even in the inorganic materials synthesis, by choosing supercritical fluids, ionic liquids, and even deep eutectic solvents, as reaction mediums. Although some basic differences in the flow methodologies for material synthesis and organic synthesis exist, the overall picture aims at sustainable process development.

This chapter mainly discusses about the principles and practices of flow synthesis of various functional materials like metals, metal oxides, micro-, nano-, and ultra-small NPs, porous materials, followed by high-throughput flow synthesis. This chapter also offers an overview of the methods of continuous flow synthesis of various functional materials at different scales. It also discusses the challenges in flow synthesis of materials and the possible methods to overcome the same.

3.1.1.1 Classifications: size, shape, and form of materials

The micro- and nanomaterials in solid phase that are synthesized in flow are analogous to the batch mode. These materials can be classified in different ways owing to their size ranges, shapes, and forms of material. In general, based on the sizes, they have characteristic length scales, that is, particle size, aspect ratio, pore size, film thickness, and so on. The other approach to classify these materials is based on shapes, that is, the particles/materials can be spherical, cylindrical, cubical, ellipsoidal, toroidal, and so on. This classification can also happen on the basis of pore specifications for porous materials. Classification of various functional materials based on their form and shape is described in Fig. 3.1.

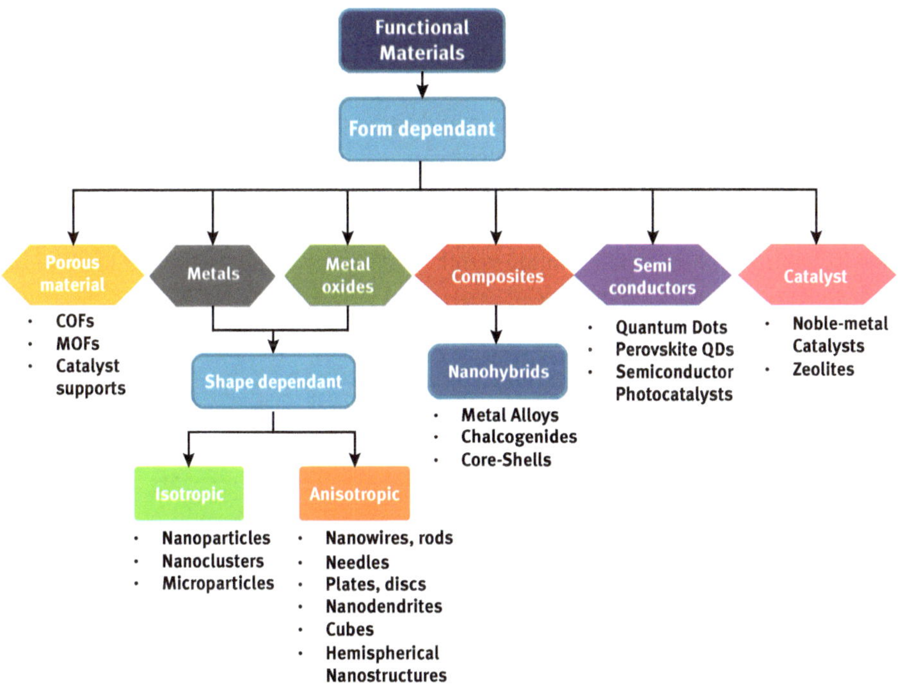

Fig. 3.1: Overall classification of functional materials.

3.1.2 Material synthesis approach in flow

The approach for flow synthesis of materials primarily depends on the material of choice and its properties, synthesis procedure, kinetics and thermodynamic data, derived for the batch process. Here, we briefly provide an overview of the strategies for flow synthesis of materials. In the case of ultra-small NPs/metal nanoclusters, arresting the particles growth is an important criterion, thereby leading to the incorporation of in-line quenching in the synthesis protocol, in addition to successful separation of nucleation and growth phases in the flow process. Whereas synthesis of metal oxide NPs, namely silica, needs separate growth period, which is due to the slow sol–gel process. In addition to the sol–gel processes, hydrothermal/solvothermal methods also need longer reaction times for the synthesis of zeolites and MOFs. Hence, for successful flow transformation of the batch hydrothermal synthesis, the thermal lag has to be minimized without compromising the quality of the crystallization process as well as without letting any evaporation in that section. Moreover, the crystallization, resulting from supersaturation, is high in solids content, which is capable of blocking the channels, and it is important to handle the crystallization meticulously. In shape- and size-controlled synthesis to produce metal NPs, axial dispersion that occurs due to the fluid-velocity difference at the wall and center of the channel has to be eliminated. This limitation can be successfully overcome by converting the continuous flow process to "batch in flow" synthesis, that is, droplet-based synthesis or by introducing a modification in the channel geometry. The droplet flow strategy also addresses the fouling of channel, occurring due to particle deposition. Similar strategy can also be used to generate emulsion and curing it by UV/thermal treatment to produce polymer microparticles. Additionally, core–shell structures and semiconductor QD synthesis often involves multistep synthesis/heating. Multistep microfluidic synthesis strategy could be the apt method for the production of such materials.

3.2 Protocols for flow synthesis of materials and various examples

Flow synthesis enables fine-tuning the shape and size of materials in desired form. Reactions can be flexible with respect to a number of factors including but not limited to precursors, solvents, and additives, and aqueous and organic environments, which sometimes are not compatible with batch. The objective of this section is to give an overview regarding the methods for synthesizing various high-value materials in continuous flow. We discuss a variety of experimental protocols and reactors used for the synthesis of comprehensively studied materials in their nano/micro/atomic cluster forms (with examples listed from literature). Furthermore, we also discuss the flow synthesis of semiconductor materials such as QDs and core–shells via multistep synthesis route or hydrothermal flow synthesis route (for graphene QD (GQD) and carbon QD (CQD)), Semiconductor photocatalysts, and perovskite QDs. An overview of nanohybrids is also given to elaborate the synthesis of other materials such as 2D materials, catalysts (noble metals, zeolites) and porous materials, namely, catalyst supports, COFs, and MOFs at the end of this section.

3.2.1 Flow synthesis of metal, metal oxides, and silica particles

Utility of flow reactors for material synthesis yields a versatile range of products such as metal NPs, non-metals, metal oxides, semiconductors, and salt nanoparticles, as well as core–shells, dioxides, and composites. Among these, the most extensively studied materials are metals, metal oxides in their nano- and microforms, and a special focus on silica, primarily because of its ability to remain functional in different physical forms.

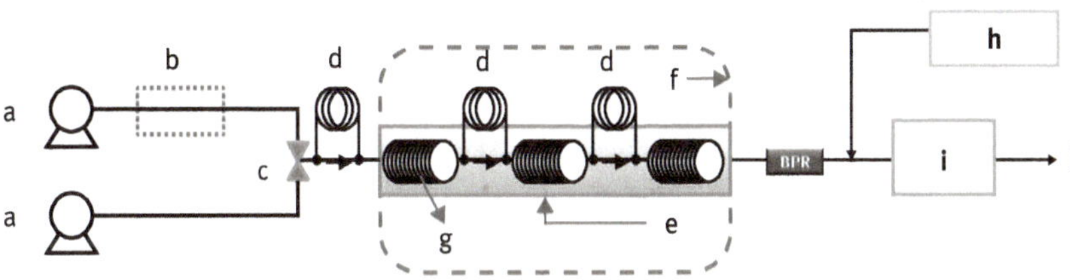

a. **Pumps** (Syringe/Peristatic/Piston/Diaphragm)
b. **Preheater** (If required)
c. **Micro-mixer** (T/Y/Cross)
d. **Reaction loop** (For addition of reagents/multistep synthesis)
e. **Flow Reactor** (helical coil/microchannel reactor/Oscillatory baffled reactor/coiled flow inverter)
f. **Reaction medium** (Heating/cooling/Hydrothermal/Microwave/oscillatory/Electrochemistry/Photochemistry/Light/Ultrasound)
g. **Coil Reactor**
h. **Quenching** (Seperation/Work-up)
i. **Inline analysis**
i. **Product Collection**

Fig. 3.2: A representative flow chemistry setup for material synthesis.

3.2.1.1 Metals

Continuous flow systems are employed for the synthesis of metal NPs with different sizes or shapes due to their precise control, rapid heat and mass transfer, high mixing efficacy, large reaction interfaces, and compatibility with online analysis. Due to their size-dependent applications, it is necessary to achieve uniform size distribution of particles, which can be achieved in flow synthesis. Figure 3.2 represents a typical flow chemistry setup utilized for synthesis of various organic and inorganic compounds. A number of metals and nanomaterials, namely, Ag, Au, Pt, Pd, Ni, Cu, Co, Bi, Ru, Rh, Fe, Zn, and Ti have been synthesized using the continuous flow technology. For noble metal NPs, the ionic precursor such as $HAuCl_4$, $AgNO_3$, or Pt(acac)$_2$ is reduced by the chemical reduction of a metal-salt solution with the aid of commonly used reducing agents such as $NaBH_4$, sodium citrate, anti-cyclic acid, a mixture of lithium hydrotriethyl borate (Li [Bet3H]) with 3 (*N*,*N*-dimethyldodecylammonia) propane sulfonate (SB12), N_2H_4, and so on [8]. Additionally, stabilizers such as CTAB, PVP of different

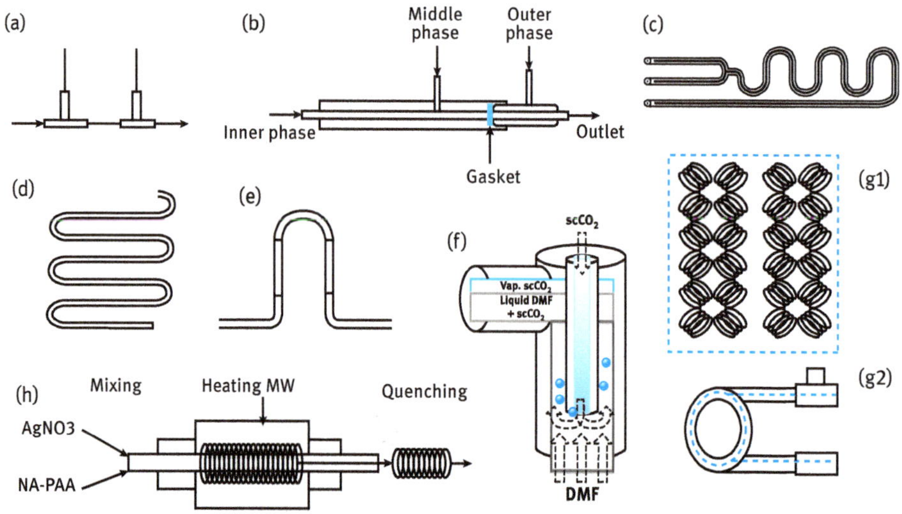

Fig. 3.3: Schematics of reactors and processes for synthesis of various materials; (a) and (b) schematics of the plug-and-play microfluidic reactors [1]; (c) microfluidic reactor used for the synthesis of ultra-small copper nanoclusters [2]; (d) oscillatory baffled flow reactor used for the synthesis of zeolite NaX [3]; (e) oscillatory baffled reactor [4]; (f) schematic of counter-current micromixing in continuous flow microreactor for the synthesis of MOF using supercritical CO_2 [5]; (g1) and (g2) tube in tube contactor and coiled flow inverter reactor used for the synthesis of gold nanoclusters [6]; and (h) schematic representation of a reactor used in experimental setup for the continuous flow synthesis of Ag nanoclusters [7].

molecular weights, tetraoctyl ammonium bromide, oleic acid, oleylamine, functionalized thiols, sodium dodecyl sulfate, and fatty amines are used in the reaction in order to increase the stability and monodispersity of NPs. It is also possible to obtain ultra-small NPs without using strong capping agents with a very narrow size distribution in continuous flow. For example, Huang et al. [9] obtained AuNPs of the size 1.9 ± 0.2 nm in 30 min using a capillary-based continuous flow system (PTFE capillary tubing ($d = 0.3$ mm)). They also demonstrated the effect of capillary wall surface on the diameter by synthesizing citrate-capped larger diameter AuNPs (17.4 ± 1.4 nm) in a segmented flow, using octane to isolate the reactants from the tubing wall. Segmented flow is often used in nanomaterial synthesis in order to achieve monodispersity [10]. Cabeza et al. [10] synthesized AuNPs of diameter 2.8 nm ($\sigma = 0.2$ nm) in merely 10s using chloroauric acid ($HAuCl_4$) with rapid reduction by sodium borohydride ($NaBH_4$) and air as the segmenting fluid. Influence of channel wettability has been demonstrated such that segmented flow microfluidic reactors achieve an excellent control of reactant diffusion, thus generating monodisperse distribution of NPs. This is also important as segmented flow reduces the axial dispersion, which otherwise gives a wider particle size distribution. Similarly, Cela et al. [11] were able to synthesize monodisperse-branched AuNPs in a droplet-based microfluidics platform, in a highly controllable and reproducible manner. Their automated process used microdroplets as microreactors for epitaxial growth of gold nanostars and was capable of adapting to a surfactant-free as well as surfactant-assisted (PVP-based) synthesis. The fine control of reagent mixing and local concentrations during particle formation led to a very low PDI (polydispersity index), in the range of 0.0005–0.002, thus indicating a nearly monodispersed product.

Tao et al. [12] synthesized spherical AgNPs (particle size (1.5–5.6 nm)) over reduced graphene oxide by carrying out the co-reduction of $AgNO_3$ and graphene oxide with $NaBH_4$, confined inside the dispersed aqueous plugs, segmented by octane. This microfluidics-based strategy was further extended for the continuous synthesis of Pt- (1.7 ± 1.0 nm) and Pd- (3.2 ± 1.1 nm) graphene oxide composites as well. On the other hand, Wojnicki et al. [13] employed sequential flow in droplet reactors (different from segmented flow, since the droplets do not attach to the channel surface) for synthesis of AgNPs. This led to a reduction in diameter of AgNPs from 4.8 ± 1.3 nm (batch) to 2.5 ± 0.5 nm (flow) and a narrower size distribution as well. Karim et al. [14] were able to produce PdNPs with diameter as low as 1 nm, by combining microfluidics with insitu analysis using small-angle X-ray scattering and X-ray absorption fine structure spectroscopy. The authors also discuss the role of a strong binding ligand and the change in metal and capping agent coordination that leads to change in the growth rate during synthesis, thus leading to narrow size distribution.

Wang et al. [15] demonstrated shape-controlled synthesis of Au nanostructures under controlled electron influence, liquid flow rate, and Au^{3+} ion supply, using liquid-cell transmission electron microscopy (LCTEM). They emphasize the importance of liquid layer thickness in affecting Au^{3+} mass transfer/diffusion and diffusion-limited growth into branched Au structures, by showing that the multitwinned nascent Au seed particles formed branched structures in thin liquid cells (100 and 250 nm) and faceted structures (e.g., spheres, rods, and prisms) in a 1 μm thick liquid cell. Similarly, Duraiswamy et al. [16] manipulated the shape of Au nanostructures into spherical–spheroidal, rod-shaped, and sharp-edged, along with excellent size and spectral tunability, by influencing reagent concentrations and feed rates in the microfluidic device. Trzcinski et al. [17] have reported a systematic study to precisely engineer AuNPs.

Predictive continuous production of shape-controlled complex NPs for bimetallic structures is also possible using microfluidic systems. For example, Au–Pd NPs, ranging from sharp branched octopods to core–shell octahedral were produced by Santana et al. [18] by manipulating the reagent flow rates in a droplet microreactor. Furthermore, monodisperse Pd NPs are produced in two-phase systems in a variety of shapes such as cubes [19], octahedral [20], plates, rods, and even dendritic particles of bimetallic Pd–Pt [21]. Alternatively, single-phase microfluidic reactor can also be used to synthesize Pd nanocubes and Pd–Pt core–shell NPs [22] in order to eliminate the time-consuming separation of the two phases (required in segmented/two-phase systems).

Other unique reactors such as the continuous Couette–Taylor reactor have been used for controlling the morphology of CuS NPs, where Tang et al. [23] were able to synthesize a range of shapes from nanofibers to hexagonal nanoplates by controlling the mean residence time and the feed solution concentration. Roberts et al. [24] synthesized NiNPs in a millifluidic reactor in higher yields and a narrow size distribution, with its catalytic activity equivalent to those produced in batch. Researchers have even used continuous flow stirred-tank reactors for synthesis of silver plates [25] as well as Ag NP [26] and studied the influences of agitation speed and molar ratio of $AgNO_3$ to the etching agent to demonstrate the possibility of mass production of other anisotropic nanostructures. Similarly, continuous stirred tank reactor (CSTR) is employed for the production of Cu nanopowder with minimal production costs and the resulting NPs are purified by a controlled flocculation process [27]. The authors found that in the case of increased particle loadings, sodium citrate is an effective stabilizer as compared to polymeric stabilizers. Metallic bismuth NPs have been synthesized in continuous flow, coupled with sonochemistry [28]. The authors have found that using green ultrasound assisted synthesis not only improves productivity, but also a smaller hydrodynamic size is obtained when compared to the batch process. Other novel methods such as ultraspray pyrolysis (USP) have also been used to produce Pt NPs in continuous flow [29]. The authors claim that preparation of materials similar to complex, nanostructured Pt/C can be scaled up effectively using USP in flow.

3.2.1.2 Microparticles to atomic clusters

Microparticle synthesis in a continuous flow system involves the generation of microdroplets from the emulsion formed by two/more immiscible liquids, either by dripping or flow focusing, followed by chemical processes (chemical/thermal/UV cross-linking/hydrolysis/oxidation). The microfluidic synthesis of microparticles results in highly monodispersed particles with a variation coefficient <3%. In the synthesis of microparticles, calculation of two important dimensionless numbers, that is, capillary number (*Ca*) and Damköhler number (*Da*) give an idea about the flow regime, the relative rates of reaction, and the mass transfer. $Ca < 1$ results in the controlled break up of droplets, while in the case of $Ca > 1$, there is a formation of extended jetting due to the dominating viscous forces. Similarly, *Da* compares the relative values of residence time and reaction times, which are to be used to select the flow regime, based on the material to be synthesized. For example, $Da < 1$ is suitable for the synthesis of TiO_2 microparticles, which avoids channel clogging and in the reactions that involve cross-linking and photopolymerization, while for fast mixing mass transfer-limited regime, $Da > 1$ is preferred. Compared to conventional microfluidic reactors whose fabrication is costly and a time-consuming process, a simple "plug-and-play" kind of tubular coil (in various forms, namely, helical coil, expanding or contracting spiral shape, coil flow inverter, serpentine, etc.) can be used as a suitable system for the synthesis of microparticles. Detailed tutorial of plug-and-play reactor construction and the various microparticles synthesized using this can be found in the literature. Figure 3.3(a) and (b) represents the schematics of the commonly used plug-and-play reactors.

Among metal oxides, TiO_2 microparticles were synthesized using such simple systems, where titanium(IV) *n*-butoxide in different organic solvents such as octanoic acid were used as dispersed phase and water as the continuous phase. Titanium precursors undergo hydrolysis at the oil–water interface and the TiO_2 shell forms as a result of hydrolysis. The generated microparticles need enough time for complete hydrolysis and they are then subjected to calcination to crystallize. Other strategies such as replacing water with polar aprotic formamide as the continuous phase were also explored to synthesize york–shell TiO_2 microparticles. In this strategy, the microparticles generated from the reactor were subjected to particle hydrolysis and then cured via photopolymerization. Controlling the titanium composition, a library of hollow york–shell macroporous TiO_2 microparticles with high surface area were generated. In addition to TiO_2, other metal oxide microparticles such as ZnO and copper sulfide were also explored in flow methods. This plug-and-play microreaction strategy can also be used to synthesize MOF particles as well as liquid metal microparticles.

(For further details on Damköhler number please see Volume 1, Chapter 1, Title: Fundamentals of Flow Chemistry) i

---

3.2.1.3 Nanoclusters/ultra-small nanoparticles

In addition to microparticles, it is also possible to synthesize ultra-small NPs/nanoclusters by microfluidic system. Earlier report on the ultra-small nanoparticle synthesis in microfluidic approach demonstrated the synthesis of Cu nanoclusters of size ~2 nm. Synthesis was carried out in PMMA microfluidic chip at different flow rates (i.e., 6.8, 14.3, 32.7, and 51.4 mL h^{-1}) and the corresponding residence times (47.49, 24.44, 16.56, and 9.02 s). The Cu nanoclusters were obtained by reacting Cu (II) nitrate solution, polymeric surfactant O-[2-(3-mercaptopropionyl-amino)ethyl]-O′-methylpolyethylene glycol (MPEG), and $NaBH_4$ solutions at higher pH, which were stable for more than three months under inert conditions.

Synthesis of gold NPs (AuNPs) is carried out by Turkevich method using the slow reducing agent citrate. The slow nucleation rate cannot facilitate ultra-small AuNPs in the batch process. However, an interesting electrostatic play between the positively charged Au(III) complex and the negatively charged wall of microreactor makes the synthesis of ultra-small AuNPs feasible in the microfluidic system by using the Turkevich method. It is important to note that the rate of nucleation (and hence the reduction in the size of NPs) increases if the local concentration of Au(III) complex concentration increases. It can be done by passing a pre-mixed solution of Au(III)-citrate solution to the microreactor tubing with negative surface potential on the wall surface. Hence, an electrostatic attraction can be expected that leads to Au (III) film formation on the surface of wall, which increases the local concentration of Au(III) because of high surface-to-volume ratio of the microreactor. This leads to an increase in the rate of nucleation and ultimately results in the formation of ultra-small AuNPs [91].

Moreover, synthesis of PVP-stabilized Au, Pd, and bimetallic Au–Pd ultra-small NPs of size 1–2 nm was demonstrated by microfluidic method by Tofighi et al. [30], where the reactants were delivered with a flow rate of 2.6 L h^{-1} from a highly pressurized fluid delivery racks to three cyclone micromixers, that were integrated in a microfluidic chip for rapid mixing. The above mentioned flow rate resulted in a *Re* of 2400 with turbulent mixing within 2 ms. $HAuCl_4$, K_2PdCl_4 were used as metal precursors and $NaBH_4$ was used as a reducing agent along with PVP as the capping agent.

Furthermore, a combination of microwave and continuous flow technologies was adopted to synthesize Ag nanoclusters. In this method, $AgNO_3$ was used as Ag precursor and sodium salt of poly (acrylic acid) [Na-PAA] was used as a reducing agent as well as a capping agent [7]. Microwave heating technology provided rapid heating rate, resulting in homogeneous nucleation, in which the heating time reduced and uniform cluster sizes was obtained. The improvement in the temperature gradients by a combination of microwave technology with continuous flow also resulted in an increase in the rate of heating, mixing, and quenching. The whole reactor setup contained three different zones, all connected by PTFE tubing. The first mixing zone comprised of a simple Y-micromixer that was connected to quartz helical coil, which was placed in the center of microwave cavity that acts as second zone, that is, heating/reaction zone. PTFE coil, immersed in the ice bath, was used as the quenching zone (third zone). This methodology successfully produced Ag nanoclusters of size <2 nm, which were further supported by SBA-15 matrix.

Atomically precise Au nanoclusters with defined number of gold atoms were synthesized using microfluidic system by Haung et al. [6] using CO. Interestingly, CO-saturated heptane was used and liquid–liquid segmented flow was generated instead of bubbling CO directly into the reaction system, to generate gas–liquid segmented flow. This strategy prevents the leakage of CO, ensures the safety of the process, and also helps contain the toxicity from getting exposed to CO. Gold precursor was used in the aqueous system (which also contained cysteine to cap/stabilize the clusters) and the transfer of CO from heptane to aqueous phase happened through the contact of liquid slugs generated in the liquid–liquid segmented flow. The experimental setup shown in Fig. 3.3g1–g2 consists of a combination of tube-in-tube contractor at room temperature (a 2 m long Teflon tube of 0.8 mm inner diameter (ID), and 1.0 mm outer diameter (OD) as the contacting interface placed inside a polytetrafluoroethylene (PTFE) tube (2.4 mm ID, 3.2 mm OD)) and coiled flow inverter as the reactor (fluorinated ethylene propylene tubing with 1 mm ID with 100 coils, each coil with 1 cm diameter), which was submerged in glycerol bath plated on hot plate. Contractor tube is used to saturate the heptane by CO, which is pressurized at the annulus of the inner tube and the outer tube. The product is precipitated with acetone and washed, which is stable up to 1 month.

3.2.1.4 Metal oxides and chalcogenides

Metal oxide NPs are usually synthesized via sol–gel methods, hydrothermal, or microwave assisted synthesis. For example, for the production of very fine nanoscale TiO_2 NPs (~6 nm), Anwar et al. [9] used continuous microwave-assisted flow synthesis, which yielded smaller NPs than that obtained by the sol–gel (~9 nm) and the chemical precipitation (~15 nm) methods. Similarly, CuO NPs [31] with controlled size were produced in a microwave-assisted continuous flow synthesis. Methods such as thermal decomposition have also been adapted to flow for the preparation of magnetite (Fe_3O_4) [32] and iron oxide NPs [33]. A comprehensive study over the years has been carried out to synthesize ZnO nano/microstructures in a variety of forms, such as wire, rod, sphere, flower, sheet, flake, spindle, and ellipsoid [34]. These are synthesized in a controlled manner, in terms of shape and size, employing a number of reactors, namely microchamber [35] Y-mixer [36], T-mixer [37], co-flowing reactor [38], membrane dispersion reactor [39], straight microchannel [40], and spiral channel [41], etc.

Recently, hydrothermal flow synthesis method is also being used to synthesize and fine-tune specific morphologies of metal oxide NPs. For example, Beyer et al. [42] were able to synthesize titania (rod-like rutile crystallites) within a specific aspect ratio of ~3.5 using hydrothermal synthesis in a supercritical continuous flow reactor, whereas Fabra et al. [43] used a pipe in a pipe-continuous reactor system to obtain nanoplates of tungsten(VI) oxide (WO_3). Another example where hydrothermal flow synthesis was carried out in a supercritical system was the production of SnO_2 NPs, where Mamakhel et al. [44] reduced the batch time of several hours to a mere few seconds in flow. Moreover, they observed no metal impurities due to reactor corrosion and were able to maintain the continuous reactor operation without clogging.

Elevated temperatures, pressure conditions, and slurry generation associated with hydro/solvothermal processes make the transition from batch to flow, a tedious task [32]. Another alternative to hydrothermal synthesis, as reported by Licona et al. [45], is the coiled flow inverter reactor, which offers flexibility in terms of production and product quality, provided it meets the mixing requirement of the reaction. The reactor turned out to be an excellent choice to obtain monodisperse ZnO NPs (PDI < 0.21). In other cases such as $BaSO_4$ NPs [46], segmented microfluidics is used to obtain a monodisperse product and avoid reactor choking. Modified segmented microreactors are employed for the production of shape-controlled and monodisperse metal chalcogenides such as CdSe NPs, where the nucleation and growth stage is separated [47], whereas confined impinging jet microreactors are used for obtaining stable monodisperse solutions of Ta-doped tin dioxides and cadmium sulfide NPs and MgO [48]. Thus, reactions should be considered owing to their unique requirements and the pros and cons of adapting the specialized reactions to flow should be carefully weighed.

3.2.1.5 Silica

Porous and solid silica materials offer good biocompatibility, high thermal and mechanical stability, and ease of surface functionalization. These properties have led to its numerous applications, ranging from drug delivery and catalysis to water treatment.

A simple laminar flow microreactor with two inlets is capable of synthesizing colloidal solid silica spheres, given the proper mean residence time and reaction conditions. A number of microreactors such as winding [49], Y junction [50], coflow [51], spiral [52], cross-junction baffled mixer for efficient mixing at high flow rates [53] and slit interdigital [54] have been used for the synthesis of solid SiO_2 spheres in the range of 30–825 nm in laminar flow reactors, whereas in the case of discrete segment and droplet microreactors, the diameter of solid spheres that are synthesized can be as low as 50 nm or as large as ~3 mm, just by varying the flowrates of continuous and dispersed phases [55]. Uniform-sized porous spherical silica are synthesized in laminar flow reactors by realizing nucleation and growth, a two-step process, whereas in the case of nonspherical silica such as hollow ellipsoids, fibers, flowers, sheets, and triangles, spiral-shaped microreactors first fabricated by the Hao group [55] with two-inlet flows are used. These spiral shaped reactors are flexible enough for the synthesis of anisotropic multifunctional hierarchical structures and mesoporous silica nanofibers [56] with tunable aspect ratios as well.

Furthermore, a series of non-spherical silica materials, including doughnut [57], raspberry [58], filbert shape [59], and disk shape [60] can be synthesized in discrete segment and droplet flow reactors. For example, Tamtaji et al. [61] used a droplet reactor for the flow synthesis of amorphous plate-like silica particles, using the sol–gel method. For the fabrication of hierarchical nanocomposites such as TiO_2–SiO_2, Fe_3O_4–SiO_2–Pt, and Co_3O_4–SiO_2, a multistep nucleation-controlled growth in laminar flow is employed, whereas in the case of droplet flow synthesis, inorganic–organic and inorganic–inorganic composites such as silica-magnetic composites and silica-titania core–shells can be fabricated by precisely controlling the emulsification process. Moreover, using gas–liquid segmented reactors, a number of core–shells, namely, SiO_2@TiO_2, SiO_2@Au, and Au@SiO_2 can also be produced.

3.2.2 Nanohybrids

A composite of two or more materials with at least one of the composite materials in the nanometer dimension is known as a nanohybrid. In the synthesis of nanohybrids, controlling the morphology and the component composition are very important. Although the choice and methodology of the microfluidic system depend on the material to be synthesized, for the synthesis of nanohybrids, multiphase/integrated microfluidic systems are identified as favorable choices.

In the case of oxides and other chalcogenide-based nanohybrids, co-axial flow reactors and plug flow reactors coupled with magnetic field, multistep synthesis methods were opted to mainly synthesize γ-Fe_2O_3–SiO_2, TiO_2–SiO_2. It is important to mention that silica is the common oxide present in these hybrids. Due to the passive surface nature of silica, surface coverage and shell thickness of the other metal oxide can be easily controlled. For chalcogenide-based nanohybrids such as CdS–ZnS, CdSe–ZnS, ZnSe–ZnS, and InP–ZnS, multistep microfluidic synthesis was chosen. A hybrid system of microfluidic reactor with a batch cooling system was used to fabricate hybrid materials such as $FeAl@Al_{(1-x)}Fe_xO_y$, $CoZn@Zn_{(1-x)}Co_xOy$, $AgAl@Al_{(1-x)}Ag_xO_y$, and $AuZn@Zn_{(1-x)}Au_xO_y$.

3.2.3 Two-dimensional materials

Two-dimensional (2D) materials are materials with ultrathin atomic thickness (1–100 nm) and consist of a single or few layers. A classification of 2D materials is given in Fig. 3.4. Two-dimensional materials are functionalized by covalent and noncovalent improvement techniques in order to explore their physicochemical properties and make them fit for processing in solvent-assisted techniques such as spin coating, filtration, and layer-by-layer assembly [62]. CHFS (continuous hydrothermal flow synthesis) is used for covalent functionalization since it uses water-soluble precursors at critical conditions and follows a greener route. A typical CHFS reactor has three feeds, namely water (preheated under set reaction temperature and pressure), precursor, and other required precursors (surface stabilizers, alkaline solution) that meet at a mixing point, that is, the reactor. The reaction precursors are heated up rapidly in the reactor and the resulting product is cooled at the base of the mixing point. A back-pressure regulator (BPR) connected before the outlet tubing, where the product is collected, ensures that an optimal pressure is maintained in the system [62]. Kellici et al. [63] functionalized a 2D-reduced graphene oxide with great controllability over oxygen functionalities and particle size using CHFS. Moreover, CHFS also allows uniform deposition of highly crystalline nanostructures into graphene and less explored 2D materials such as 2D nitrides. For instance, Saada et al. synthesized tin-doped zirconia/graphene [64] and ceria–zirconia oxide/graphene nanocomposite [65] by CHFS as catalysts for the synthesis of Dimethyl Carbonate. Promising results of materials synthesized via CHFS, such as increased activity of graphene-based CeZrLa nanocatalysts

[66], controlled and enhanced optical properties of GQDs [67], and rare blue luminescent graphitic core CQDs [68] show that CHFS is a reliable method for synthesis and surface functionalization of complex 2D materials. Although, CHFS is also used for production of other classes of 2D materials such as the layered double hydroxides (LDH) [69], different routes such as continuous flow co-precipitation technique [70], counter-current flow [71], column technique [72] and a number of continuous reactors namely stirred tank reactor [70], t-shaped micromixers [73] have also been employed to prepare LDH and its analogues in continuous flow. For example, Yaseneva et al. [74] developed doped LDH with higher surface areas in a meso-scale flow reactor using co-precipitation in flow. Furthermore, other reactors such as microdroplet reactors were used to produce 2D Zr MOF nanosheets [75] in flow and even reactors operating in intense micromixing such as vortex fluidic device have been used to synthesize spheroidal composites of C60@graphene with controlled diameters [76]. In spite of the excellent control over particle size in continuous methods, challenges such as high crystallinity and multicomponent hybrids synthesis persist.

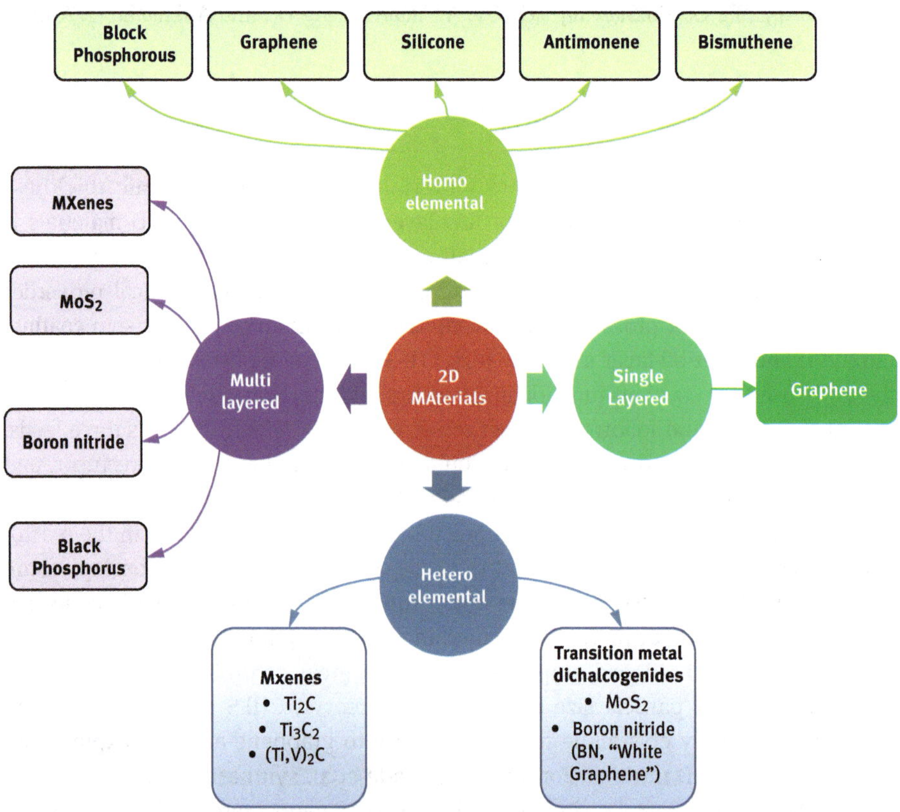

Fig. 3.4: Classification of 2D materials [62].

3.2.4 Catalysts

Synthesis of catalysts in continuous flow is one of the most important applications of flow synthesis, since a consistent product quality that can be obtained from continuous flow will ensure that further reactions carried out using the as-synthesized catalyst will have necessary controls. Of late, there have been significant efforts in the synthesis of various types of catalysts, primarily zeolites and NPs. In this section, we will focus more on how to achieve flow synthesis of zeolites in different forms.

Zeolites are crystalline aluminosilicates composed of alumina and silica tetrahedrons that form 3D porous structures. Conventional synthesis of zeolites follows hydrothermal treatment of a reactive gel composed of a structure-directing agent such that the sources of aluminum and silica are aged at low temperatures, followed by a high temperature reaction, where the products get synthesized in the form of a crystalline solid. For synthesis of catalysts or any other materials, a complete homogeneity of the reactants or a high-intensity mixing of reactants is an absolute necessity. Liu et al. [77] demonstrated it in the seed-free synthesis of ZSM-5 in a continuous flow reactor, where the reaction takes place in a few seconds using pressurized hot water as the heating medium. Direct mixing of a well-tuned precursor (90 °C) with the pressurized water preheated to extremely high temperature (370 °C) in the millimeter-sized continuous flow reactor resulted in immediate heating to high temperatures (240–300 °C) (Fig. 3.5a); consequently, the crystallization of ZSM-5 in a seed-free system proceeded to completion within tens of or several seconds. The direct contact heat transfer achieved by mixing of pressurized water at an extremely high temperature with the precursors has helped achieve instantaneous temperature increase that allowed extremely rapid kinetics and synthesis time, in the order of seconds. This implies that in addition to the conventional hydrothermal method or even the microwave heating of precursors, it is possible to use a third mode of direct contact heat transfer for facilitating catalyst synthesis in continuous manner. The crystallization rate of ZSM-5 (and also the crystallinity) was observed to be a strong function of reaction temperature and it was studied using [27] Al MAS NMR spectra of the products synthesized over 240–300 °C. However, the fastest crystallization rate was observed at a synthesis temperature of 260 °C and higher as well as at temperatures lower than 260 °C, the overall crystallization rate was slower.

It is not necessary that the zeolite is produced only in the particle form. In an interesting approach, Huang et al. [78] demonstrated the hydrothermal flow synthesis of Zeolite Socony Mobil-5 (ZSM-5) film in a quartz capillary microchannel (inner diameter of 0.53 and 0.32 mm). This was achieved by pre-coating the channel surface using ZSM-5 seed layer as active sites to induce film growth and the extent of growth was manipulated by changing the channel aspect ratio, seed coating, residence time, and the synthesis time (Fig. 3.5b). This is an amazing application of flow synthesis as

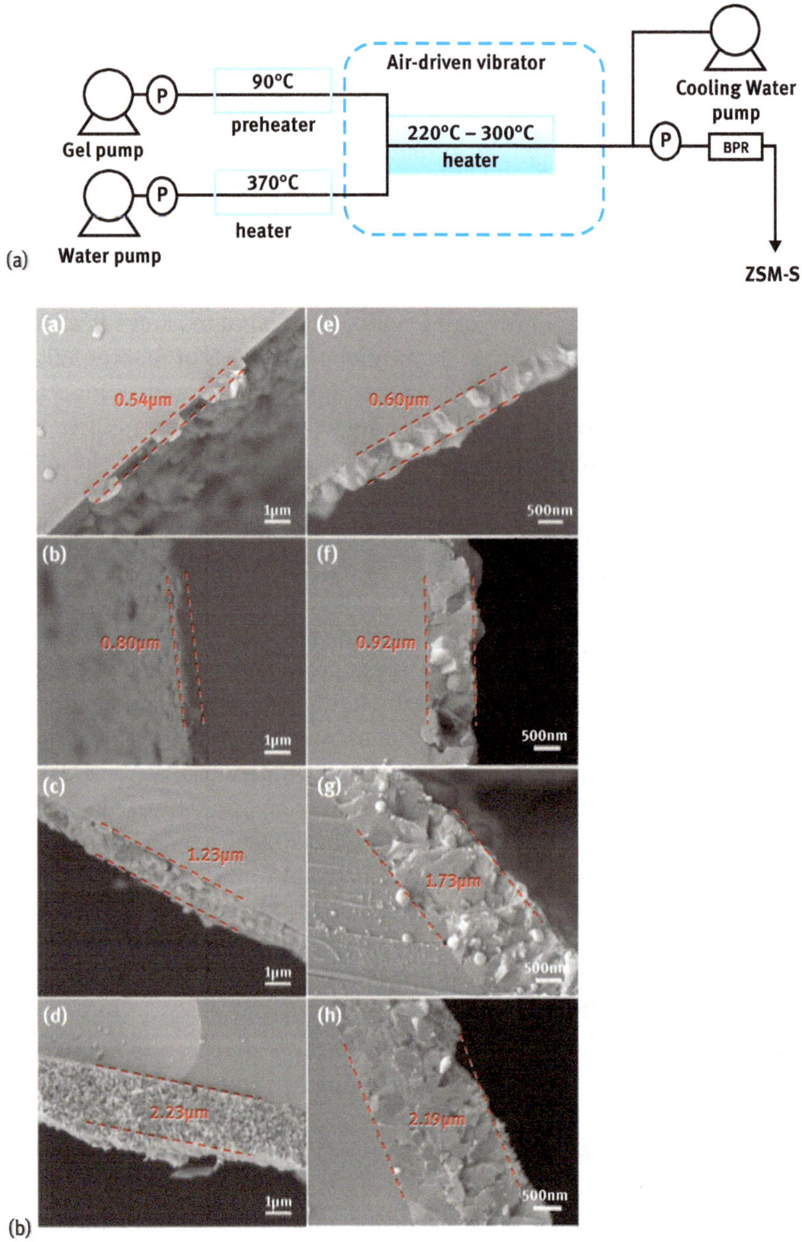

Fig. 3.5: (a) Setup for flow synthesis of ZSM-5 [77], (b) cross-sectional view of the SEM images of ZSM-5 film synthesized inside of a 0.53 mm capillary at different residence time (20, 24, 36, and 48 h) and in 0.32 mm capillary for (24, 36, 54, and 60 h) (reproduced from [78]).

it can be extended for catalyst impregnation on monoliths. The film thickness could be controlled by adjusting the feed flow rate and/or synthesis time.

Another approach for making zeolites is to subject the precursors to rapid heating using microwaves. Bukhari and Rohani (2016) demonstrated the conversion of coal fly ash into zeolite using a continuous tubular microwave reactor, which helps in drastically reducing the wastewater from the process [79]. The concentration of the cations was studied using inductively coupled plasma analysis method during and after each run to observe the concentration of the cations. Higher microwave irradiation resulted in a faster crystal growth; however, the product crystallization was seen to be a combined effect of microwave irradiation power (in W) and the residence time. For example, the authors reported identical crystallization at 810 and 335 W of microwave irradiation for 60 and 120 min, respectively. The system was seen to work very well even with recycled water and the product is shown in Fig. 3.6.

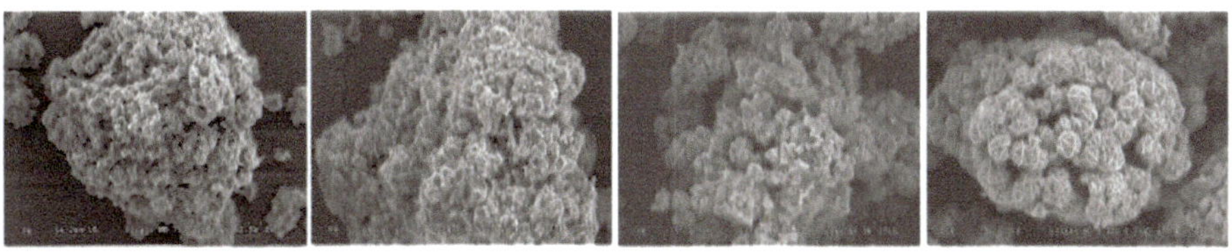

Fig. 3.6: SEM images of the zeolite coal fly ash using recycled water. Sequence from left to right: deionized water, first recycled water, second recycled water, and third recycled water (reproduced with permission from the author) [79].

Yoshioka et al. [80] have shown the flow synthesis of small-pore zeolites that feature eight-membered rings as the largest pore. Ultrafast synthesis was realized within 10 min through interzeolite conversion. FAU zeolite is used as a source of silicon and aluminum, and acid-leached seeds are added for promoting the crystallization of AFX zeolite. Such a method helps to eliminate formation of by-products, thereby facilitating the catalyst material. Interestingly, the approach helps to control the primary and secondary nucleation effects, thereby controlling the pore sizes. Beyond the conventional proof of concept, catalyst synthesis can also be achieved at larger scale using oscillatory baffle reactor. Mendoza et al. [3] demonstrated the continuous flow synthesis of zeolite NaX successfully at a pilot scale (up to 100 L h^{-1}) using continuous oscillatory baffled reactor (OBR) by an in-line mixing of reactants and seeds at high temperature [3]. Such systems help to have excellent control on spatial mixing, where one can change the number of baffles and their design spatially and also apply temperature profile, if needed. Such systems also avoid any fouling on the walls of the reactor and this proposes an excellent way to produce high density catalytic materials. Analogous approaches will also include compressed air-driven vibrator, or impinging jet reactors [81], or using simple CSTR.

A continuous flow reaction system with a number of components, namely, high pressure syringe pumps, BPR, SS tubular reactor (immersed in oil bath of 190/210 °C), was used for the rapid synthesis of industrially important silicoaluminophosphate zeolites SAPO-CHA, SAPO-AFI [82]. The precursor solution was passed through the continuous flow system and the residence time was varied by varying the flow-rates and the length of the tubular reactor. Here, it is important to mention that the precursor solution was prepared separately (Al and Si sources were added to phosphoric acid with organic ligands and stirred for few hours). It is interesting to note that the final zeolite product obtained in the flow process was dependent on the rate of crystallization that was dependent on the heating rate and temperature.

Slow and time-taking hydrothermal synthesis of zeolites can be successfully replaced by continuous flow methods in which the reaction rate is rapid and the reaction time decreases many folds as compared to the typical hydrothermal synthesis. Our discussion on the continuous flow synthesis of zeolites extends to another example, that is, ZSM-5 zeolite. In continuous flow, the reaction time was decreased by an instantaneous increase in temperature, which was facilitated by mixing preheated precursor solution (aluminoslilicate) with pressurized water heated at very high temperatures (370 °C) [77]. These solutions were dispensed to a millimeter-sized continuous flow reactor with the help of pumps. In this method, the crystallization of the zeolite (in the present case, ZSM-15) occurs immediately as soon as both the solutions mix without any thermal lag (which is the main limitation of hydrothermal synthesis) [77]. Another important challenge that is to be addressed is the hydrodynamic failure of continuous flow systems with high solid content. This limitation of the flow reactors, which finally leads to clogging of the reactor, was overcome by subjecting the continuous flow reactor to compressed air-driven vibrator, maintaining stable hydrodynamic conditions.

Handling the highly viscous nature of the precursor solution used for zeolite synthesis is another important challenge in the continuous flow synthesis of zeolites. It can be overcome by opting for emulsion-based continuous flow system in which the immiscible liquid is introduced to the emulsion. This emulsion-based continuous flow system was used in the synthesis of Erionite (ERI) zeolite, where the precursor emulsion solution containing hexamethonium bromide, alumina, silica, and KOH was injected into the Teflon-lined pipe system and cyclohexane acted as the continuous phase in the emulsion. It is subjected to heating by a feed of pressurized hot water from the inlet. Fully crystalline ERI zeolite of particle size 600 nm was successfully obtained at 240 °C after 7 min reaction time.

3.2.5 Porous materials

Porous materials are another important class of materials. In the synthesis of porous materials like zeolites, MOF, and other allied materials, the mechanism was less understood even in batch processes. The following examples discuss the novel methodologies/strategies that have been successful in implementing the flow synthesis of porous materials.

A combination of continuous flow OBR (CF-OBR) with microwave heating was demonstrated as an efficient method for the process intensification for the synthesis of MOF (HKUST-1), with a production rate of ~97 g h^{-1}. CF-OBR is generally used for the process intensification of many laboratory reactions. CF-OBR is a tubular reactor with transversely placed baffle plates to oscillated flow. The baffle plates interrupt the boundary wall and the oscillated flow generated by the pump (used on one side or both the sides) creates vortices to improve the mixing. When CF-OBR is integrated with microwave heating, it could create an ideal reactor for process intensification. A typical CF-OBR is shown in Fig. 3.3e. The experimental setup rig is used to synthesize the well-studied MOF, that is, HKUST-1. Two feed stock solutions (one consists of Cu salt solution and another of linker, made from trimesic acid) were dispensed at equal flow rates, thereby achieving a residence time of 2.2–44.4 s in the microwave heating. A third feed stock was used to generate oscillation, which results in the highest space–time yield (STY) of 6.32×10^5 kg m^{-3} day^{-1}.

Similar to the above example, the concept of integrating microwave heating with continuous flow synthesis was also used for the synthesis of MOF MIL-100 (Fe). Coiled PTFE tubular reactor of length 20 m and 2 mm inner diameter was used as a continuous flow reactor, and it was placed in the microwave oven. Using syringe pumps, the reactant solutions: (i) metal precursor $FeCl_3$ (ii) linker, that is, 1,3,5-benzenetricarboxylic acid (H_3BTC) in DI water, and DMF and HNO_3 was used for the mineralization were dispensed to the coiled tubular reactor. The obtained MOF, synthesized by the continuous flow method, exhibited high crystallinity, porosity, and STY (771 kg m^{-3} day^{-1}) than the batch sample synthesized using conventional heating.

Integration of flow systems with supercritical fluids is the right model for the greener process intensification. One such combination of slow synthesis and supercritical CO_2 ($scCO_2$) was chosen for process intensification in the synthesis of MOF. To address the solubility between $scCO_2$ and metal precursor, mass transfer and heat transfer counter-current mixing (CCM) section was chosen. The schematic diagram of the CCM flow reactor is shown in Fig. 3.3(f). The reactor was operated between 298–873 K and 0.101–30 MPa. Once the reactor was pressurized, CO_2 gas was supplied to the reactor by the HPLC pump, by condensing CO_2 to liquid form before it reaches the pump. Then, the metal precursor materials were fed to the reactor at a flow rate of 10 mL min^{-1} without any pre-heating. T-micromixer was used for mixing the metal precursors at ambient temperatures, then mixed with $scCO_2$, and finally passed to the reactor section. Under optimized experimental conditions, the reactor consumed only 65% of the energy requirement of the hydrothermal reactor and produced UiO-66 MOF at a rate of 104 g h^{-1}.

Covalent organic frameworks were also synthesized in the flow system consisting of a PTFE tubular reactor with a length of 0.1 m and ID 0.3 mm [92]. The feedstock solutions 1,3,5-triformylbenzene (TFB) *p*-phenylenediamine (PDA) in a mixture of dia-oxane, acetic acid solutions were injected into the T-micromixer, followed by the flow reactor, to give a residence time of 11 s. The collected COF was purified and confirmed by various characterization techniques. The produced COF, at a rate of 41 mg h^{-1} in the flow synthesis, possessed high BET surface than the sample synthesized in batch. This could be due to the improved crystallization under flow conditions.

3.2.6 Mesoporous materials

Evaporation-induced self-assembly (EISA) is a method in which silica precursor and surfactants are mixed in ethanol/water solution. It undergoes evaporation, followed by self-assembly into mesophases. Combining the EISA method with emulsion would make the synthesis of mesoporous materials in the microfluidic platform. Droplets of silica precursor were generated in a T-micromixer and EISA was carried out at the downstream of microfluidic channel. Fast diffusion of ethanol from droplet-to-oil phase resulted in an accelerated self-assembly with spherical silica particles of mesoporosity. This method was successful in tuning the size and morphology of mesoporous materials by tuning the flow conditions.

3.2.7 Quantum dots

Flow synthesis of QDs is mainly opted as high temperature synthesis in organic solvents and in the presence of capping agents. deMello and coworkers explored the novel scale-out strategy using multichannel droplet reactors and produced ~145 g day^{-1} of CdTe QDs, without any fouling of the PTFE tube, which was used as the droplet flow reactor [83]. Such high temperature routes were also explored for the continuous flow synthesis of hybrid/core–shell QDs such as CdSe/CdS/ZnS, CdS/ZnS, CdSeS/ZnS nanocrystals [84]. Hybrid programmed microfluidic-batch approach was demonstrated by Song and coworkers to successfully synthesize CdSe nanocrystals with controlled shapes such as angular, tripodia, near-spherical, and spherical nanostructures [85]. Lead halide perovskite QDs (PQDs) that got wide attention because of energy-related applications were also synthesized in the continuous flow methods via single- and two-phase flows. Similar to Cd-/Zn-based QDs for Pb PQDs, high temperature synthesis method was also the most opted for flow synthesis. However, the flow synthesis of Pb PQDs was also extended to low temperature synthesis (e.g., $CsPbBr_3$ synthesis at room temperature; $MAPbX_3$ PQDs synthesis in single phase in PTFE tuning at 30 °C) and post processing of PQDs in the flow for halide-exchange reactions. Detailed summary on the synthesis of the abovementioned QDs can be found in the recent reviews [86, 87].

Nonmetallic QDs such as GQDs, CQDs, doped CQDs were also synthesized in continuous flow hydrothermal reactors. Kellici and coworkers synthesized N-doped CQDs, in which, super critical water was used as one feed, and citric acid and ammonia as the other two feeds. Synthesis was carried out at 450 °C and 24.8 MPa [88]. Similarly, for the synthesis of S-doped CQDs, glucose and *p*-sulfonic acid calix[4] arene were used as feed stocks, in addition to supercritical water [88]. In all such cases, temperature plays an important role and extremely fast nucleation is the key to restrict the growth.

3.3 High-throughput continuous flow synthesis of materials

Microfluidic technology can also be implemented for the high-throughput synthesis of materials. Van Nguyen et al. [89] developed a novel centrifugal microfluidic device for the high-throughput synthesis of Pd@Au–Pt core–shell NPs. The device consists of a PMMA sheet on which micropatterns are generated, both on the top and bottom of the PMMA with CNC machine. This micropatterned PMMA sheet is sealed in between two PSA films. This device consists of 60 reaction chambers, three reagent loading channels, and the device is operated by a simple rotational operation. Sample injection and loading into the 60 reaction chambers would be initiated by simple automated 10 min rotation. By varying the concentrations of the reactants and their molar ratio, up to 60 different core–shell structures can be generated in a high-throughput manner.

Other than the above discussed complex and novel designs, an inexpensive and simple PTFE coil can also be used as a microfluidic reaction system for high-throughput synthesis. A simple helical tubular PTFE coil is used for the high-throughput synthesis of SiO_2 NPs by basic hydrolysis of tetraethyl orthosilicate (TEOS) using ammonia. Dean flow in the helical coil helps to mix the TEOS in non-aqueous phase and NH_3 in the aqueous phase to produce silica NPs of sizes that range from 100 to 600 nm, and produce an yield of up to 0.234 g h^{-1}.

Such PTFE coils are also used for the high-throughput synthesis of metal NPs (Ni) in a continuous flow manner. Precursors solution that are used for Ni NPs synthesis consist of $Ni(acac)_2$, oleylamine, oactadecene, and trioactyl phosphine are preheated to 80 °C using a heating tape. The preheated solution is passed to the PTFE reaction channel, which is placed in the preheated convection oven at 220 °C. The weight loss in the preheated solution is continuously measured by an analytical balance. High reaction temperature causes a faster nucleation, and at higher flow rates, the plug formation is observed in the reaction due to generation of gas. This plug flow helps to reduce the axial dispersion. At a flow rate of 133 mL h^{-1}, >27 g of Ni NPs are produced with a relative yield of 62%. In addition to NiNPs, the above method is also used to produce >0.5 kg SiO_2-supported NiNPs with 5 wt.% SiO_2.

Furthermore, AuNPs are also produced with excellent reproducibility by the combination of precursor passivation and well-designed reactor model, to give a highly reproducible, high yield, AuNPs without any fouling of reactor.

3.4 Challenges and future directions

3.4.1 Challenges associated with separation and purification of the materials (and recent developments in this direction)

The most conventional ways of separation and isolation (rather recovery) of materials from solutions/suspensions is filtration. In general, the suspension concentrations vary over a range of 0.01–5% wt./vol. At the laboratory scale, such a range of concentrations can be isolated using membrane filtration, or centrifugation, or by using an antisolvent. The former works in the continuous mode and there are very few reports of using continuous centrifugation at laboratory scale. In general, while membrane filtration is quite popular, it is important to choose the right membrane to ensure that the membrane pore size is at least 5 times smaller than the particle size. This indicates that once the particles start getting accumulated on the membrane pores, subsequent filtration experiences a lot of pressure drop for the filtration to happen. This is very common and one may need to increase the suction pressure or applied pressure with time within the range of mechanical strength of the membrane used for filtration, however.

On the other hand, using continuous centrifugation is possible, but it needs certain operational protocols as discussed by Deshpande and Kulkarni, where an annular centrifugal extractor is used [90]. The centrifugation speed needs to be increased significantly for very dilute suspensions as well as for very small extractor sizes.

The last approach is by using an antisolvent, where a low density and highly soluble antisolvent is added to the reaction mass. This significantly reduces the density of the liquid phase of the suspension. In general, using an antisolvent needs subsequent settling of particles, which can be recovered either by settling (which takes a few minutes to hours, depending on the particle size and material density) or by accelerated settling by batch centrifugation. For making this approach continuous, one has to continuously inject the antisolvent in the outlet stream from the continuous flow synthesis stream (only after bringing the stream at temperatures lower than the boiling point of the antisolvent), followed by a settling chamber having a calming and liquid discharge sections that allow for the settling of the material and a continuous discharge of the liquid. For achieving a truly continuous operation, one has to optimize the protocol to identify the most suitable volume ratios of solvent to antisolvent, and mixing the length and size of the calming and discharge sections of the settler.

3.4.2 Immobilization of materials

From an application point of view, most of the functional materials synthesized using continuous flow approaches are used in powder form or as a coating or as a film made using a formulation. This specific aspect of using materials is important as it requires the material to have an approach for transforming into an application. Among these, the most important feature is the immobilization of catalyst on an adsorbent-like carbon or alumina, etc. Thus, it is necessary that a material needs to have an ability to get adsorbed without losing its functionality.

3.4.3 Process control

This is the most important challenge among all the operational protocols of flow synthesis of materials. It deals with the consistency in the product quality, quantifying the real time deviation, and identification of the corresponding control parameter. Recently, few excellent efforts in that direction are seen from the Gavriilidis' group from UCL, London. An exceptional ability of the control algorithm to identify the deviations in the wavelength of maximum absorbance in the UV–vis spectra for AuNPs and correcting the flow rates or temperature of the system to eliminate deviation is reported in an excellent manner. More such efforts are needed for almost all kinds of materials as continuous flow synthesis is expected to produce consistent quality. This depends on several parameters that can make changes at the nanometer or micrometer length scales.

3.4.4 Cleaning of systems

Deposition of material on the reactor wall or fouling is a common observation. In general, it is known that fouling can eventually lead to clogging, thereby forcing the shutdown of the system. Such situations are encountered very commonly and should be avoided either by using a suitable dispersed phase system or by periodic washing of the system by monitoring the pressure drop in the reactor. This leads to loss of time and loss of precious material and hence, the design of continuous flow synthesis setup should have a built-in cleaning protocol or should not need cleaning at all.

3.5 Summary and recommendations

Continuous flow synthesis of materials will be the future of manufacturing of functional materials. On-site-on-demand nature of synthesis kits and plants will make it more realistic and reliable. The vast body of literature shows that it is possible to synthesize almost all kinds of nanomaterials in flow, with more focus on complex asymmetric as well as unidirectional materials. All these efforts can be put in the periodic table so as to get a glimpse of the range of elements that have been used so far for the flow synthesis of functional materials (Fig. 3.7). Synthesis of core–shell materials, Janus particles, unidirectional metal rods/wires, nanomaterials in the form of flakes or dots, and others have been demonstrated using flow synthesis, using simple systems.

An important aspect that remains to be explored at large is the detailed reaction kinetics of individual materials. Following the conventional batch protocol gives restricted information on kinetics, and hence, flow synthesis needs to be exhaustively used for the same. This will help allow monitor growth along certain crystal planes or the development of certain shapes.

Although several materials, their oxides, composites with other inorganic and organic substrates have been synthesized in flow, it just opens a vast set of opportunities. The use of impinging jet reactors, bubble column reactors, fluidized bed reactors, oscillating baffle reactors, and so on actually facilitate taking the size of operation beyond the laboratory scale. Synthesis labs have to go beyond the conventional tubular systems.

Using the laboratory-scale synthesis systems for understanding transient effects on materials is still an unexplored area. Conventional batch protocols do not allow it to that extent, which flow synthesis can definitely help. Such studies will help understand the nucleation and growth mechanisms with more accuracy, and also allow some sensitivity studies to be carried out that can lead to deviations from the particle properties. This also means that there lies a huge opportunity to develop miniaturized versions of electron microscopes and in-line scattering measurements so as to do real time monitoring of material synthesis/phasing out.

Many new flow synthesis concepts that can handle suspensions with higher solid percentages will need to be explored. Including these simple techniques in the academic experiments will help students view the synthesis of complex materials from a different angle and will use flow synthesis for making new materials and fill more blocks in the periodic table in the time to come.

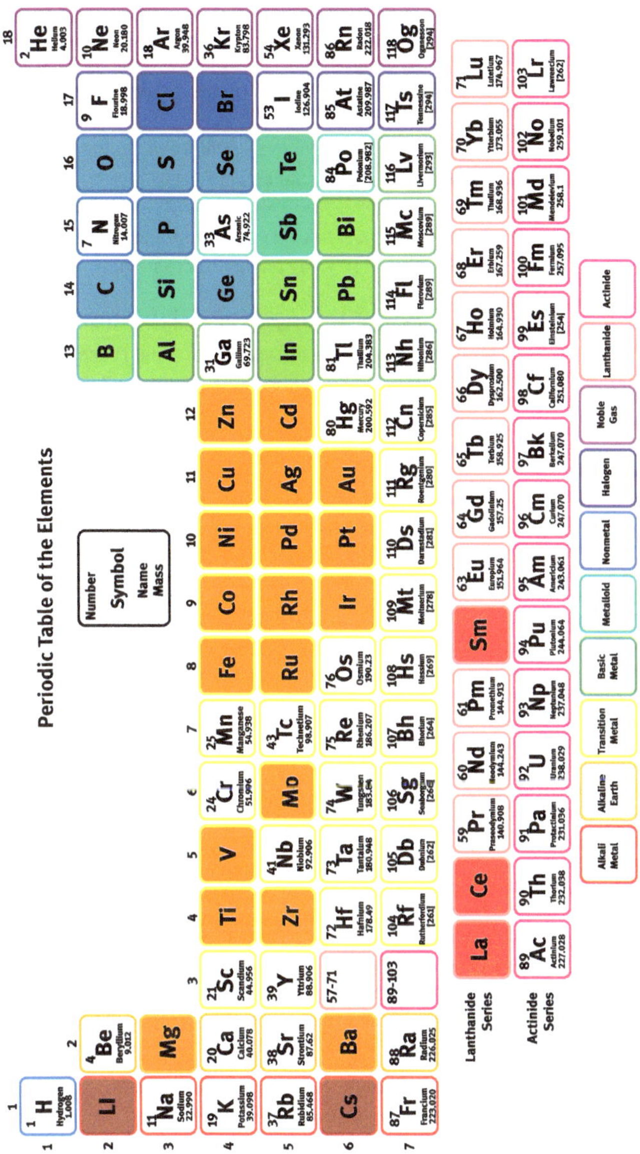

Fig. 3.7: Occupancy in the periodic table by the materials that are synthesized in flow.

References

[1] Campbell ZS, Abolhasani M. Facile synthesis of anhydrous microparticles using plug-and-play microfluidic reactors. React. Chem. Eng., 2020, 5, 1198-1211

[2] Biswas S, Miller JT, Li Y, Nandakumar K, Kumar CSSR. Developing a millifluidic platform for the synthesis of ultrasmall nanoclusters: Ultrasmall copper nanoclusters as a case study. Small, 2012, 8 (5), 688–698

[3] Ramirez Mendoza H, Valdez Lancinha Pereira M, Van Gerven T, Lutz C. Continuous flow synthesis of zeolite FAU in an oscillatory baffled reactor. J Adv Manuf Process. 2020; 2: e10038

[4] Laybourn A, López-Fernández AM, Thomas-Hillman I, Katrib J, Lewis W, Dodds C, et al. Combining continuous flow oscillatory baffled reactors and microwave heating: Process intensification and accelerated synthesis of metal-organic frameworks. Chem Eng J. 2019; 356, 170-177

[5] Bayliss PA, Ibarra IA, Pérez E, Yang S, Tang CC, Poliakoff M, et al. Synthesis of metal-organic frameworks by continuous flow. Green Chem. 2014; 16, 3796-3802

[6] Huang H, Hwang GB, Wu G, Karu K, Du Toit H, Wu H, et al. Rapid synthesis of [Au25(Cys)18] nanoclusters via carbon monoxide in microfluidic liquid-liquid segmented flow system and their antimicrobial performance. Chem Eng J. 2020; 383, 123176

[7] Manno R, Ranjan P, Sebastian V, Mallada R, Irusta S, Sharma UK, et al. Continuous Microwave-Assisted Synthesis of Silver Nanoclusters Confined in Mesoporous SBA-15: Application in Alkyne Cyclizations. Chem Mater. 2020;

[8] Tian Z, Ge X, Wang Y, Xu J. Nanoparticles and Nanocomposites with Microfluidic Technology [Internet]. Polymer-based Multifunctional Nanocomposites and their Applications. Elsevier Inc.; 2019. 1–33 p.

[9] Anwar A, Akbar S, Kazmi M, Sadiqa A, Gilani SR. Novel synthesis and antimicrobial studies of nanoscale titania particles. Ceram Int [Internet]. 2018; 44(17):21170–5.

[10] Sebastian Cabeza V, Kuhn S, Kulkarni AA, Jensen KF. Size-controlled flow synthesis of gold nanoparticles using a segmented flow microfluidic platform. Langmuir. 2012; 28, 17, 7007–7013

[11] Abalde-Cela S, Taladriz-Blanco P, De Oliveira MG, Abell C. Droplet microfluidics for the highly controlled synthesis of branched gold nanoparticles. Sci Rep [Internet]. 2018; 8(1):1–6.

[12] Tao S, Yang M, Chen H, Chen G. Continuous Synthesis of Highly Uniform Noble Metal Nanoparticles over Reduced Graphene Oxide Using Microreactor Technology. ACS Sustain Chem Eng. 2018; 6(7):8719–26.

[13] Wojnicki M, Tokarski T, Hessel V, Fitzner K, Luty-Błocho M. Continuous, monodisperse silver nanoparticles synthesis using microdroplets as a reactor. J Flow Chem. 2019; 9(1):1–7.

[14] Karim AM, Al Hasan N, Ivanov S, Siefert S, Kelly RT, Hallfors NG, et al. Synthesis of 1 nm Pd nanoparticles in a microfluidic reactor: Insights from in situ X-ray absorption fine structure spectroscopy and small-angle X-ray scattering. J Phys Chem C. 2015; 119, 23, 13257–13267

[15] Wang ST, Lin Y, Nielsen MH, Song CY, Thomas MR, Spicer CD, et al. Shape-controlled synthesis and: In situ characterisation of anisotropic Au nanomaterials using liquid cell transmission electron microscopy. Nanoscale. 2019; 11(36):16801–9.

[16] Duraiswamy S, Khan SA. Droplet-based microfluidic synthesis of anisotropic metal nanocrystals. Small. 2009; 5, No. 24, 2828–2834

[17] Trzciński JW, Panariello L, Besenhard MO, Yang Y, Gavriilidis A, Guldin S. Synthetic guidelines for the precision engineering of gold nanoparticles. Curr Opin Chem Eng. 2020; 29:59–66.

[18] Santana JS, Koczkur KM, Skrabalak SE. Kinetically controlled synthesis of bimetallic nanostructures by flowrate manipulation in a continuous flow droplet reactor. React Chem Eng. 2018; 3(4):437–41.

[19] Kim YH, Zhang L, Yu T, Jin M, Qin D, Xia Y. Droplet-based microreactors for continuous production of palladium nanocrystals with controlled sizes and shapes. Small. 2013; 9(20):3462-7
[20] Zhang L, Niu G, Lu N, Wang J, Tong L, Wang L, et al. Continuous and scalable production of well-controlled noble-metal nanocrystals in milliliter-sized droplet reactors. Nano Lett. 2014; 14, 11, 6626–6631
[21] Sebastian V, Smith CD, Jensen KF. Shape-controlled continuous synthesis of metal nanostructures. Nanoscale. 2016; 8, 7534-7543
[22] Pekkari A, Say Z, Susarrey-Arce A, Langhammer C, Härelind H, Sebastian V, et al. Continuous Microfluidic Synthesis of Pd Nanocubes and PdPt Core-Shell Nanoparticles and Their Catalysis of NO2 Reduction. ACS Appl Mater Interfaces. 2019; 11(39):36196–204.
[23] Tang Z, Kim WS, Yu T. Studies on morphology changes of copper sulfide nanoparticles in a continuous Couette-Taylor reactor. Chem Eng J. 2019; 359(September 2018):1436–41.
[24] Roberts EJ, Habas SE, Wang L, Ruddy DA, White EA, Baddour FG, et al. High-Throughput Continuous Flow Synthesis of Nickel Nanoparticles for the Catalytic Hydrodeoxygenation of Guaiacol. ACS Sustain Chem Eng. 2017; 5(1):632–9.
[25] Tang Z, Kim WS, Yu T. Continuous synthesis of silver plates in a continuous stirring tank reactor (CSTR). J Ind Eng Chem [Internet]. 2018; 66:411–8.
[26] Deshpande JB, Kulkarni AA. Reaction Engineering for Continuous Production of Silver Nanoparticles. Chem Eng Technol. 2018; 41(1):157–67.
[27] Maji NC, Krishna HP, Chakraborty J. Low-cost and high-throughput synthesis of copper nanopowder for nanofluid applications. Chem Eng J [Internet]. 2018; 353(May):34–45.
[28] Gomez C, Hallot G, Pastor A, Laurent S, Brun E, Sicard-Roselli C, et al. Metallic bismuth nanoparticles: Towards a robust, productive and ultrasound assisted synthesis from batch to flow-continuous chemistry. Ultrason Sonochem [Internet]. 2019; 56(April):167–73.
[29] Domínguez C, Metz KM, Hoque MK, Browne MP, Esteban-Tejeda L, Livingston CK, et al. Continuous Flow Synthesis of Platinum Nanoparticles in Porous Carbon as Durable and Methanol-Tolerant Electrocatalysts for the Oxygen Reduction Reaction. Chem Electro Chem. 2018; 5(1):62–70.
[30] Tofighi G, Gaur A, Doronkin DE, Lichtenberg H, Wang W, Wang D, et al. Microfluidic Synthesis of Ultrasmall AuPd Nanoparticles with a Homogeneously Mixed Alloy Structure in Fast Continuous Flow for Catalytic Applications. J Phys Chem C. 2018; 122(3):1721–31.
[31] Nikam A V., Dadwal AH. Scalable microwave-assisted continuous flow synthesis of CuO nanoparticles and their thermal conductivity applications as nanofluids. Adv Powder Technol [Internet]. 2019; 30(1):13–7.
[32] Fong EJ, Han J, Cornell CC, Han TYJ. Opportunities and Challenges for Nanoparticle Synthesis using Continuous Flow Systems: A Magnetite Nanocluster Case Study. ChemRxiv. 2019; 1–50.
[33] Vangijzegem T, Stanicki D, Panepinto A, Socoliuc V, Vekas L, Muller RN, et al. Influence of experimental parameters of a continuous flow process on the properties of very small iron oxide nanoparticles (VSION) designed for T1-weighted magnetic resonance imaging (MRI). Nanomaterials. 2020; 10(4), 757
[34] Hao N, Zhang M, Zhang JXJ. Microfluidics for ZnO micro-/nanomaterials development: Rational design, controllable synthesis, and on-chip bioapplications. Biomater Sci. 2020; 8(7):1783–801.
[35] Xie Y, Yang S, Mao Z, Li P, Zhao C, Cohick Z, et al. In situ fabrication of 3D Ag@ZnO nanostructures for microfluidic surface-enhanced Raman scattering systems. ACS Nano. 2014; 8, 12, 12175–12184
[36] He Z, Li Y, Zhang Q, Wang H. Capillary microchannel-based microreactors with highly durable ZnO/TiO2 nanorod arrays for rapid, high efficiency and continuous-flow photocatalysis. Appl Catal B Environ. 93 2010; 376–382

[37] Sue K, Kimura K, Arai K. Hydrothermal synthesis of ZnO nanocrystals using microreactor. Mater Lett. 2004; 58 (25); 3229-3231
[38] Roig Y, Marre S, Cardinal T, Aymonier C. Synthesis of exciton luminescent ZnO nanocrystals using continuous supercritical microfluidics. Angew Chemie - Int Ed. 2011; 50, 12071 –12074
[39] Huang C, Wang Y, Luo G. Preparation of highly dispersed and small-sized ZnO nanoparticles in a membrane dispersion microreactor and their photocatalytic degradation. Ind Eng Chem Res. 2013; 52, 16, 5683–5690
[40] Lee J, Choi KH, Min J, Kim HJ, Jee JP, Park BJ. Functionalized ZNO nanoparticles with gallic acid for antioxidant and antibacterial activity against methicillin-resistant S. aureus. Nanomaterials. 2017; 7(11): 365.
[41] Hao N, Xu Z, Nie Y, Jin C, Closson AB, Zhang M, et al. Microfluidics-enabled rational design of ZnO micro-/nanoparticles with enhanced photocatalysis, cytotoxicity, and piezoelectric properties. Chem Eng J. 2019; 378: 122222.
[42] Beyer J, Mamakhel A, Søndergaard-Pedersen F, Yu J, Iversen BB. Continuous flow hydrothermal synthesis of phase pure rutile TiO2 nanoparticles with a rod-like morphology. Nanoscale. 2020; 12(4):2695–702.
[43] Gimeno-Fabra M, Dunne P, Grant D, Gooden P, Lester E. Continuous flow synthesis of tungsten oxide (WO3) nanoplates from tungsten (VI) ethoxide. Chem Eng J. 2013; 226:22–9.
[44] Mamakhel A, Søndergaard M, Borup K, Brummerstedt Iversen B. Continuous flow hydrothermal synthesis of rutile SnO2 nanoparticles: Exploration of pH and temperature effects. J Supercrit Fluids. 2020; 166, 105029
[45] Delgado-Licona F, López-Guajardo EA, González-García J, Nigam KDP, Montesinos-Castellanos A. Intensified tailoring of ZnO particles in a continuous flow reactor via hydrothermal synthesis. Chem Eng J. 2020; 396, 125281
[46] Sen N, Koli V, Singh KK, Panicker L, Sirsam R, Mukhopadhyay S, et al. Segmented microfluidics for synthesis of BaSO4 nanoparticles. Chem Eng Process - Process Intensif. [Internet]. 2018; 125:197–206.
[47] Tian ZH, Shao M, Zhao XY, Wang YJ, Wang K, Xu JH. Morphology Control of CdSe Nanoparticles via Two-Step Segmented Microreactors. Cryst Growth Des. 2018; 18(7):3953–8.
[48] Hiemer J. MicroJet Reactor Technology : An Automated, Continuous Approach for Nanoparticle Syntheses. 2019; (10):2018–27.
[49] Khan SA, Günther A, Schmidt MA, Jensen KF. Microfluidic synthesis of colloidal silica. Langmuir. 2004; 20, 20, 8604–8611
[50] Shiba K, Kambara K, Ogawa M. Size-controlled syntheses of nanoporous silica spherical particles through a microfluidic approach. Ind Eng Chem Res. 2010; 49, 17, 8180–8183
[51] Abou-Hassan A, Bazzi R, Cabuil V. Multistep continuous-flow microsynthesis of magnetic and fluorescent γ-Fe2O3OSiO2 core/shell nanoparticles. Angew Chemie - Int Ed. 2009; 48, 7180 –7183
[52] Hao N, Nie Y, Shen T, Zhang JXJ. Microfluidics-enabled rational design of immunomagnetic nanomaterials and their shape effect on liquid biopsy. Lab Chip. 2018; 18, 1997-2002
[53] Chung CK, Shih TR, Chang CK, Lai CW, Wu BH. Design and experiments of a short-mixing-length baffled microreactor and its application to microfluidic synthesis of nanoparticles. Chem Eng J. 2011; 168 (2), 790-798
[54] Gomez L, Arruebo M, Sebastian V, Gutierrez L, Santamaria J. Facile synthesis of SiO 2-Au nanoshells in a three-stage microfluidic system. J Mater Chem. 2012; 22, 21420-21425
[55] Hao N, Nie Y, Zhang JXJ. Microfluidics for silica biomaterials synthesis: Opportunities and challenges. Biomater Sci. 2019; 7(6):2218–40.
[56] Hao N, Nie Y, Zhang JXJ. Microfluidic Flow Synthesis of Functional Mesoporous Silica Nanofibers with Tunable Aspect Ratios. ACS Sustain Chem Eng. 2018; 6(2):1522–6.

[57] Duraiswamy S, Khan SA. Plasmonic nanoshell synthesis in microfluidic composite foams. Nano Lett. 2010; 10, 9, 3757–3763
[58] Zhao CX, Middelberg APJ. Microfluidic synthesis of monodisperse hierarchical silica particles with raspberry-like morphology. RSC Adv. 2013; 3, 21227-21230
[59] Ju M, Ji X, Wang C, Shen R, Zhang L. Preparation of solid, hollow, hole-shell and asymmetric silica microspheres by microfluidic-assisted solvent extraction process. Chem Eng J. 2014; 10, 1177
[60] Yan H, Kim C. Formation of monodisperse silica microparticles with various shapes and surface morphologies using double emulsion templates. Colloids Surfaces A Physicochem Eng Asp. 2014; 8, 22
[61] Tamtaji M, Mohammadi A. Continuous synthesis of plate-like silica microparticles using microfluidics. J Flow Chem. 2019; 9(3):161–74.
[62] Alli U, Hettiarachchi SJ, Kellici S. Chemical Functionalisation of 2D Materials by Batch and Continuous Hydrothermal Flow Synthesis. Chemistry - A European Journal. 2020; 26, 6447 –6460
[63] Kellici S, Acord J, Power NP, Morgan DJ, Coppo P, Heil T, et al. Rapid synthesis of graphene quantum dots using a continuous hydrothermal flow synthesis approach. RSC Adv. 2017; 7, 14716-14720
[64] Saada R, AboElazayem O, Kellici S, Heil T, Morgan D, Lampronti GI, et al. Greener synthesis of dimethyl carbonate using a novel tin-zirconia/graphene nanocomposite catalyst. Appl Catal B Environ. 2018; 226, 451-462
[65] Saada R, Kellici S, Heil T, Morgan D, Saha B. Greener synthesis of dimethyl carbonate using a novel ceria-zirconia oxide/graphene nanocomposite catalyst. Appl Catal B Environ. 2015; 226, 353-362
[66] Middelkoop V, Slater T, Florea M, Neaţu F, Danaci S, Onyenkeadi V, et al. Next frontiers in cleaner synthesis: 3D printed graphene-supported CeZrLa mixed-oxide nanocatalyst for CO 2 utilisation and direct propylene carbonate production. J Clean Prod. 2019; 214, 606-614
[67] Kellici S, Acord J, Moore KE, Power NP, Middelkoop V, Morgan DJ, et al. Continuous hydrothermal flow synthesis of graphene quantum dots. React Chem Eng. 2018; 3, 949-958
[68] Baragau IA, Power NP, Morgan DJ, Heil T, Lobo RA, Roberts CS, et al. Continuous hydrothermal flow synthesis of blue-luminescent, excitation-independent nitrogen-doped carbon quantum dots as nanosensors. J Mater Chem A. 2020; 8, 3270-3279
[69] Luo X, Yuan S, Pan X, Zhang C, Du S, Liu Y. Synthesis and Enhanced Corrosion Protection Performance of Reduced Graphene Oxide Nanosheet/ZnAl Layered Double Hydroxide Composite Films by Hydrothermal Continuous Flow Method. ACS Appl Mater Interfaces. 2017; 9, 21, 18263–18275
[70] Chang Z, Evans DG, Duan X, Vial C, Ghanbaja J, Prevot V, et al. Synthesis of [Zn-Al-CO3] layered double hydroxides by a coprecipitation method under steady-state conditions. J Solid State Chem. 2005; 178 (9), 2766-2777
[71] Elbasuney S. Surface engineering of layered double hydroxide (LDH) nanoparticles for polymer flame retardancy. Powder Technol. 2015; 277, 63-73
[72] Forticaux A, Dang L, Liang H, Jin S. Controlled synthesis of layered double hydroxide nanoplates driven by screw dislocations. Nano Lett. 2015; 5, 3403–3409
[73] Pang X, Liu Y, Chen L, Zhong Y, Li Z, Liu M, et al. Preparation and formation mechanism of pure phase Ca2Al-layered double hydroxides nanosheets synthesized by a T-type microchannel reactor: Application as hardening accelerator for mortar. Appl Clay Sci. 2018; 166, 174-180
[74] Yaseneva P, An N, Finn M, Tidemann N, Jose N, Voutchkova-Kostal A, et al. Continuous synthesis of doped layered double hydroxides in a meso-scale flow reactor. Chem Eng J. [Internet]. 2019; 360(August 2018):190–9.

[75] Wang Y, Li L, Yan L, Gu X, Dai P, Liu D, et al. Bottom-Up Fabrication of Ultrathin 2D Zr Metal-Organic Framework Nanosheets through a Facile Continuous Microdroplet Flow Reaction. Chem Mater. 2018; 30(9):3048–59.
[76] Alsulam IK, Alharbi TMD, Moussa M, Raston CL. High-Yield Continuous-Flow Synthesis of Spheroidal C60@Graphene Composites as Supercapacitors. ACS Omega. 2019; 4(21):19279–86.
[77] Liu Z, Okabe K, Anand C, Yonezawa Y, Zhu J, Yamada H, et al. Continuous flow synthesis of ZSM-5 zeolite on the order of seconds. Proc Natl Acad Sci U S A. 2016; 113 (50) 14267-14271
[78] Huang X, Zhang G, Zhang L, Zhang Q. Continuous flow synthesis of a ZSM-5 film in capillary microchannel for efficient production of solketal. ACS Omega. 2020; 5, 20784–20791
[79] Bukhari S, Rohani S. Continuous flow synthesis of zeolite - A from coal fly ash utilizing microwave irradiation with recycled liquid stream. In: Materials Engineering and Sciences Division 2016 - Core Programming Area at the 2016 AIChE Annual Meeting. 2016; 13 (3), 2017, 233-244
[80] Yoshioka T, Liu Z, Iyoki K, Chokkalingam A, Yonezawa Y, Hotta Y, et al. Ultrafast and continuous-flow synthesis of AFX zeolite via interzeolite conversion of FAU zeolite. React Chem Eng. 2021; 6, 74-81
[81] Shukla CA, Atapalkar RS, Kulkarni AA. Efficient Processing of Reactions Involving Diazonium Salts: Meerwein Arylation in an Impinging-Jet Reactor. Org Process Res Dev. 2020; 24, 9, 1658–1664
[82] Liu Z, Wakihara T, Nomura N, Matsuo T, Anand C, Elangovan SP, et al. Ultrafast and continuous flow synthesis of silicoaluminophosphates. Chem Mater. 2016; 28, 13, 4840–4847
[83] Nightingale AM, Bannock JH, Krishnadasan SH, O'Mahony FTF, Haque SA, Sloan J, et al. Large-scale synthesis of nanocrystals in a multichannel droplet reactor. J Mater Chem A. 2013; 1, 4067-4076
[84] Naughton MS, Kumar V, Bonita Y, Deshpande K, Kenis PJA. High temperature continuous flow synthesis of CdSe/CdS/ZnS, CdS/ ZnS,and CdSeS/ZnS nanocrystals. Nanoscale. 2015; 7, 15895-15903
[85] Wang J, Zhao H, Zhu Y, Song Y. Shape-Controlled Synthesis of CdSe Nanocrystals via a Programmed Microfluidic Process. J Phys Chem C. 2017; 121, 6, 3567–3572
[86] Abdel-Latif K, Bateni F, Crouse S, Abolhasani M. Flow Synthesis of Metal Halide Perovskite Quantum Dots: From Rapid Parameter Space Mapping to AI-Guided Modular Manufacturing. Matter. 2020; 3(4),1053-1086
[87] Kubendhiran S, Bao Z, Dave K, Liu RS. Microfluidic Synthesis of Semiconducting Colloidal Quantum Dots and Their Applications. ACS Applied Nano Materials. 2019, 2, 4, 1773–1790
[88] Baragau IA, Lu Z, Power NP, Morgan DJ, Bowen J, Diaz P, et al. Continuous hydrothermal flow synthesis of S-functionalised carbon quantum dots for enhanced oil recovery. Chem Eng J. 2021; 405, 126631
[89] Nguyen H Van, Kim KY, Nam H, Lee SY, Yu T, Seo TS. Centrifugal microfluidic device for the high-throughput synthesis of Pd@AuPt core-shell nanoparticles to evaluate the performance of hydrogen peroxide generation. Lab Chip. 2020; 20, 3293-3301
[90] Deshpande JB, Navale GR, Dharne MS, Kulkarni AA. Continuous Interfacial Centrifugal Separation and Recovery of Silver Nanoparticles. Chem Eng Technol. 2020; 43(3),582–592
[91] Huang H, Toit H du, Besenhard MO, Ben-Jaber S, Dobson P, Parkin I, et al. Continuous flow synthesis of ultrasmall gold nanoparticles in a microreactor using trisodium citrate and their SERS performance. Chem Eng Sci. 2018; 189:422–30
[92] Peng Y, Wong WK, Hu Z, Cheng Y, Yuan D, Khan SA, et al. Room Temperature Batch and Continuous Flow Synthesis of Water-Stable Covalent Organic Frameworks (COFs). Chem Mater. 2016; 28(14):5095–101.

Tanja Junkers

4 Polymer synthesis in continuous flow

4.1 Introduction

Industrially, polymer materials make for one of the largest markets in terms of production volume and turnover. Commodity plastics such as low-density polyethylene or polystyrene are produced on enormous scale. Mostly, this production is done in continuous flow reactors or continuous stirred tank reactors. On smaller scale, batch processing prevails. The rise of continuous flow reactors on small and intermediate scale has, however, soon also raised the interest of polymer chemists. A clear disadvantage of polymer reactions in flow is the high viscosities that are automatically reached. While this is an obstacle that needs to be considered, it is, however, rarely truly prohibitive to carry a reaction out in flow.

First, flow polymerizations on lab scale have already been reported decades ago [1], and the concept of polymer on a tap has intrigued researchers ever since. The last ten years have then seen a remarkable evolution in continuous flow polymer reaction toward highly integrated reactor systems, precision polymer synthesis on demand and product intensification on lab scale. Practically all areas of polymer synthesis have to date been applied to continuous reactors. As will be discussed, flow reactors show tremendous advantages over batch-based chemistry. Not only do reactions become scalable, and proceed with less batch-to-batch variation, reactions also become more precise. This means that key parameters such as average molecular weight, dispersity (hence, the broadness of a polymer distributions), and end-group distributions can be controlled in much finer increments. This is of high importance, as only with precise control over these attributes, polymer materials with pre-determined physical properties can become accessible. The increase in precision in continuous flow reactors for almost all types of polymerization stems from the better isothermicity found in flow reactors, and as for most other flow application the improved mass transfer and excellent mixing properties. Temperature is, however, one of the most important factors. Polymerizations are inherently highly exothermic, making an adequate temperature control in batch even on small scale a challenge. By providing much more stable temperature conditions, automatically fewer side reactions occur, and a much more uniform kinetic profile with respect to the overall rate of polymerization is given. In contrast to most small molecule synthesis, kinetics of a polymerization directly influences not only the yield of a reaction, but also the composition, average molecular weight, and uniformity of the polymer product. An outline over the most important polymerizations and polymer modification reactions is given in this chapter, showing how all microstructural features as shown in Figure 4.1 can be accessed in continuous reactions, and eventually improved by switching from batch to continuous flow processing.

https://doi.org/10.1515/9783110693690-004

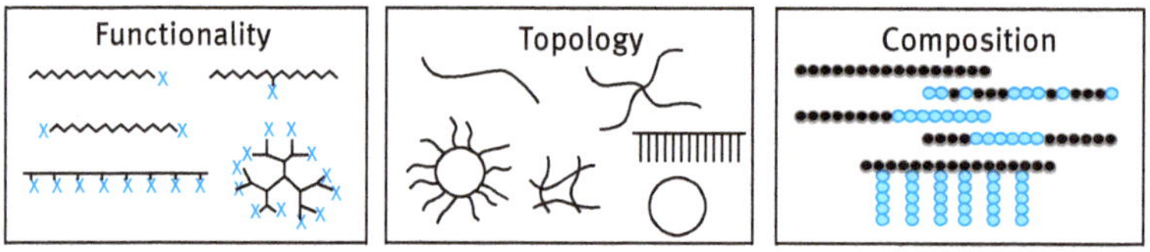

Figure 4.1: Microstructural features determining physical properties of a polymer besides the average molecular weight and dispersity of a polymer.

4.2 Anionic polymerization

The first type of polymerization that was translated to continuous flow technology was ionic polymerizations [3]. Flow technology displayed a major advantage for such reactions due to the necessity for ionic reactions to be performed under demanding reaction conditions. Oxygen and water-free conditions are more easily realized in flow reactor compared to batch; any residual water adsorbed on the reactor walls is removed during conditioning of the reactor when flushing the entire system with solvent, and does not influence the polymerization after an initialization period. Additionally, ionic polymerizations are rapid, requiring the use of cryogenic conditions in batch reactors. The improved heat exchange in flow reactors also allows these reactions to be performed closer to room temperature (or higher), and at the same time provides conditions where very fast reactions can be efficiently controlled [2]. Interestingly, the first lab-scale flow reactors for anionic polymerization were already introduced in the 1970s by Schulz and coworkers. They employed what they then coined tube reactors to study the kinetic of polymerizations with high precision [3]. In many ways, this work already demonstrated the potential of flow chemistry long before the technology made comeback decades later. Müller followed the early work in a series of investigations into anionic polymerization kinetics [4–6]. Much later, Frey and coworkers used the same principle to produce styrene- and 4-tert-butoxystyrene homo- and block copolymers in microflow reactors with molecular weights of up to 70 kDa with a narrow dispersity [7]. Yoshida and coworkers also demonstrated anionic polymerization of styrene in flow, allowing for synthesis of low dispersity materials at 0 °C, thus impressively underpinning the sophistication that flow chemistry can provide for such reactions (see Figure 4.2). Next to styrene, also methacrylate polymerization was successfully carried out [8]. Polymerizations could be carried out at 0 °C, yet for methacrylates slightly higher dispersities were obtained in the range of 1.2–1.3. Still, short reaction times on the scale of seconds to reach full monomer conversions were only required. Mixed block copolymers made from styrene and methacrylates were also achievable under similar conditions [9, 10], as well as mixed acrylate–methacrylate block copolymers [11]. Hence, a broad variety of (block) copolymers could be obtained in these flow reactors.

Since reactions are fast and reach high conversions easily, block copolymers are achievable with relative ease. Further, reactions are also well scalable, as was also shown by Yoshida and coworkers, when the anionic styrene polymerization was scaled to 1 kg production in 3 h, without any loss of definition of the product [12].

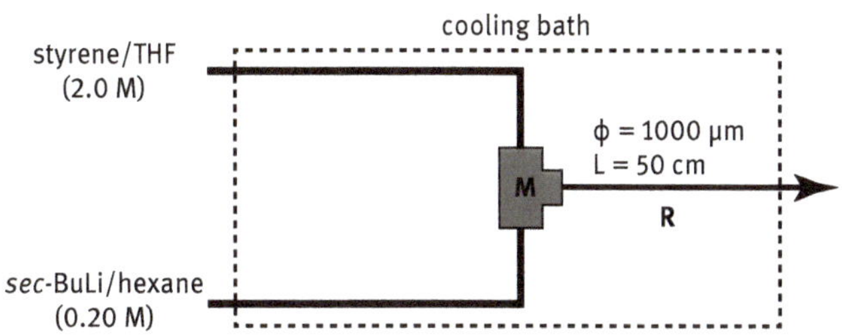

Figure 4.2: Example of an anionic polymerization carried out in a flow reactor [9]. Copyright (2010). Reproduced with permission from American Chemical Society.

(For further details on these approaches please see Volume 1, Chapter 2, Title: Principles of controlling reactions in flow chemistry)

4.3 Homogeneous radical polymerization

Polymerizations in flow can either be carried out in homogeneous or heterogeneous reaction conditions. While heterogeneous, slug flow type polymerizations are somewhat more difficult to realize, they are more tolerable for viscosity influences. Hence, homogeneous polymerizations are mostly carried out on controlled polymerization methods, where molecular weight, and hence the overall viscosity of solutions stays comparatively low. For the same reason, branched polymers can be obtained well in flow reactor, as they are also associated with lower viscosities. Conventional radical polymerizations typically result in high-molecular-weight polymers, and hence are found less in micro- or millisized continuous flow reactors. This does not mean though that they cannot be carried out in flow. In fact, most high-volume industrial production of radical polymerization products is obtained in continuous reactors. These are them, however, typically specifically designed toward their purpose. Yet, some work has been carried out on free radical solution polymerizations. Yoshida and coworkers studied various conventional (meth) acrylate polymerizations initiated by azoisobutyronitrile AIBN. Unsurprisingly, they found that flow reactors yield more consistent molecular weight distributions throughout their studies, closer to ideal polymerization behavior [13]. This can be explained by the much more stable reaction condition, and the often cited history dependence of conventional radical polymerizations; small changes during the initial (highly exothermic) period of the reaction influence the viscosity, and in consequence the kinetics of the following reaction. High molecular weights could be obtained generally, up to several 100 kDa, in small scale microreactors [14]. As a further interesting result, Ito and coworkers tested uncontrolled free radical polymerizations in laboratory and 100 t per year plant-sized reactors, showing that independent of production volume, very consistent products can be obtained, with variations below 5% in molecular weight and dispersity [15]. The accuracy of flow reactors was further used by Baxendale and coworkers who polymerized acrylic acid [16]. They were able to interpolate the conditions required to reach specific product characteristics and to scale the process with high accuracy.

Controlled radical polymerization, also referred to as reversible deactivation radical polymerization (RDRP), has been studied in much more detail for flow reactions. Practically all major RDRP routes have been studied in this way, including reversible addition fragmentation chain transfer (RAFT), polymerization [17–19], nitroxide-mediated polymerization (NMP) [20, 21], atom transfer radical polymerization (ATRP) [22–24], or Cu (0)-mediated single electron transfer polymerization [25–28]. Next to a higher consistency with respect to reaction rate, RDRP reactions in flow yield better results with respect to end-group chemistry and polymer dispersity [29]. Monte Carlo simulations of the flow polymerization kinetics demonstrated that this improvement can be mostly traced to the isothermicity in microflow devices [30]. Yet, while the improved isothermicity allows for a higher precision in the polymerization outcome, the laminar flow profile and the residence time distribution (RTD) connected to this also have a broadening effect on the dispersity of products. In fact, it was shown that if ideal plug flow conditions are achieved via Taylor flow conditions, the lowest dispersities can be accessed [31–33]. This should, however, not be read as a general advantage of Taylor flow over homogeneous laminar flow. The choice of mode of flow profile must be made with respect to overall viscosity and temperature. It may also not always be feasible to create a stable Taylor flow. More important than the flow profile is, particularly for fast reactions, mixing of initiator and monomer. For polymerizations that occur on the timescale of mixing (such as some anionic polymerizations), the product dispersity becomes a function of the mixing time [34].

4.3.1 Atom transfer radical polymerization

ATRP is one of the most widespread RDRP methods. While it requires generally copper or iron catalysts, it can be carried out in homogeneous conditions, or can employ metal reactor walls as catalyst. Already soon after the discovery of ATRP in the middle of the 1990s, first flow ATRPs were carried out [23, 35, 36]. These reactors used immobilized catalyst on a silica bed and were yet not overly efficient, but displayed already the typical advantages of flow reactions as described above.

Following these packed column continuous flow protocols, more classical laminar flow in homogeneous conditions was used in later work [37]. For instance, polymerization of 2-hydroxypropyl methacrylate (HPMA), St, and *n*BA was achieved at temperatures between 60 and 90 °C, resulting in polymers with molecular weights up to ~13 kg/mol with dispersities lower than 1.3. This is typically seen as a well-controlled polymerization with a good outcome. Other examples displayed even lower dispersities, and unambiguously demonstrated the high level of control that can be achieved in tubular reactors [38]. This marks also one of the first instances where the flow process was shown to be superior to classical batch approaches with respect to reaction outcome. Until this date, flow chemistry was seen more as an engineering tool allowing for scaling of these reactions, but with the advantage in precision being evident, flow polymerizations became slowly tools for novel synthesis and polymer reaction engineering in general.

ATRP has several subvarieties that work with different types of activators and modes of initiation. More or less all of these variants have been shown to be applicable to flow conditions. Activator regenerated by electron transfer (ARGET) is one of the main variants of ATRP that is interesting because it reduces the catalyst load to the ppm range. In consequence, this means that polymer products do not need to be purified and freed from the metal catalyst, which is attractive especially for continuous processes [39].

Copper tubing as both catalyst source and continuous flow reactor was also introduced as a facile ATRP-variant-type flow polymerization [40, 41]. The high surface area by the copper tubing reduced the amount of ligand required in the reaction, while retaining short reaction times, and well-controlled products. Next to using the reactor wall as catalyst source, also copper wire inserts to PTFE (polytetrafluroréthylene) tube flow reactors were investigated (see Figure 4.3) [42]. In such reactions, monomers could be polymerized up to 95% monomer conversion with dispersities under 1.2.

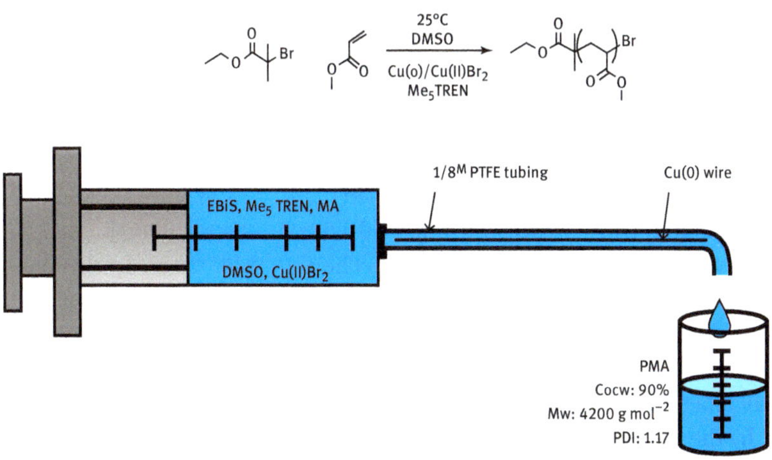

Figure 4.3: Example of a copper-mediated flow polymerization [42]. Copyright (2013). Reproduced with permission from Royal Society of Chemistry.

4.3.2 Nitroxide-mediated polymerization

In contrast to ATRP, the somewhat older method of NMP proceeds without use of metal catalysts. NMP requires though relatively high temperatures in order to establish the required reversible deactivation equilibrium. While high temperatures are a disadvantage in batch-wise processing, this can be comparatively easily realized in continuous flow reactors. Overpressure of the reactors gives further access to performing NMP in benign solvents. Serra and coworkers had shown that styrene and *n*-butyl acrylate can be well polymerized and controlled in flow using stainless steel reactors [43]. As advantageous as high temperature is for flow application, no direct improvements of the polymerization product were seen though when the same reaction temperature was used. However, flow processing allows to increase the overall reaction temperature to 140 °C, giving rise to product intensification and an overall acceleration without loss of reaction control. This effect was later confirmed in other investigations as well for a variety of vinyl monomers (see Figure 4.4) [44]. As NMP is generally less versatile compared to RAFT and ATRP, it is however, not as much used in practice despite its probably high industrial potential.

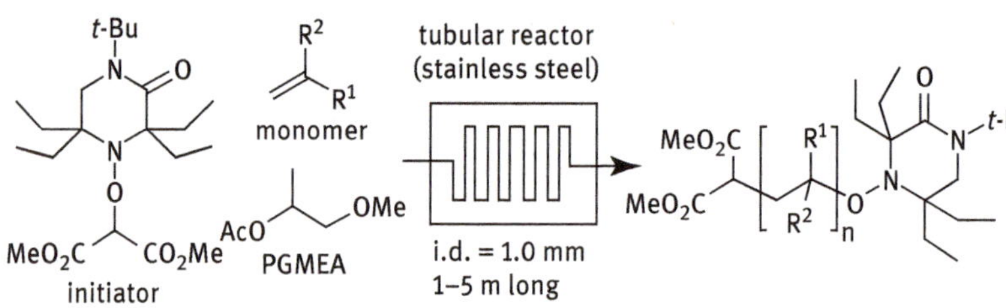

Figure 4.4: Example of a nitroxide-mediated polymerization in continuous flow [44]. Copyright (2012). Reproduced with permission from Royal Society of Chemistry.

4.3.3 RAFT polymerization

Homogeneous RAFT polymerization was first carried out in flow reactors in 2010 [45]. *N*-Isopropyl acrylamide (NIPAM) was polymerized with 2-dodecylsulfanylthiocarbonylsulfanyl-2-dimethylpropionic acid (DDMAT) as chain transfer agent. In this early work, PNIPAM was obtained at 88% monomer conversion after 1 h reaction time at 90 °C with a M_n of 21.5 kDa and a $Đ$ 1.4. Under similar experimental conditions only 40% monomer conversion is, however, reached in a batch reactor, demonstrating a very direct advantage of using flow polymerization. This significant improvement can be explained by the better mixing and heat transfer that is achieved in a flow reactor. Later, Hornung and team [46] not only polymerized NIPAM but also dimethyl acrylamide (DMAA), *n*BA, and vinyl acetate (VAc) using various thermal initiators and various RAFT agents. In all cases, RAFT polymers with narrow molecular weight distributions (Đ 1.1–1.3) and high conversions of 80–100% were synthesized with relative ease. The RAFT polymerization of DMAA was upscaled to a total daily output more than a kg without loss of precision of the polymerization with respect to dispersity and obtained average molecular weight.

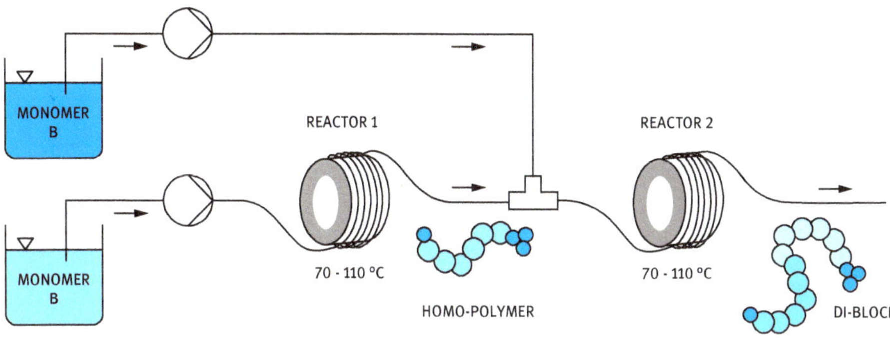

Figure 4.5: Experimental setup for the two-stage continuous flow process toward RAFT di-block copolymers [47]. Copyright (2013). Reproduced with permission from CSIRO publishing.

Further, the same authors explored monomer conversion profiles for reactors with different geometries [48]. Changing the reactor length while maintaining reaction time (i.e., increasing flow rate) appeared to have little effect on the conversion profiles. Increasing the inner diameter from 1 to 2.2 mm, though, increased the stabilization time. These changes were attributed to increased axial dispersion in the thicker tubing, which decreased the homogeneity of the polymerization mixture during the early stages of the reaction. Yet, overall RAFT polymerization can be scaled in flow reactors quite well, and product intensification can be reached by applying elevated temperatures. Yet, the stability of the RAFT end group (prerequisite for good control) limits the upper temperature range that is accessible, and typically from about 110–120 °C degradation of the RAFT group is observed. Today, the majority of RDRP reaction carried out in flow is of the RAFT type, as this technique gives access to a broad range of sophisticated macromolecular structures, caters for the broadest monomer compatibility and is carried out with ease in flow as no metal compounds are required to catalyze the reactions.

4.4 Ring-opening (metathesis) polymerization

Ring-opening polymerization (ROP) gives facile access to degradable poly(esters), poly(carbonates), poly(phosphoesters), poly(2-oxazoline) (POx), poly(ethers) and poly(amino acids), and this type of polymerization has found rapidly increasing interest in the past years in polymer science.

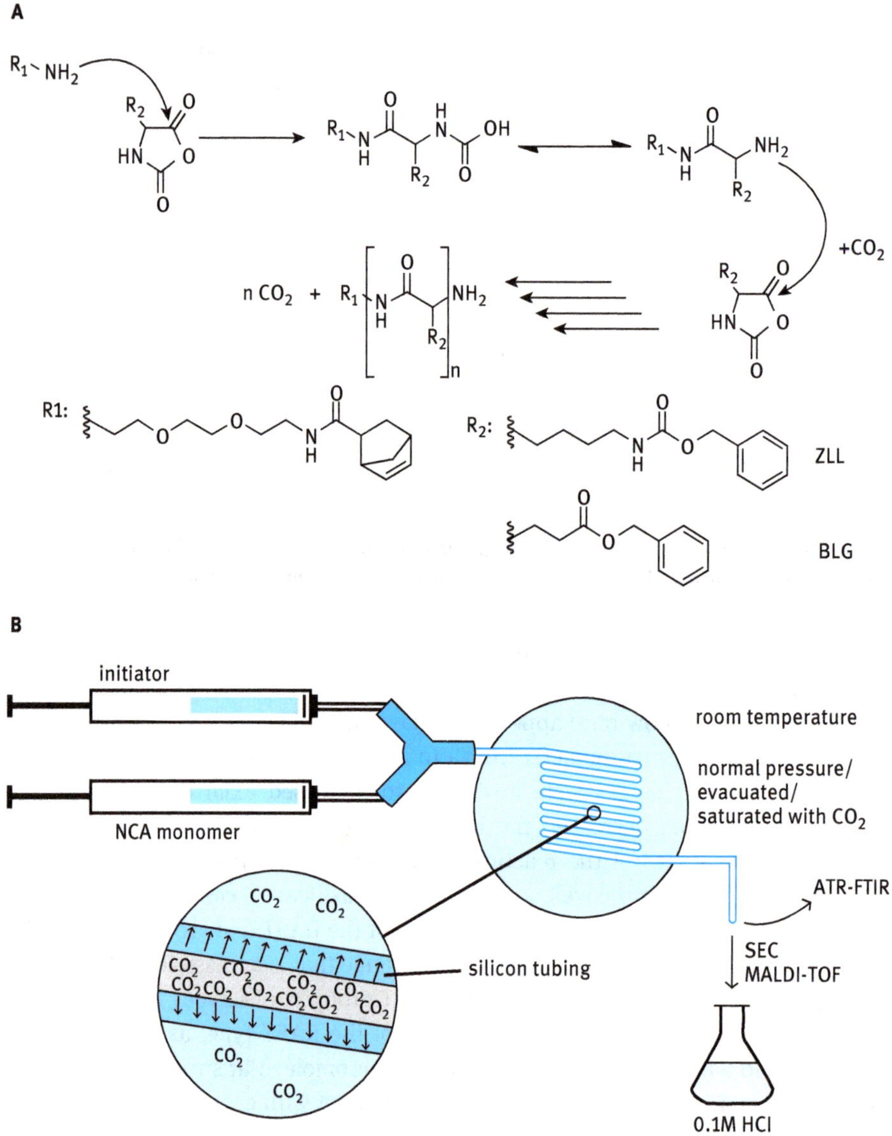

Figure 4.6: N-Carboxyanhydride (NCA) polymerization in continuous flow [49]. Copyright (2020). Reproduced with permission from Wiley.

The first example of a ROP in continuous flow was already demonstrated in 2005 [50] on the example of *n*-carboxyanhydride (NCA) polymerization, which form polypeptides. NCAs are cyclic monomers that become initiated by tertiary amines as initiator, followed by ring opening and decarboxylation (see Figure 4.6). These reactions are very sensitive to any impurity, and hence are ideal to be carried out in flow to improve the quality of polymerization. Also, they produce carbon dioxide as a side product, which needs to be removed to push the reaction equilibrium to the product site [49]. In flow, narrow molecular weight distributions and a significant increase in reaction yield were observed. As an example, polylysine with an M_n of 20 kDa and $Đ$ of 1.17 was obtained in this way. Hyper branched polyglycerols with high-molecular-weight fractions (M_n of 150 kDa) and narrow molecular weight distributions were synthesized by Frey and coworkers through the ROP of glycidol in microreactors [51]. Schubert and coworkers polymerized 2-ethyl-2-oxazoline (EtOx) in microwave-assisted continuous flow reactor. However, the residence time distribution in flow had a negative influence and broader molecular weight distributions were obtained when compared to batch, a problem that was later solved [52].

Aliphatic polyesters such as poly(lactide) (PLA) and poly(ε-caprolactone) (PCL) are important biopolymers due to their excellent biocompatibility, biofeedstock, and biodegradability [53]. Classically these polymer are obtained employing $Sn(OTf)_2$ as catalyst. While tin-catalyzed ROP is possible and demonstrates the typical advantages over classical batch reactions, more focus was given in recent years to organocatalyzed ROP, which again is much better performed in flow reactors due to the absence of metal. Further, also the obtained polymers are of better quality and biocompatibility when organocatalysts are used. Commercially available 1,5,7-triazabicyclo[4.4.0]dec-5-ene (TBD) is, for example, a highly active organocatalyst for the ROP of lactides and lactones [54].

Guo and coworkers investigated the polymerization of TBD-catalyzed CL and δ-valerolactone (VL) in continuous flow [55, 56]. Apparent rate constants were doubled in flow compared to batch for both CL and VL. Further, when coupling two flow reactors in sequence, well-defined block copolymers of PVL-*b*-PCL and PCL-*b*-PVL could be obtained directly. The TBD-catalyzed polymerization of L-lactide was also optimized for continuous flow, allowing to polymerize this monomer fully within seconds of reaction time [57]. PLA with molecular weights up to 44 kDa and narrow dispersities of less than 1.3 were obtained even with only moderate to low catalysts loadings of around 1%. This exceptionally fast reaction was later subject to creation of a programmable and automated continuous flow system by Waymouth and coworkers (see Figure 4.7). Using automation and this fast reaction, 100 different distinct polymers were obtained in merely 9 min. In their reactions, they used even more active urea catalysts to rapidly polymerize LA, VL, CL, and trimethylene carbonate (TMC) [58]. Impressively, full conversion was practically reached in just 6 ms to 2 s reaction time with these catalysts, while keeping dispersity low ($Đ < 1.14$) and molecular weights high (25,000 Da).

ROP can also be carried out using enzymes at catalyst, for example, *Candida antarctica* Lipase B (CALB) in the form of Novozym 435 (N435). Continuous flow ROPs using immobilized enzymes were used consequently also carried out (see Figure 4.8) [59, 60]. Due to the high surface area to volume ratio and better mass transport, the apparent polymerization rate was increased by a factor 24 when compared to classical reactions. Viscosity increased steeply with monomer conversion and higher molecular weights, which complicates the use of a catalyst packed bed. Hence, ultrasound assisted continuous was carried out [61], to enhance the high viscosity processing of PCL.

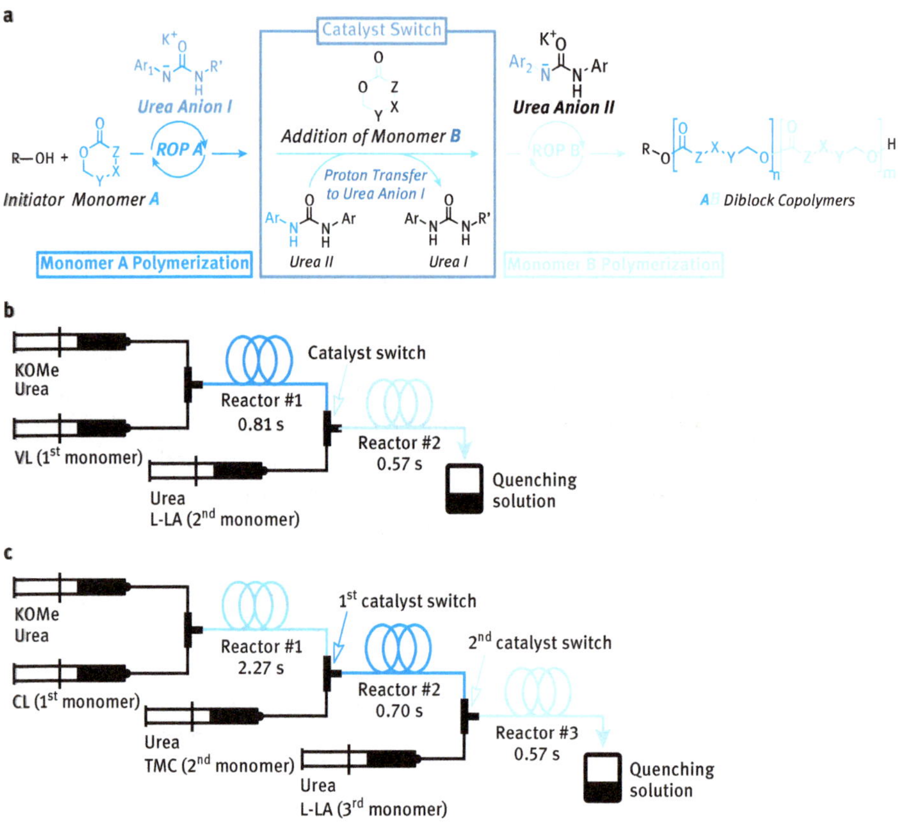

Figure 4.7: (A) Block copolymer synthesis with catalyst switch. (B) Catalyst switch based on proton transfer. (C) Reactor setup for synthesis of ABC triblock copolymer with sequential catalyst switches [58]. Copyright (2019). Reproduced with permission from American Chemical Society.

POx is a class of biocompatible polymers with tunable smart responses. POx is widely considered as a promising alternative to poly(ethylene glycol) (PEG) in biomedical application. POx triblock copolymers were synthesized in one pass using microreactor cascades in cationic ROP of EtOx and 2-*n*-propyl-2-oxazoline (nPropOx) [62]. Full monomer conversion could be reached for each block in 5 min at 160 °C and various sets of triblock copolymers EtOx-*b*-*n*PropOx-*b*-EtOx and *n*PropOx-*b*-EtOx-*b*-*n*PropOx were synthesized (see figure 4.9).

A last type of ROP that has been studied for flow polymerization is the synthesis of poly(phosphoesters). Poly(phosphoesters) are obtained via transesterification and

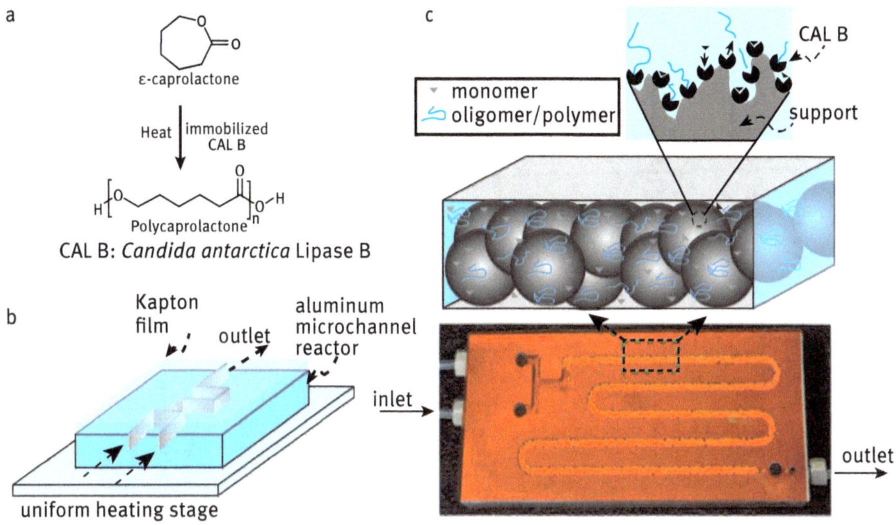

Figure 4.8: (a) Reaction scheme for ring-opening polymerization of ε-caprolactone to polycaprolactone. (b) Schematic of the microreactor setup. The microreactor was made of aluminum and was covered with Kapton film using a thermally cured epoxy. The microreactor was placed on a uniform heating stage for temperature control. (c) Image shows photograph of a typical microreactor used in this study. CAL B immobilized solid beads (macroporous poly(methyl methacrylate)) were filled in the channel [59]. Copyright (2011). Reproduced with permission from American Chemical Society.

polycondensation, or more commonly via metal- and organocatalyzed ROP to produce polymers with controlled molecular weight and narrow molecular weight distributions. 2-Isobutyoxy-2-oxo-1,3,2-dioxaphospholane (iBP) and 2-butenoxy-2-oxo-1,3,2-dioxaphospholane (BP) were polymerized within rapidly to full conversion in continuous flow, again using TBD as catalyst. In addition, in a telescoped flow process, the alkene functionality of PBP was modified in a post-polymerization polymer analogous conversion using UV-induced thiol-ene chemistry to yield highly functionalized poly(phosphoesters) polymers [63].

Ring-opening reactions can also form purely carbon–carbon polymer backbones via metathesis polymerization. Ring-opening metathesis polymerization (ROMP) is a widely used technique to synthesize poly(olefins) with similar level of control as RDRP and ROP from norbonene monomers. In 2019, Hobbs and coworkers demonstrated the first ROMP carried out in continuous flow [64]. A fast initiating 3rd generation Grubbs initiator was used to obtain poly(norbornene) in 95% monomer conversion in 22.5 s reaction time. While the outcome of the polymerizations was a little less favorable compared to its batch counterparts, a coupled flow process was further set up, giving access to poly(norbornene) block copolymers.

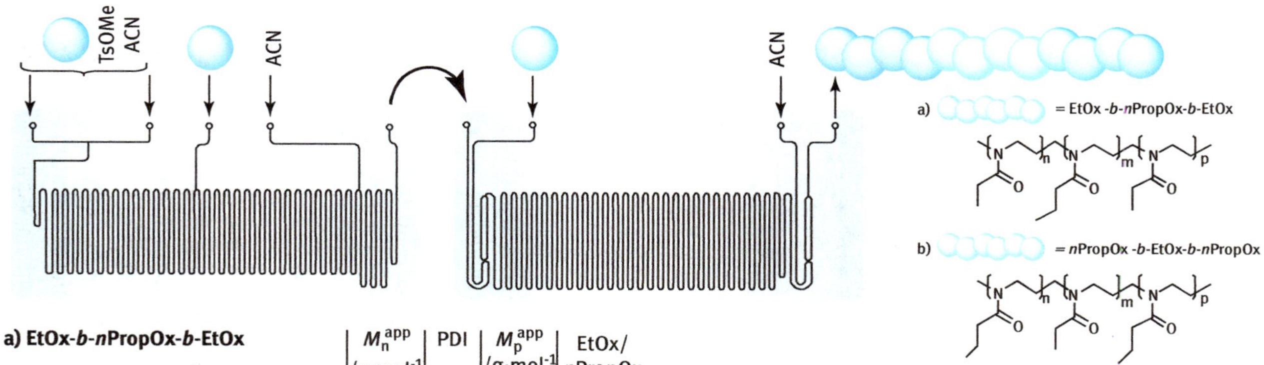

a) EtOx-*b*-*n*PropOx-*b*-EtOx

	M_n^{app} /g·mol^{-1}	PDI	M_p^{app} /g·mol^{-1}	EtOx/ *n*PropOx
EtOx homopolymer	3510	1.10	3740	-
EtOx-*b*-*n*PropOx diblock	6140	1.12	7130	1 / 1.18
EtOx-*b*-*n*PropOx-*b*-EtOx triblock	7400	1.25	10910	2 / 1.02

b) *n*PropOx-*b*-EtOx-*b*-*n*PropOx

	M_n^{app} /g·mol^{-1}	PDI	M_p^{app} /g·mol^{-1}	*n*PropOx EtOx/
*n*PropOx homopolymer	3940	1.09	4210	-
*n*PropOx-*b*-EtOx diblock	5540	1.17	6490	1 / 0.85
*n*PropOx-*b*-EtOx-*b*-*n*PropOx triblock	7320	1.21	9660	2 / 1.11

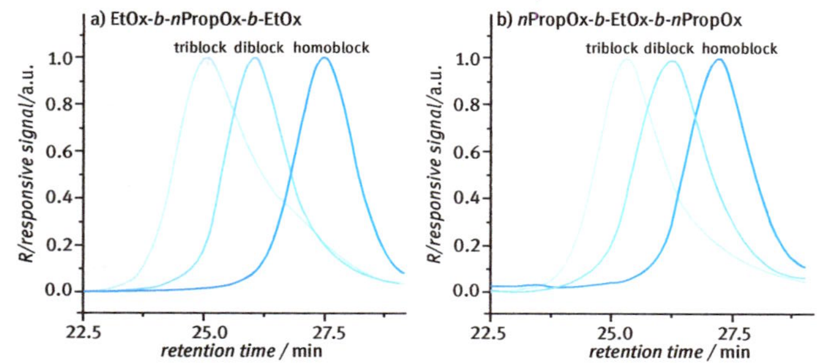

Figure 4.9: Schematic representation of the microfluidic cascade consisting of a two-stage 15 μL and a 19.5 μL microreactor. Injection of pure monomer (in a 1/1 ratio to the first monomer) via the additional inlet in the 15 μL reactor enables the formation of diblock copolymers, while the injection of monomer in the 19.5 μL reactor leads to triblock copolymers [62]. Copyright (2015). Reproduced with permission from Royal Society of Chemistry.

4.5 Photopolymerization

As described above, thermal polymerizations can greatly benefit from continuous flow applications. Generally, polymerizations are better behaved in flow, yield polymers with lower dispersity and better control over average molecular weight. Further, the applicable temperature window is increased, allowing for facile product intensification. While the majority of polymers in research and in industrial application are produced via thermal polymerization, photopolymerization is well known since decades and used in curing applications, surface modifications and also for solution polymerization. Of course, it is very worthwhile to explore the scope of flow chemistry toward photopolymerization, as the typical advantages should apply. The low optical pathlengths in micro- and millireactor allow for efficient illumination of all reactant, giving rise potentially to very high polymerization rates (as in thin film curing devices). The difference in illumination between a batch reactor and a tubular reactor wrapped around a light source is shown in Figure 4.10 for a fluorescent polythiophene solution. At the same time, lower energy light sources can in principle be used, making processes more economic, and avoiding formation of degradation products. Interestingly, almost all photopolymerizations can be initiated by light, yet propagation is exclusively thermally activated. This allows to decouple the initiation from the propagation step and gives rise to interesting possibilities toward process intensification as will be discussed below. At the same time, this distinction sets photopolymerizations apart from small molecule reactions, where light is continuously needed to drive a reaction.

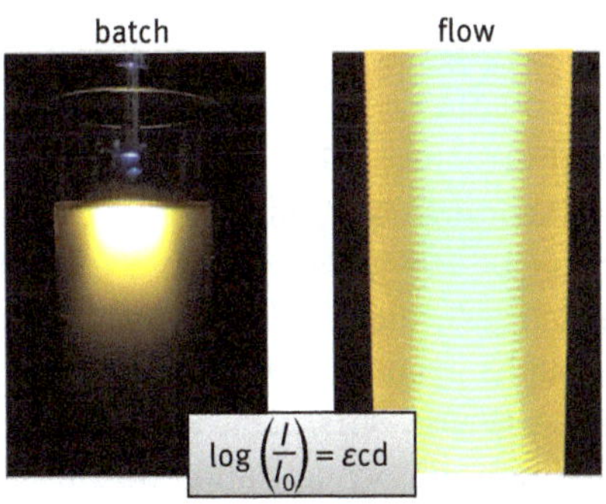

Figure 4.10: Light intensity profile in a batch reactor (left) and a tubular flow reactor (right). Copyright (2015). Reproduced with permission from Royal Society of Chemistry.

In order to perform light-induced polymerization, transparent reactors need to be employed. Typical wavelengths for initiation are UV light, but in recent years also the full spectrum of visible light is used. While glass chips are convenient and wavelength transparent, they often feature high refraction at the glass surface, which can lead to imperfect use of photons. Plastic tubing, on the other hand, can be permeable to oxygen, and hence lead to inhibition of polymerizations [65, 66]. Hawker and coworkers investigated which type of tubing is most ideal for polymerizations, on the example of a photo-controlled ATRP polymerization. In photo-ATRP, a photoelectron transfer catalyst is used to activate the bromine-terminated chains, and thus to initiate chain growth [67]. Out of four different tubing materials, namely PFA (which is also mostly used for thermal polymerizations), FEP, Tefzel and Halar were tested and compared. Polymerizations proceeded faster for tubing with lower oxygen permeability. Halar showed best performance. PFA was found to be a reasonable tubing material as well.

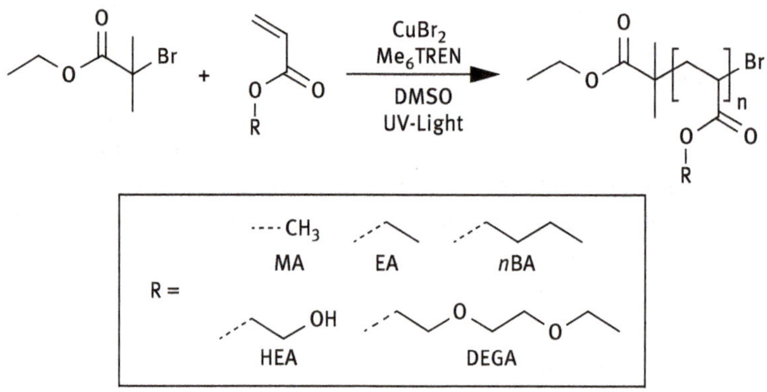

Figure 4.11: Example of photo-ATRP polymerizations carried out in continuous flow [68]. Copyright (2016) Reproduced with permission from Springer.

The first photomediated RDRP was presented in 2014 on photo-ATRP polymerization in tubular reactors as well as for microchip reactors [69]. Using UV light and active cooling of the reactors allowed to avoid the formation of midchain radicals in acrylate polymerization, which would otherwise be inevitable to disturb the reaction. Reactions were highly efficient, proceeded fast, and chain extensions were possible using other acrylates (and later also methacrylates). Polymerizations were scalable to hundreds of grams on lab scale (see Figure 4.11) [68]. Following this early work, also photoinduced cobalt-mediated radical polymerization was investigated, giving access to PVAc and PVAc-*co*-P(1-octene) polymers which were unobtainable in same quality in batch reactions [70]. A remarkable rate increase was observed for these reactions in continuous flow, and the involved researchers realized soon that photopolymerization can greatly benefit from elevated temperatures.

This temperature effect was first investigated in depth for RAFT polymerizations that were initiated by classical photoinitiators. Obviously, when the light intensity was decreased, the polymerization slowed down [71]. Various photoinitiators were also examined to find the optimal initiator for the reaction. Benzoin was identified as the best choice, yet the overall dispersity of the obtained polymer depended on the initial initiator concentration. Interesting, absence of any initiator still gave good results in terms of reaction rate and product specification. By increasing temperature, the lack of initiator (and hence low radical flux) could be compensated to reach similar overall polymerization rates. In the absence of photoinitiators, the RAFT agent (or the macro-RAFT agent for that matter) cleaves itself upon irradiation of a suitable wavelength, and forms radicals ready to propagate. The mechanism is then superimposed with a photoiniferter mechanism. Chen and Johnson also conducted an investigation into photoiniferter RAFT polymerization at room temperature [72]. Using symmetrical trithiocarbonate RAFT agents, triblock copolymers were synthesized from acrylates and acrylamides in tubular reactors. Using this strategy, high-molecular-weight polymers of up to 100,000 g/mol were achievable (see Figure 4.12).

Returning to the question of elevated temperatures, this combination can be applied to sluggishly propagating monomers, which under purely thermal conditions lead to unsatisfyingly long reaction times (and hence remain largely out of reach for continuous flow reactors) became processable. First, they showed that methacrylate polymerization could be performed to high monomer conversion within 10–30 min. The high temperatures used for polymerization increased the propagation and hence overall polymerization rate and had no significant effect on the photocleavage of the RAFT agent itself. The polymerizations remained well controlled and delivered polymers with low dispersity even at higher temperatures. Various methacrylates were polymerized with success, showing the versatility of the approach. Also block copolymers were obtained in a telescoped reactor approach (see Figures 4.13 and 4.14). After that work, also isoprene and styrene were polymerized with great success through so-called RAFT photoiniferter polymerization at elevated temperatures (120° C) [73] (see Figure 4.15). For isoprene, which typically required more than a day to reach reasonable monomer conversions, the UV-induced flow protocol allowed polymerization to reach completion in 30–60 min, representing a ~ 50 fold increase in reaction rate. The use of RAFT photoiniferter polymerization yields polymers with high end-group fidelity and hence allows the formation of well-defined PI-*b*-PSt diblock copolymers and PSt-*b*-PI-*b*-PSt triblock copolymers. The authors demonstrated that the product quality and reaction rates were largely unaffected by increases to the reactor volume, hence providing an attractive route for fast upscaling of the polymerizations.

Next to photoiniferter RAFT, the so-called photoelectron transfer RAFT (PET-RAFT) can also be used in flow reactors. PET-RAFT can be activated by a broad wavelength range up to green light, and is also oxygen tolerant, removing the need for deoxygenation or nonpermeable tubing [75]. This oxygen tolerance was demonstrated using ZnTPP catalyst for the homopolymerization of acrylamides. As for the other described reactions, also here block copolymers could be accessed with ease by connecting two reactors together. Since reactions are fast, polymerizations can be achieved in as little as 15 min of polymerization time [76, 77]. Interestingly, these reactions were not only be used to produce linear polymers of narrow dispersity,

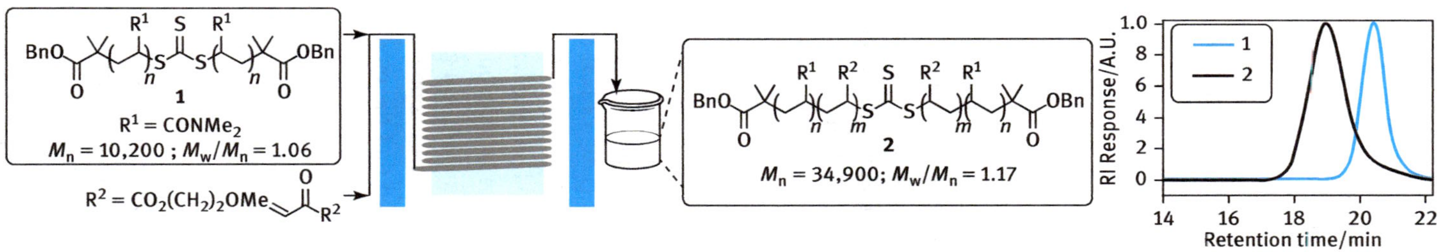

Figure 4.12: Triblock copolymer synthesis using a telechelic RAFT agent in flow under light irradiation [72]. Copyright (2015). Reproduced with permission from Royal Society of Chemistry.

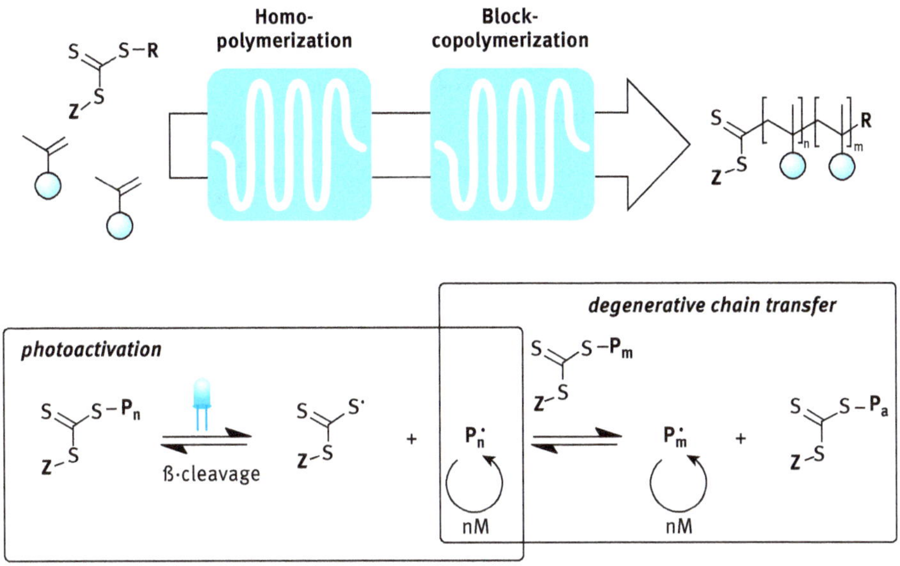

Figure 4.13: Mechanism of the photo-RAFT polymerization via an iniferter mechanism [74]. Copyright (2017). Reproduced with permission from Royal Society of Chemistry.

but also for the semiautomated generation of tailor-made molecular weight distributions by mixing of different polymers, a strategy that can also be used for thermal polymerizations [78, 79]. In this way, polymers with varying physical properties, but identical microstructure and overall average molecular weight can be obtained.

Other applications of photo-RAFT to polymerizations in flow reactors extend nanoparticle synthesis. In polymerization-induced self-assembly (PISA) different nanomorphologies can be obtained for block copolymer nanoparticles. As an example, PISA can be done in a tubular flow reactor using a trithiocarbonate macro-RAFT agent PEG, which was in-situ polymerized with HPMA in water at 37 °C under blue light irradiation. HPMA is water soluble, yet upon polymerization becomes insoluble, triggering the self-assembly during the polymerization process. A wide range of morphologies including spheres, worms, and vesicles could be accessed (see Figure 4.16) [80]. Generally, nanoparticle formation is of high interest for flow application [81–83]. While the above PISA example remained in thermodynamic equilibrium, one can also use flow mixing to kinetically trap specific morphologies, and hence create different particles from the same block copolymer. This was demonstrated by dissolving a polystyrene-b-poly(hydroxyethyl acrylate) polymer in THF and mixing it at different flow rates and mixer geometries with water [84]. As elucidated with neutron scattering, the shape and size of the obtainable nanoaggregates vary with the flow properties. Since the same parameters cannot be tuned in the same way in batch processing, this approach allows to explore new pathways in block copolymer self-assembly, and the application of the residual nanoparticles in applications.

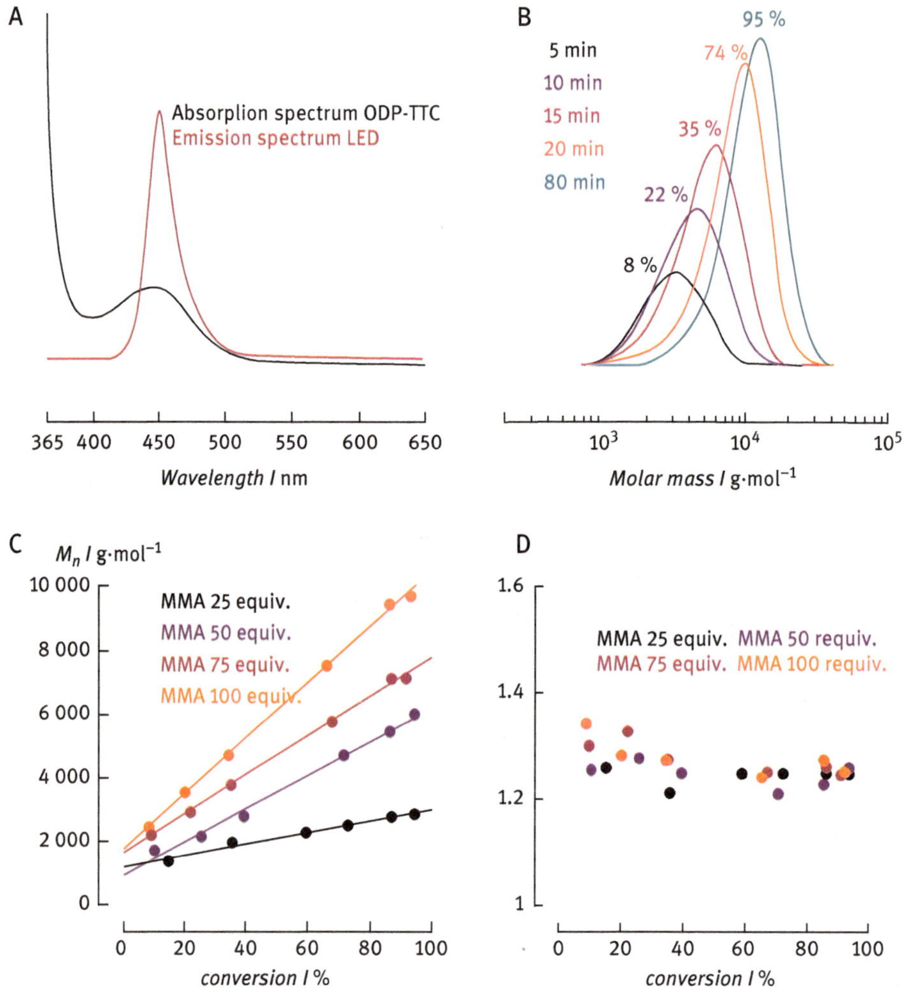

Figure 4.14: Typical results as obtained from photo-RAFT polymerization in continuous flow. (A) The relevant light absorption and emission spectra, (B) the evolution of molecular weight distributions during polymerization, (C) average molecular weight as a function of conversion, and (D) the change in dispersity with conversion [74]. Copyright (2017). Reproduced with permission from Royal Society of Chemistry.

As mentioned above, photo-ATRP was one of the earliest examples of continuous flow photopolymerizations. Yet, these reactions still required metal catalysts. Interestingly, photo-ATRP can also be induced using organocatalysts, then termed O-ATRP. Also these reactions can be performed well in flow [86]. Miyake and team polymerized a range of methacrylates using O-ATRP in continuous reactors [87]. As for most flow polymerizations, also here a greater precision in the polymerization in

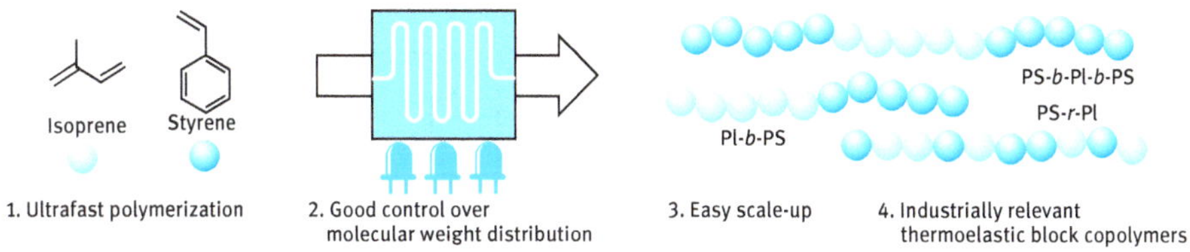

Figure 4.15: Reaction of isoprene and styrene ion photoiniferter polymerization [73]. Copyright (2019). Reproduced with permission from Wiley.

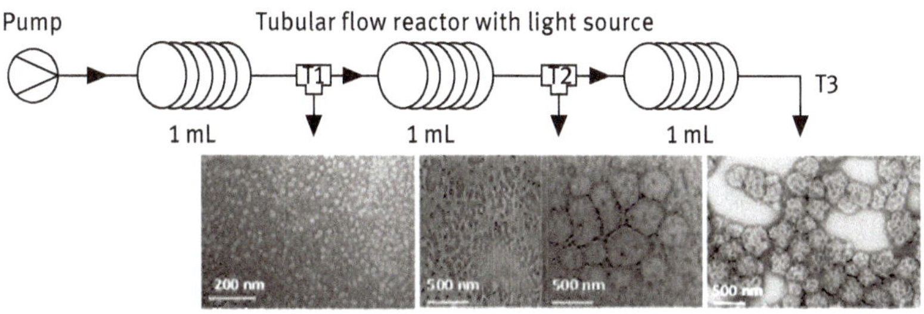

Figure 4.16: Reactor setup for the synthesis of a wide range of nanoparticle morphologies via polymerization-induced self-assembly in continuous flow [85]. Copyright (2018). Reproduced with permission from American Chemical Society.

terms of dispersity and average molecular weight was observed, alongside a reaction rate increase. Since O-ATRP does not use any basic ligands, it also makes polymerization of acidic monomers accessible [86].

(For further details on flow photochemistry please see Volume 2, Chapter 1, Title: Photochemical transformations in continuous-flow reactors)

4.6 Polymer modification in continuous flow

Next to polymerizations, also post-polymerization modification reactions can be interested for flow application. Many different examples exist for successful transformations of polymer end groups, or polymer-analogous reactions where a chemical moiety per monomer repeat unit is reacted. Some of the reactions have already been discussed above. For example, for the removal of RAFT end groups, amines can be added to quench the polymerization, reducing the end group to a thiol. These thiols can then be used for Michael addition modification in thiol-ene reactions (see Figure 4.17) [88, 89].

Seeberger and coworkers have utilized continuous flow systems for functionalizing sequence defined poly(amidoamine) (PAA) through thiol-ene click chemistry

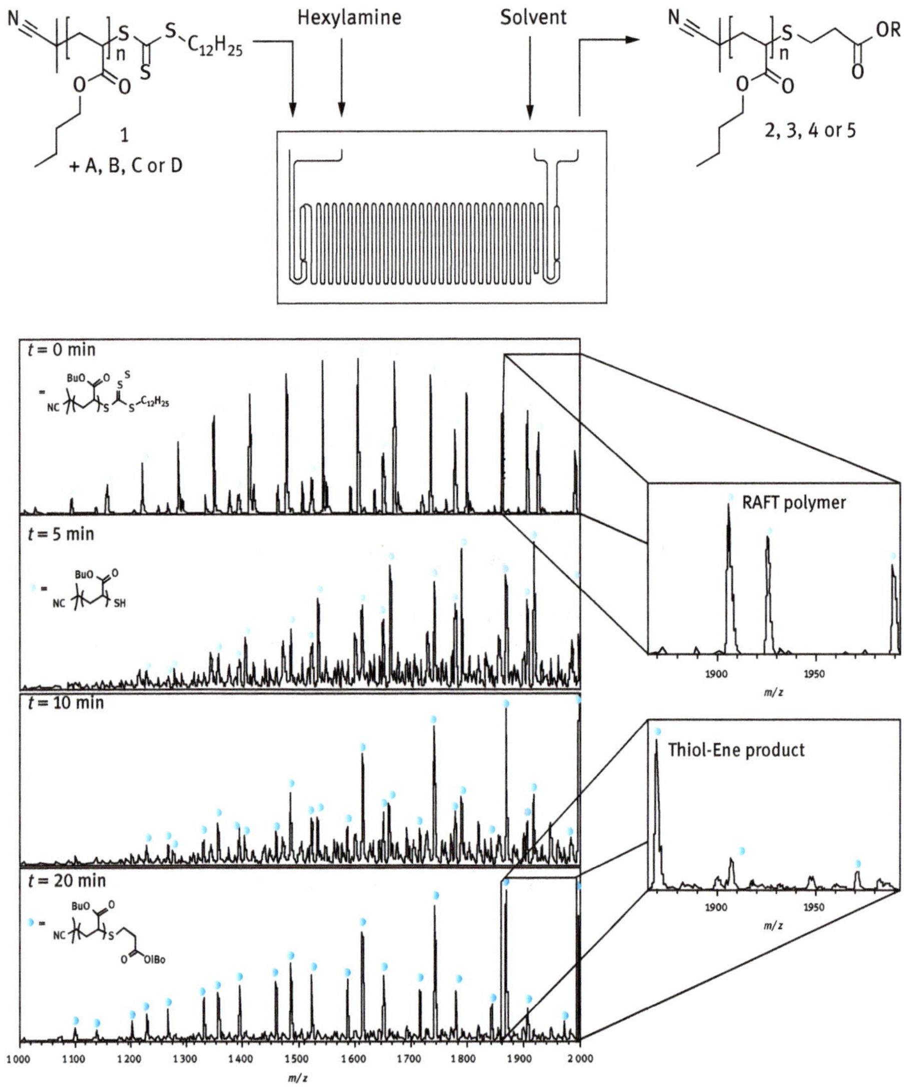

Figure 4.17: Thiol-ene end-group modification of RAFT polymer in continuous flow. The mass spectra show the progress of the reaction in time [90]. Copyright (2014). Reproduced with permission from Royal Society of Chemistry.

starting from solid-phase peptide synthesis followed by addition of β-thioglucose or α-thioethyl-mannose under UV irradiation in continuous flow [91]. Other examples include the end-group modification of ATRP and RAFT polymers and the coupling of such end-group modified chains via a CuAAC click reaction to obtain diblock copolymers [90]. In such an example, homopolymers were first obtained by ATRP and RAFT polymerization following the protocols described above; then the ATRP-derived chains feature their characteristic bromine end group, which can be substituted by an azide. The RAFT polymers can be reacted in a concomitant aminolysis and Michael addition with propargyl acrylate to obtain alkyne-terminated polymers. As is the case for polymerizations, also end group transformations are increased in rate when being performed in flow. Lastly, the azide-terminal and alkyne-terminal polymers are then clicked together in a copper-catalyzed azide–alkyne cycloaddition quantitatively. Interestingly, after optimization this reaction only needed 40 min to proceed to completion, compared to hours in batch.

Another interesting application is the use of UV light to transform end groups. [2 + 2] cycloadditions can be used in facile manner to functionalize ATRP polymers [92]. In this case, polymers featuring aldehyde end groups (introduced via the ATRP initiator) can be activated via UV, and with thioxanthane as photosensitizer, very fast oxetane formation is observed (~1 min), allowing to introduce new functional groups via the choice of alkene used in the [2 + 2] cycloaddition. Reaction in continuous flow did thereby not only lead to an increase in reaction rate, but also allowed to reduce the amount of ene component used in the reaction from 20-fold excess being required in batch versus equimolar rations being permitted in flow (see Figure 4.18).

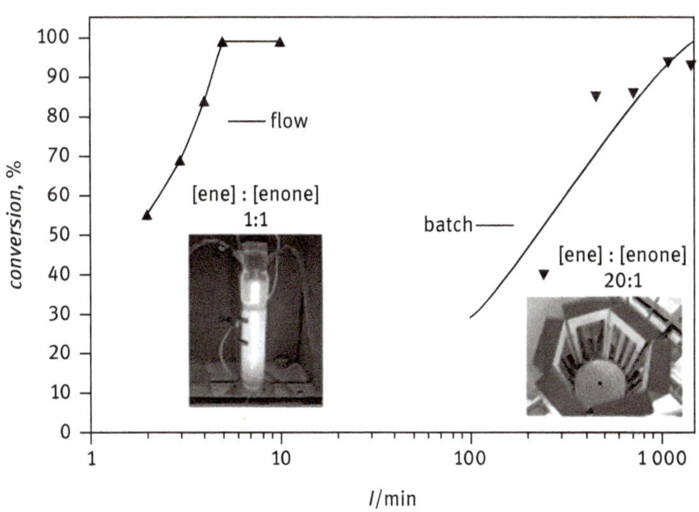

Figure 4.18: Comparison of reaction rate for the alkene-enone cycloaddition between maleimide and 1-octene in a flow reactor compared to a batch reactor, demonstrating the higher efficiency achieved in continuous flow reactors [93]. Copyright (2015). Reproduced with permission from Elsevier.

4.7 Online monitoring of continuous flow polymerizations

As shown in the previous sections, flow polymerization and end-group modification are attractive reactions to be carried out in flow. They unfold their potentially specifically when being combined in telescoped reactors, and many examples can be found where two or more stages are combined to yield complex polymers. A recent example of such synthesis is the star polymer formation via an arm-first polymerization approach. In such a reaction, linear polymers are independently polymerized, and then upon addition of specific amounts of difunctional monomers, they were cross-linked to yield star-shaped macromolecules featuring several dozens of individual arms. Junkers and coworkers used this principle to obtain to acrylate polymer in parallel via photo-ATRP, and then to mix the product streams of those both reactants together with a diacrylate to produce said star polymers, where two different polymer types are mixed in the material. Such telescoping is highly attractive as it allows to obtain these high value-added materials directly from monomers without requiring any purification (see Figure 4.19).

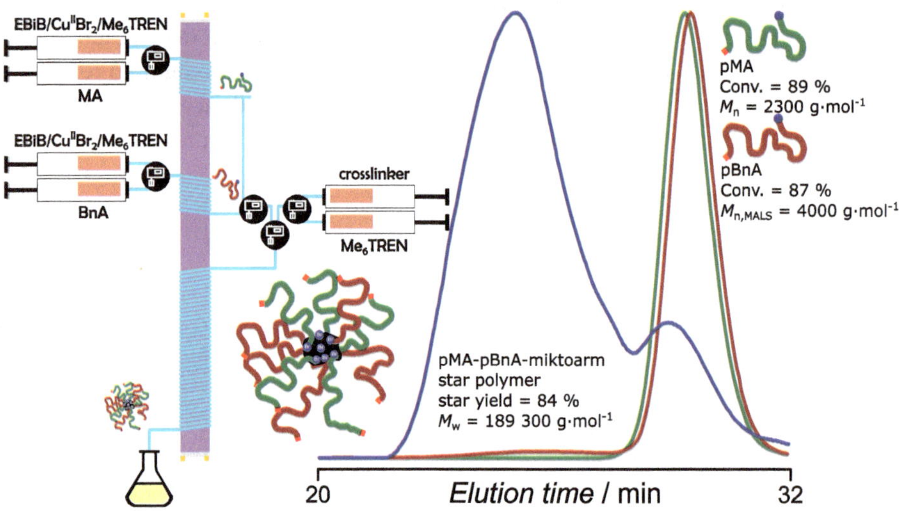

Figure 4.19: Telescoped synthesis of miktoarm core cross-linked star (CCS) polymers in a flow reactor cascade. Left, schematic visualization of the reactor. Right, SEC elugram of the MIs and resulting miktoarm CCS polymer [94]. Copyright (2019). Reproduced with permission from Royal Society of Chemistry.

As interesting as telescoping is (mostly used for sequences of polymerization followed by end-group medication, or for the synthesis of sequence-controlled multiblock copolymers), it suffers from the drawback that each individual stage must be monitored precisely. High conversions at each stage are obligatory, as residual monomer can have a detrimental effect on the following reactions. For example, in block copolymer synthesis, presence of residual monomer will lead to formation of a statistical copolymer in the second block rather than a pure block copolymer. Hence, the introduction of online monitoring tools that provide direct feedback with high time resolution to the operator of a reactor is of high interest. Monitoring can be carried out in-line, thus directly observing the mass flow in the reactor, or online via a sampling method, in which a side stream of the product stream is continuously analyzed. Only few methods are directly applicable to in-line monitoring, as product concentrations and volumes are typically high.

Online analysis is also very useful in cases where reactions cannot be easily quenched, or where toxic and reactive intermediates are formed. Further, online monitoring gives access to very systematic kinetic datasets which allow for a precise prediction of reaction parameters. At the same time, it can also be used as process control feature [95–98]. This is an attractive feature especially for industry where constant product output is required without significant variation of product features.

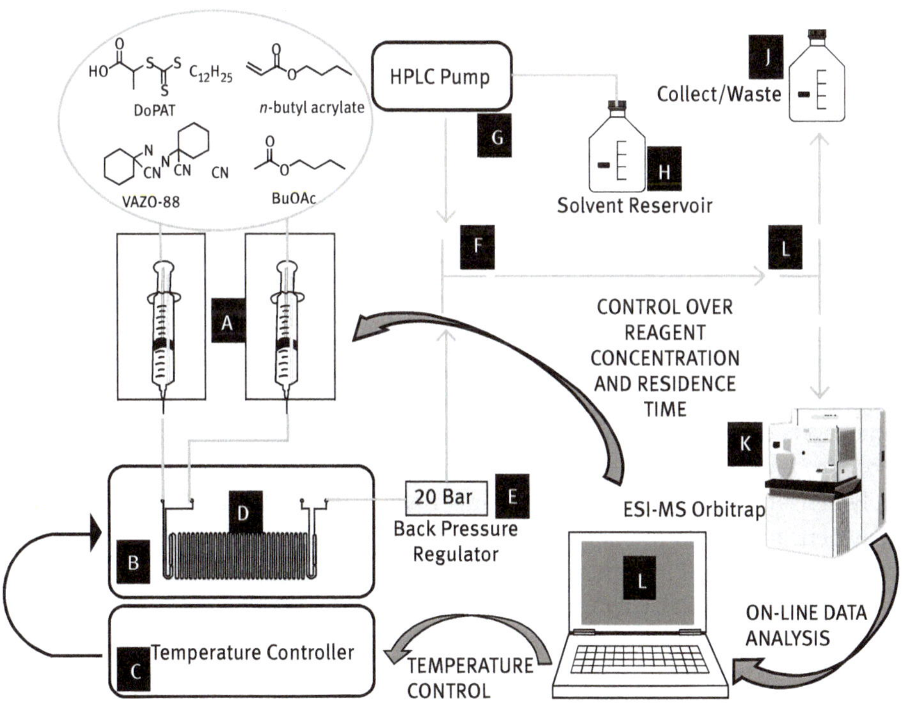

Figure 4.20: Schematic representation of online microreactor ESI-MS coupling [100]. Copyright (2015). Reproduced with permission from Royal Society of Chemistry.

For flow reactions in general, the analytical techniques applied online are typically HPLC [100], MS [101, 102], fluorescence spectroscopy [103], IR [104], Raman [105], UV-Vis [106], NMR [107], and SEC [108]. For flow polymerizations, mostly SEC, NMR, FT-IR, and MS methods are relevant to probe the progress of polymerization, end-group fidelity, and chain length distribution. An early example for online mass spectrometry in the field of polymerizations is the monitoring of the early stages of the Brookhart alkene and Ziegler-Natta ethene polymerization [109, 110]. This technique was later fully developed by Junkers and coworkers and applied to various polymer reactions using glass chip microreactors [99]. One example is the monitoring of the end-group modification via Passerini three-component reactions to obtain diblock copolymers [101]. Another is the synthesis of multifunctional photocaged dienes [111]. A further interesting work is the screening of the RAFT polymerizations [112]. In there, also the concept of transient monitoring, so-called time-sweeps, was introduced. In such time-sweeps, the flow rate in the microreactor is suddenly changes, and the change in reactor output is then monitored as a function of time. In this way, a full reaction can be monitored within a single residence time of the reactor, giving access to high speed, and fully transient data recording.

Molecular weights and molecular weight distributions can be accessed via online SEC. This is somewhat more difficult to achieve than MS monitoring as SSE requires injection of sample aliquots at defined points in time rather than continuous infusion of an analyte stream. Hadzijoannou and coworkers were the first to develop a coupled SEC system that allowed for automated injections of polymer samples coming from a flow reactor [43]. They used this system to follow NMP polymerizations online, and later added a feature that also allowed for recovery of product via nanoprecipitation [113]. Later other groups followed this example, and online SEC has become a common tool in many flow polymerization labs.

Next to MS and SEC, FT-IR is an established spectroscopic technique to follow reactions. It can be used well to follow monomer conversions; yet more in-depth analysis is usually hampered by the complex spectra obtained. Online NMR spectroscopy is more straightforward to gather chemical structure analysis data during polymerization. Both high field and low field NMR can be used to monitor flow polymerizations, yet the newer generation of low-filed benchtop NMR has proven to be very powerful for this matter. Various examples of flow NMR monitoring of polymerizations exist, giving access to direct in-line data. Warren and coworkers demonstrated first the application of benchtop NMRs to RAFT polymerization, giving access to transient data acquisition and rapid optimization of flow polymerizations [114].

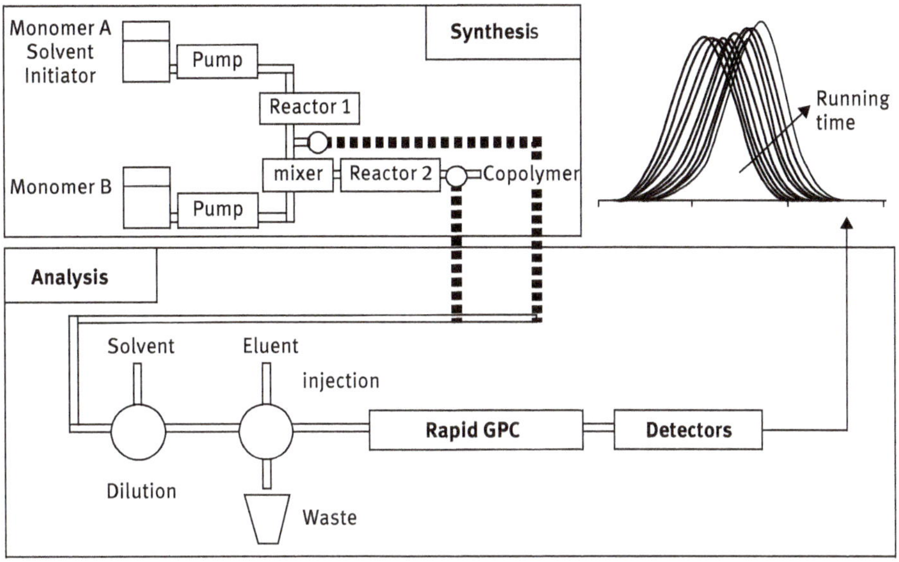

Figure 4.21: Example for the first implementation of size exclusion chromatography online monitoring of polymerization reactions [43]. Copyright (2007). Reproduced with permission from Wiley.

4.8 Machine learning in polymer flow synthesis

Once online monitoring is established for flow polymerizations, a multitude of kinetic data becomes available for reactions. This allows for more than only optimization of flow rates for best product yields, but also for the control of parameters such as molecular weight, dispersity, and end-group fidelity. As polymerizations are inherently chain reactions, control over kinetics means control over product structure. With this, an automatic control over product structures becomes available when the online monitoring data is connected with machine learning algorithms [115, 116]. Also, polymerizations are highly sensitive to changes in reagent concentrations, temperature, and presence of small amounts of impurities. The early stage of a polymerization, and slight variations in that time period, can influence the overall outcome of reactions at longer residence times.

Automation in chemical synthesis can be differentiated in three distinct areas [117]: (i) *autonomous high throughput screening*, (ii) *autonomous optimization*, and (iii) *enhanced process control*. For *autonomous high throughput* screening predefined experiments with iterative variation in reaction conditions and reactant concentrations are screened stepwise in an automated fashion. This type of automation is particularly useful for the development of kinetic models and mechanistic investigations. *Autonomous optimization* refers to online monitoring and self-optimization of the reactor with the use of smart computer programs. Last, but not least, *enhanced process* control allows to keep the reaction system in steady state and can correct for small changes in reaction conditions throughout longer runtimes of a reactor.

In recent years, autonomous self-optimizing reactors were described for numerous organic reactions, with many examples given by the groups of Rueping [118.119]. Ley [120–122], Jensen [98, 123], Bourne [124, 125], Poliakoff [120, 126], and Cronin [127, 128]. For polymer chemistry, autonomous self-optimizing reactors are to date still much less developed. A first example of such automated polymer synthesis platform is given by

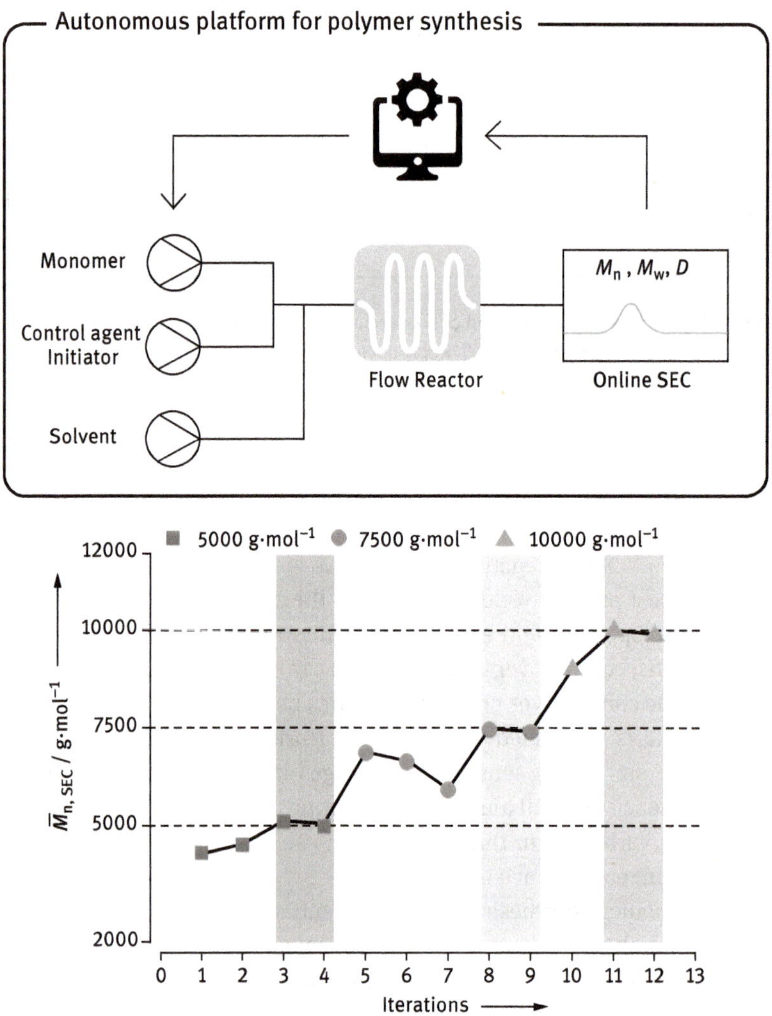

Figure 4.22: Autonomous synthesis platform for polymer with pre-programmed molecular weight via online monitoring and machine-learning algorithm guidance of flow rates. Left: schematic of the setup, right: evolution of molecular weight toward the desired molecular weights with each iteration [129]. Copyright (2019). Reproduced with permission from Wiley.

coupling a SEC analytical system to a continuous flow reactor for the controlled RAFT polymerization of (meth)acrylic monomers. Via optimization algorithms, preprogrammed molecular weights could in this fashion be obtained with an accuracy of less than 2.5%. This marks a remarkable advantage over classical batch wise synthesis, where deviations of 20% or higher from target are commonly observed [129].

In a follow up, Junkers and coworkers developed an autonomous flow platform using lowfield NMR for targeting specific monomer conversions in a radical polymerization. Transient NMR provides thereby real-time information on the monomer conversion of a RAFT-mediated polymerization and comprehensive kinetic data is obtained from time-sweep experiments. With use of this transient data, algorithms can precisely predict the reactor residence time needed to achieve a specific conversion. Again, such task is not easily carried out by hand, and only becomes possible via the increased stability of flow processes and automatic flow rate control. For both high-molecular-weight polymerizations and low-molecular-weight oligomerization, the system targeted monomer conversions with discrepancies of less than 3% from targeted values [130].

4.9 Conclusion and outlook

Within the past 10 years, flow polymerizations and polymer reactions in continuous flow in general have evolved from an oddity to standard methodology. The advantage of flow chemistry in the field is significant. Reactions show less batch to batch variation, are more directly scalable, and display increased precision. Via telescoping, processes can be coupled that allow to obtain complex products in unmatched sophistication directly from monomers. It can be safely assumed that the number of flow applications will further increase in the next years. Especially for photoinduced polymer reactions the advantage is so large that flow processing is likely to replace the existing batch technology entirely. For the newly emerging electrochemistry field, which also finds first application on polymers a similar outlook can be given. It may also be expected that the synthesis of polymer nanoparticles will gain more traction in the next years. Some examples have already shown that flow chemistry is very useful in facilitating self-assembly of polymers into a variety of nanostructures. Also emulsion and miniemulsion polymerization will be interesting to perform in continuous flow.

The most profound change is, however, the application of machine learning [86, 131]. The development of fully autonomous reactor systems removes the need of presence of a trained chemist in the synthesis process, enabling research groups or companies to perform synthesis themselves where it is today still not possible. Further, product libraries become available that are unseen today. The full elucidation of (micro)structure–property relationships is finally in reach.

Further reading

- Daniel Wilms, Johannes Klos, Holger Frey, Microstructured Reactors for Polymer Synthesis: A Renaissance of Continuous Flow Processes for Tailor-Made Macromolecules?, Macromol. Chem. Phys. 2008, 209, 343–356.
- Christoph Tonhauser, Adrian Natalello, Holger Löwe, Holger Frey, Microflow Technology in Polymer Synthesis, Macromolecules 2012, 45, 24, 9551–9570.
- Neomy Zaquen, Maarten Rubens, Nathaniel Corrigan, Jason Xu, Per. Zetterlund, Cyril Boyer, Tanja Junkers, Polymer Synthesis in Continuous Flow Reactors, Prog. Polym. Sci. 2020, 107, 101256.

Study questions

4.1 What are the main advantages of carrying out a polymerization in continuous flow?

4.2 Is mixing efficiency crucial to the outcome of a flow polymerization?

4.3 Are photopolymerizations best carried out at room temperature?

4.4 What are the challenges in setting up a telescoped flow process that involves at least one polymerization step?

4.5 Which detectors are most important to follow a polymer reaction online, and which product feature do they each characterize?

References

[1] Löhr G, Schulz GV, Der einfluß der turbulenz auf die molekulargewichtsverteilung von im strömungsrohr hergestellten polymeren, Z Physik Chem NF, 1969, 65(1–4), 170–180.

[2] Mastan E, He J, Continuous production of multiblock copolymers in a loop reactor: when living polymerization meets flow chemistry, Macromolecules, 2017, 50(23), 9173–9187.

[3] Barnikol VWKR, Schulz GV, Zur Kinetik der anionischen Polymerisation von Styrol in Tetrahydropyran, kurzmitteilung, Macromol Chem Phys, 1963, 68(1), 211–215.

[4] Königsmann H, Jüngling S, Müller AHE, Metal-free anionic polymerization of methyl methacrylate in tetrahydrofuran using bis(triphenylphosphoranilydene)ammonium (PNP+) as counterion, Macromol Rapid Commun, 2000, 21(11), 758–763.

[5] Müller AHE, Kinetics of the anionic polymerization of tert-butyl methacrylate in tetrahydrofuran, Makromol Chem, 1981, 182(10), 2863–2871.

[6] Hofe T, Mauerer A, Müller AHE, Ein neuer strömungsrohrreaktor für die kinetik schneller chemischer reaktionen – demonstration der leistungsfähigkeit am beispiel schneller polymerreaktionen, GIT Labor-Fachz, 1998, 42(11), 1127–1127.

[7] Wilms D, Klos J, Frey H, Microstructured reactors for polymer synthesis: a renaissance of continuous flow processes for tailor-made macromolecules?, Macromol Chem Phys, 2008, 209(4), 343–356.

[8] Nagaki A, Tomida Y, Miyazaki A, Yoshida J-I, Microflow system controlled anionic polymerization of alkyl methacrylates, Macromolecules, 2009, 42(13), 4384–4387.

[9] Nagaki A, Miyazaki A, Tomida Y, Yoshida J-I, Anionic polymerization of alkyl methacrylates using flow microreactor systems, Chem Eng J, 2011, 167(2–3), 548–555.

[10] Nagaki A, Miyazaki A, Yoshida J-I, Synthesis of polystyrenes–poly(alkyl methacrylates) block copolymers via anionic polymerization using an integrated flow microreactor system, Macromolecules, 2010, 43(20), 8424–8429.
[11] Nagaki A, Takahashi Y, Akahori K, Yoshida J-I, Living anionic polymerization of tert-butyl acrylate in a flow microreactor system and its applications to the synthesis of block copolymers, Macromol React Eng, 2012, 6(11), 467–472.
[12] Nagaki A, Nakahara Y, Furusawa M, Sawaki T, Yamamoto T, Toukairin H, Tadokoro S, Shimazaki T, Ito T, Otake M, Arai H, Toda N, Ohtsuka K, Takahashi Y, Moriwaki Y, Tsuchihashi Y, Hirose K, Yoshida J-I, Feasibility study on continuous flow controlled/living anionic polymerization processes, Org Process Res Dev, 2016, 20(7), 1377–1382.
[13] Iwasaki T, Yoshida J-I, Free radical polymerization in microreactors. Significant improvement in molecular weight distribution control, Macromolecules, 2005, 38(4), 1159–1163.
[14] Song Y, Shang M, Zhang H, Xu W, Pu X, Lu Q, Su Y, Process characteristics and rheological properties of free radical polymerization in microreactors, Ind Eng Chem Res, 2018, 57(32), 10922–10934.
[15] Asano Y, Togashi S, Ito Y, A continuous synthesis of polymethylmethacrylate using a microreactor plant, J Chem Eng Jpn, 2014, 47(5), 429–434.
[16] Brocken L, Price PD, Whittaker J, Baxendale IR, Continuous flow synthesis of poly(acrylic acid) via free radical polymerisation, React Chem Eng, 2017, 2(5), 662–668.
[17] Mayadunne RTA, Rizzardo E, Chiefari J, Chong YK, Moad G, Thang SH, Living radical polymerization with reversible addition– fragmentation chain transfer (RAFT polymerization) using dithiocarbamates as chain transfer agents, Macromolecules, 1999, 32(21), 6977–6980.
[18] Barlow KJ, Bernabeu V, Hao X, Hughes TC, Hutt OE, Polyzos A, Turner KA, Moad G, Triphenylphosphine-grafted, RAFT-synthesised, porous monoliths as catalysts for Michael addition in flow synthesis, React Funct Polym, 2015, 96, 89–96.
[19] Moad G, RAFT Polymerization – then and now, In: Controlled Radical Polymerizations: Mechanisms, Ser AS, Ed, ACS Symp Ser, 2015, Vol. 1187, 211–246.
[20] Nicolas J, Guillaneuf Y, Lefay C, Bertin D, Gigmes D, Charleux B, Nitroxide-mediated polymerization, Prog Polym Sci, 2013, 38(1), 63–235.
[21] Hawker CJ, Bosman AW, Harth E, New polymer synthesis by nitroxide mediated living radical polymerizations, Chem Rev, 2001, 101(12), 3661–3688.
[22] Matyjaszewski K, Tsarevsky NV, Macromolecular engineering by atom transfer radical polymerization, J Am Chem Soc, 2014, 136(18), 6513–6533.
[23] Wang J-S, Matyjaszewski K, Controlled/"living" radical polymerization. atom transfer radical polymerization in the presence of transition-metal complexes, J Am Chem Soc, 1995, 117(20), 5614–5615.
[24] Kato M, Kamigaito M, Sawamoto M, Higashimura T, Polymerization of methyl methacrylate with the carbon tetrachloride/dichlorotris-(triphenylphosphine) ruthenium (II)/ methylaluminum bis (2, 6-di-tert-butylphenoxide) initiating system: possibility of living radical polymerization, Macromolecules, 1995, 28(5), 1721–1723.
[25] Percec V, Guliashvili T, Ladislaw JS, Wistrand A, Stjerndahl A, Sienkowska MJ, Monteiro MJ, Sahoo S, Ultrafast synthesis of ultrahigh molar mass polymers by metal-catalyzed living radical polymerization of acrylates, methacrylates, and vinyl chloride mediated by SET at 25 C, J Am Chem Soc, 2006, 128(43), 14156–14165.
[26] Zhang N, Samanta SR, Rosen BM, Percec V, Single electron transfer in radical ion and radical-mediated organic, materials and polymer synthesis, Chem Rev, 2014, 114(11), 5848–5958.
[27] Rosen BM, Percec V, Single-electron transfer and single-electron transfer degenerative chain transfer living radical polymerization, Chem Rev, 2009, 109(11), 5069–5119.

[28] Konkolewicz D, Wang Y, Krys P, Zhong M, Isse AA, Gennaro A, Matyjaszewski K, SARA ATRP or SET-LRP. End of controversy?, Polym Chem, 2014, 5(15), 4396–4417.
[29] Junkers T, Precision polymer design in microstructured flow reactors: improved control and first upscale at once, Macromol Chem Phys, 2017, 218(2), 1600421. 1–9.
[30] Derboven P, Van Steenberge PH, Vandenbergh J, Reyniers MF, Junkers T, D'Hooge D, R., Marin GB, Improved livingness and control over branching in RAFT polymerization of acrylates: could microflow synthesis make the difference?, Macromol Rapid Commun, 2015, 36(24), 2149–2155.
[31] Corrigan N, Zhernakov L, Hashim MH, Xu J, Boyer C, Flow mediated metal-free PET-RAFT polymerisation for upscaled and consistent polymer production, React Chem Eng, 2019, 4(7), 1216–1228.
[32] Corrigan N, Manahan R, Lew ZT, Yeow J, Xu J, Boyer C, Copolymers with controlled molecular weight distributions and compositional gradients through flow polymerization, Macromolecules, 2018, 51(12), 4553–4563.
[33] Reis MH, Varner TP, Leibfarth FA, The influence of residence time distribution on continuous-flow polymerization, Macromolecules, 2019, 52(9), 3551–3557.
[34] Morsbach J, Müller AHE, Berger-Nicoletti E, Frey H, Living polymer chains with predictable molecular weight and dispersity via carbanionic polymerization in continuous flow: mixing rate as a key parameter, Macromolecules, 2016, 49(14), 5043–5050.
[35] Ando T, Kato M, Kamigaito M, Sawamoto M, Living Radical Polymerization of Methyl Methacrylate with Ruthenium Complex: formation of Polymers with Controlled Molecular Weights and Very Narrow Distributions, Macromolecules, 1996, 29(3), 1070–1072.
[36] Wang J-S, Matyjaszewski K, Controlled/"living" radical polymerization. Halogen atom transfer radical polymerization promoted by a Cu(I)/Cu(II) redox process, Macromolecules, 1995, (28), 7901–7910.
[37] Wu T, Mei Y, Cabral JT, Xu C, Beers KL, New Synthetic A, Method for Controlled Polymerization using a Microfluidic System, J Am Chem Soc, 2004, 126(32), 9880–9881.
[38] Noda T, Grice AJ, Levere ME, Haddleton DM, Continuous process for ATRP: synthesis of homo and block copolymers, Eur Polym J, 2007, 43(6), 2321–2330.
[39] Chan N, Boutti S, Cunningham MF, Hutchinson RA, Continuous atom transfer radical polymerization with low catalyst concentration in a tubular reactor, Macromol React Eng, 2009, 3(5–6), 222–231.
[40] Chan N, Cunningham MF, Hutchinson RA, Copper-mediated controlled radical polymerization in continuous flow processes: synergy between polymer reaction engineering and innovative chemistry, J Polym Sci, Part A: Polym Chem, 2013, 51(15), 3081–3096.
[41] Chan N, Cunningham MF, Hutchinson RA, Continuous controlled radical polymerization of methyl acrylate in a copper tubular reactor, Macromol Rapid Commun, 2011, 32(7), 604–609.
[42] Burns JA, Houben C, Anastasaki A, Waldron C, Lapkin AA, Haddleton DM, Poly(acrylates) via SET-LRP in a continuous tubular reactor, Polym Chem, 2013, 4(17), 4809–4813.
[43] Rosenfeld C, Serra C, O'Donohue S, Hadziioannou G, Continuous online rapid size exclusion chromatography monitoring of polymerizations, Macromol Rapid Commun, 2007, 1(5), 547–552.
[44] Studer A, Ryu I, Fukuyama T, Kajihara Y, Nitroxide-mediated polymerization of styrene, butyl acrylate, or methyl methacrylate by microflow reactor technology, Synthesis, 2012, 44(16), 2555–2559.
[45] Diehl C, Laurino P, Azzouz N, Seeberger PH, Accelerated continuous flow RAFT polymerization, Macromolecules, 2010, 43(24), 10311–10314.

[46] Hornung CH, Guerrero-Sanchez C, Brasholz M, Saubern S, Chiefari J, Moad G, Rizzardo E, Thang SH, Controlled RAFT polymerization in a continuous flow microreactor, Org Process Res Dev, 2011, 15(3), 593–601.

[47] Hornung CH, Nguyen X, Kyi S, Chiefari J, Saubern S, Synthesis of RAFT block copolymers in a multi-stage continuous flow process inside a tubular reactor, Aust J Chem, 2013, 66(2), 192–198.

[48] Hornung CH, Nguyen X, Dumsday G, Saubern S, Integrated continuous processing and flow characterization of RAFT polymerization in tubular flow reactors, Macromol React Eng, 2012, 6(11), 458–466.

[49] Vrijsen JH, Rasines Mazo A, Junkers T, Qiao GG, Accelerated polypeptide synthesis via n-carboxyanhydride ring opening polymerization in continuous flow, Macromol Rapid Commun, 2020, 41(18), 2000071.

[50] Honda T, Miyazaki M, Nakamura H, Maeda H, Controllable polymerization of N-carboxy anhydrides in a microreaction system, Lab Chip, 2005, 5(8), 812–818.

[51] Wilms D, Nieberle J, Klos J, Löwe H, Frey H, Synthesis of hyperbranched polyglycerol in a continuous flow microreactor, Chem Eng Technol, 2007, 30(11), 1519–1524.

[52] Paulus RM, Erdmenger T, Becer CR, Hoogenboom R, Schubert US, Scale-up of microwave-assisted polymerizations in continuous-flow mode: cationic ring-opening polymerization of 2-ethyl-2-oxazoline, Macromol Rapid Commun, 2007, 28(4), 484–491.

[53] Dechy-Cabaret O, Martin-Vaca B, Bourissou D, Controlled ring-opening polymerization of lactide and glycolide, Chem Rev, 2004, 104(12), 6147–6176.

[54] Hu X, Zhu N, Fang Z, Guo K, Continuous flow ring-opening polymerizations, React Chem Eng, 2017, 2(1), 20–26.

[55] Zhu N, Huang W, Hu X, Liu Y, Fang Z, Guo K, Chemoselective polymerization platform for flow synthesis of functional polymers and nanoparticles, Chem Eng J, 2018, 333, September 2017, 43–48.

[56] Zhu N, Liu Y, Feng W, Huang W, Zhang Z, Hu X, Fang Z, Li Z, Guo K, Continuous flow protecting-group-free synthetic approach to thiol-terminated poly(ε-caprolactone), Eur Polym J, 2016, 80, 234–239.

[57] Van Den Berg SA, Zuilhof H, Wennekes T, Clickable polylactic acids by fast organocatalytic ring-opening polymerization in continuous flow, Macromolecules, 2016, 49(6), 2054–2062.

[58] Lin B, Hedrick JL, Park NH, Waymouth RM, Programmable high-throughput platform for the rapid and scalable synthesis of polyester and polycarbonate libraries, J Am Chem Soc, 2019, 141(22), 8921–8927.

[59] Bhangale AS, Beers KL, Gross RA, Enzyme-catalyzed polymerization of end functionalized polymers in a microreactor, Macromolecules, 2012, 45(17), 7000–7008.

[60] Kundu S, Bhangale AS, Wallace WE, Flynn KM, Guttman CM, Gross RA, Beers KL, Continuous flow enzyme-catalyzed polymerization in a microreactor, J Am Chem Soc, 2011, 133(15), 6006–6011.

[61] Gumel AM, Annuar MSM, Chisti Y, Ultrasound-assisted enzymatic synthesis of poly-ε-caprolactone: kinetic behavior and reactor design, Int J Chem Reactor Eng, 2014, 11(1), 609–617.

[62] Baeten E, Verbraeken B, Hoogenboom R, Junkers T, Continuous poly(2-oxazoline) triblock copolymer synthesis in a microfluidic reactor cascade, Chem Commun, 2015, 51(58), 11701–11704.

[63] Baeten E, Vanslambrouck S, Jérôme C, Lecomte P, Junkers T, Anionic flow polymerizations toward functional polyphosphoesters in microreactors: polymerization and UV-modification, Eur Polym J, 2016, 80, 208–218.

[64] Subnaik SI, Hobbs CE, Flow-facilitated ring opening metathesis polymerization (ROMP) and post-polymerization modification reactions, Polym Chem, 2019, 10(33), 4524–4528.
[65] Ligon SC, Husár B, Wutzel H, Holman R, Liska R, Strategies to reduce oxygen inhibition in photoinduced polymerization, Chem Rev, 2013, 114(1), 557–589.
[66] Yeow J, Chapman R, Gormley AJ, Boyer C, Up in the air: oxygen tolerance in controlled/living radical polymerisation, Chem Soc Rev, 2018, 47(12), 4357–4387.
[67] Melker A, Fors BP, Hawker CJ, Poelma JE, Continuous flow synthesis of poly(methyl methacrylate) via a light-mediated controlled radical polymerization, J Polym Sci, Part A: Polym Chem, 2015, 53(23), 2693–2698.
[68] Railian S, Wenn B, Junkers T, Photo-induced copper-mediated acrylate polymerization in continuous-flow reactors, J Flow Chem, 2016, 6(3), 260–267.
[69] Wenn B, Conradi M, Carreiras AD, Haddleton DM, Junkers T, Photo-induced copper-mediated polymerization of methyl acrylate in continuous flow reactors, Polym Chem, 2014, 5(8), 3053–3060.
[70] Kermagoret A, Wenn B, Debuigne A, Jérôme C, Junkers T, Detrembleur C, Improved photo-induced cobalt-mediated radical polymerization in continuous flow photoreactors, Polym Chem, 2015, 6(20), 3847–3857.
[71] Junkers T, Wenn B, Continuous photoflow synthesis of precision polymers, React Chem Eng, 2016, 1(1), 60–64.
[72] Chen M, Johnson JA, Improving photo-controlled living radical polymerization from trithiocarbonates through the use of continuous-flow techniques, Chem Commun, 2015, 51(31), 6742–6745.
[73] Lauterbach F, Rubens M, Abetz V, Junkers T, Ultrafast photoRAFT block copolymerization of isoprene and styrene facilitated through continuous flow operation felix lauterbach, Angew Chem Int Ed Engl, 2018, 57, 14260–14264.
[74] Rubens M, Latsrisaeng P, Junkers T, Visible light-induced iniferter polymerization of methacrylates enhanced by continuous flow, Polym Chem, 2017, 8(42), 6496–6505.
[75] Corrigan N, Rosli D, Jones JWJ, Xu J, Boyer C, Oxygen tolerance in living radical polymerization: investigation of mechanism and implementation in continuous flow polymerization, Macromolecules, 2016, 49(18), 6779–6789.
[76] Zaquen N, Kadir AMNBPHA, Iasa A, Corrigan N, Junkers T, Zetterlund PB, Boyer C, Rapid oxygen tolerant aqueous RAFT photopolymerization in continuous flow reactors, Macromolecules, 2019, 52(4), 1609–1619.
[77] Corrigan N, Almasri A, Taillades W, Xu J, Boyer C, Controlling molecular weight distributions through photoinduced flow polymerization, Macromolecules, 2017, 50(21), 8438–8448.
[78] Rubens M, Junkers T, A predictive framework for mixing low dispersity polymer samples to design custom molecular weight distributions, Polym Chem, 2019, 10(42), 5721–5725.
[79] Rubens M, Junkers T, Comprehensive control over molecular weight distributions through automated polymerizations, Polym Chem, 2019, 10(46), 6315–6323.
[80] Zaquen N, Azizi WAAW, Yeow J, Kuchel RP, Junkers T, Zetterlund PB, Boyer C, Alcohol-based PISA in batch and flow: exploring the role of photoinitiators, Polym Chem, 2019, 10(19), 2406–2414.
[81] Serra CA, Khan IU, Cortese B, De Croon MHJM, Hessel V, Ono T, Anton N, Vandamme T, Microfluidic Production of Micro- and Nanoparticles, Encycl Polym Sci Technol.
[82] Serra CA, Chang Z, Microfluidic-assisted synthesis of polymer particles, Chem Eng & Technol, 2008, 31(8), 1099–1115.
[83] Buckinx A-L, Verstraete K, Baeten E, Tabor RF, Sokolova A, Zaquen N, Junkers T, Kinetic control of aggregation shape in micellar self-assembly, Angew Chem Int Ed, 2019, 58(39), 13799–13802.

[84] Buckinx AL, Verstraete K, Baeten E, Tabor RF, Sokolova A, Zaquen N, Junkers T, Kinetic control of aggregation shape in micellar self-assembly, Angew Chem Int Ed, 2019, 58(39), 13799–13802.
[85] Zaquen N, Yeow J, Junkers T, Boyer C, Zetterlund PB, Visible light-mediated polymerization-induced self-assembly using continuous flow reactors, Macromolecules, 2018, 51(14), 5165–5172.
[86] Ramakers G, Krivcov A, Trouillet V, Welle A, Möbius H, Junkers T, Organocatalyzed photo-atom transfer radical polymerization of methacrylic acid in continuous flow and surface grafting, Macromol Rapid Commun, 2017, 38(21), 1700423/18.
[87] Ramsey BL, Pearson RM, Beck LR, Miyake GM, Photoinduced organocatalyzed atom transfer radical polymerization using continuous flow, Macromolecules, 2017, 50(7), 2668–2674.
[88] Hornung CH, Postma A, Saubern S, Chiefari J, A continuous flow process for the radical induced end group removal of RAFT polymers, Macromol React Eng, 2012, 6(6-7), 246–251.
[89] Vandenbergh J, Junkers T, Use of a continuous-flow microreactor for thiol–ene functionalization of RAFT-derived poly(butyl acrylate), Polym Chem, 2012, 3(10), 2739–2742.
[90] Vandenbergh J, Tura T, Baeten E, Junkers T, Polymer end group modifications and polymer conjugations via “Click” chemistry employing microreactor technology, J Polym Sci, Part A: Polym Chem, 2014, 52(9), 1263–1274.
[91] Wojcik F, O’Brien AG, Götze S, Seeberger PH, Hartmann L, Synthesis of carbohydrate-functionalised sequence-defined oligo(amidoamine)s by photochemical thiol-ene coupling in a continuous flow reactor, Chemistry, 2013, 19(9), 3090–3098.
[92] Conradi M, Junkers T, Fast and efficient [2 + 2] UV cycloaddition for polymer modification via flow synthesis, Macromolecules, 2014, 47(16), 5578–5585.
[93] Junkers T, [2+2] Photo-cycloadditions for polymer modification and surface decoration, Eur Polym J, 2015, 62, 273–280.
[94] Vrijsen JH, Osiro Medeiros C, Gruber J, Junkers T, Continuous flow synthesis of core cross-linked star polymers via photo-induced copper mediated polymerization, Polym Chem, 2019, 10(13), 1591–1598.
[95] Plutschack MB, Pieber B, Gilmore K, Seeberger PH, The hitchhiker’s guide to flow chemistry parallel, Chem Rev, 2017, 117(18), 11796–11893.
[96] Rossetti I, Compagnoni M, Chemical reaction engineering, process design and scale-up issues at the frontier of synthesis: flow chemistry, Chem Eng J, 2016, 296, 56–70.
[97] Noël T, Su Y, Hessel V, Beyond organometallic flow chemistry: the principles behind the use of continuous-flow reactors for synthesis, In: Organometallic Flow Chemistry, Noël TCCSIPA, Ed, 2015, 1–41.
[98] Reizman BJ, Jensen KF, Feedback in flow for accelerated reaction development, Acc Chem Res, 2016, 49(9), 1786–1796.
[99] Haven JJ, Vandenbergh J, Junkers T, Watching polymers grow: real time monitoring of polymerizations via an on-line ESI-MS/microreactor coupling, Chem Commun, 2015, 51(22), 4611–4614.
[100] Foley DA, Wang J, Maranzano B, Zell MT, Marquez BL, Xiang Y, Reid GL, Online NMR, HPLC as a reaction monitoring platform for pharmaceutical process development, Anal Chem, 2013, 85(19), 8928–8932.
[101] Haven JJ, Baeten E, Claes J, Vandenbergh J, Junkers T, High-throughput polymer screening in microreactors: boosting the Passerini three component reaction, Polym Chem, 2017, 8(19), 2972–2978.
[102] Browne DL, Deadman BJ, Ashe R, Baxendale IR, Ley SV, Continuous flow processing of slurries: evaluation of an agitated cell reactor, Org Process Res Dev, 2011, 15(3), 693–697.

[103] Sorensen JPR, Vivanco A, Ascott MJ, Gooddy DC, Lapworth DJ, Read DS, Rushworth CM, Bucknall J, Herbert K, Karapanos I, Gumm LP, Taylor RG, Online fluorescence spectroscopy for the real-time evaluation of the microbial quality of drinking water, Water Res, 2018, 137, 301–309.
[104] Skilton RA, Parrott AJ, George MW, Poliakoff M, Bourne RA, Real-time feedback control using online attenuated total reflection Fourier transform infrared (ATR FT-IR) spectroscopy for continuous flow optimization and process knowledge, Appl Spectrosc, 2013, 67(10), 1127–1131.
[105] Roberto MF, Dearing TI, Martin S, Marquardt BJ, Integration of continuous flow reactors and online raman spectroscopy for process optimization, J Pharm Innov, 2012, 7(2), 69–75.
[106] Li X, Mastan E, Wang W-J, Li B-G, Zhu S, Progress in reactor engineering of controlled radical polymerization: a comprehensive review, React Chem Eng, 2016, 1(1), 23–59.
[107] Haven JJ, Junkers T, Online monitoring of polymerizations: current status, Eur J Org Chem, 2017, 2017(44), 6474–6482.
[108] Levere ME, Willoughby I, O'Donohue S, De Cuendias A, Grice AJ, Fidge C, Becer CR, Haddleton DM, Assessment of SET-LRP in DMSO using online monitoring and Rapid GPC, Polym Chem, 2010, 1(7), 1086–1094.
[109] Santos LS, Metzger JO, Study of homogeneously catalyzed Ziegler-Natta polymerization of ethene by ESI-MS, Angew Chem Int Ed, 2006, 45(6), 977–981.
[110] Wang HY, Yim WL, Kluner T, Metzger JO, ESIMS studies and calculations on alkali-metal adduct ions of ruthenium olefin metathesis catalysts and their catalytic activity in metathesis reactions, Chemistry, 2009, 15(41), 10948–10959.
[111] Zaquen N, Haven JJ, Rubens M, Altintas O, Bohländer P, Offenloch JT, Barner-Kowollik C, Junkers T, Exploring the photochemical reactivity of multifunctional photocaged dienes in continuous flow, ChemPhotoChem, 2019, 3(11), 1146–1152.
[112] Haven JJ, Zaquen N, Rubens M, Junkers T, The kinetics of n-butyl acrylate radical polymerization revealed in a single experiment by real time on-line mass spectrometry monitoring, Macromol React Eng, 2017, 11, 1700016/12.
[113] Bally F, Serra CA, Hessel V, Hadziioannou G, Micromixer-assisted polymerization processes, Chem Eng Sci, 2011, 66(7), 1449–1462.
[114] Knox ST, Parkinson S, Stone R, Warren NJ, Benchtop flow-NMR for rapid online monitoring of RAFT and free radical polymerisation in batch and continuous reactors, Polym Chem, 2019, 10(35), 4774–4778.
[115] Reed WF, Alb AM, Monitoring Polymerization Reactions: From Fundamentals to Applications, John Wiley & Sons, 2014, 229–324.
[116] Frauendorfer E, Wolf A, Hergeth WD, Polymerization online monitoring, Chem Eng Technol, 2010, 33(11), 1767–1778.
[117] Shukla CA, Kulkarni AA, Automating multistep flow synthesis: approach and challenges in integrating chemistry, machines and logic, Beilstein J Org Chem, 2017, 13, 960–987.
[118] Poscharny K, Fabry DC, Heddrich S, Sugiono E, Liauw MA, Rueping M, Machine assisted reaction optimization: a self-optimizing reactor system for continuous-flow photochemical reactions, Tetrahedron, 2018, 74(25), 3171–3175.
[119] Fabry DC, Sugiono E, Rueping M, Online monitoring and analysis for autonomous continuous flow self-optimizing reactor systems, React Chem Eng, 2016, 1(2), 129–133.
[120] Amara Z, Streng ES, Skilton RA, Jin J, George MW, Poliakoff M, Automated serendipity with self-optimizing continuous-flow reactors, Eur J Org Chem, 2015, 2015(28), 6141–6145.
[121] Fitzpatrick DE, Battilocchio C, Ley SV, A novel internet-based reaction monitoring, control and autonomous self-optimization platform for chemical synthesis, Org Process Res Dev, 2015, 20(2), 386–394.

[122] Ley SV, Fitzpatrick DE, Ingham RJ, Myers RM, Organic synthesis: march of the machines, Angew Chem Int Ed Engl, 2015, 54(11), 3449–3964.
[123] McMullen JP, Jensen KF, Integrated microreactors for reaction automation: new approaches to reaction development, Annu Rev Anal Chem, 2010, 3(1), 19–42.
[124] Schweidtmann AM, Clayton AD, Holmes N, Bradford E, Bourne RA, Lapkin AA, Machine learning meets continuous flow chemistry: automated optimization towards the Pareto front of multiple objectives, Chem Eng J, 2018, 352, 277–282.
[125] Holmes N, Akien GR, Savage RJD, Stanetty C, Baxendale IR, Blacker AJ, Taylor BA, Woodward RL, Meadows RE, Bourne RA, Online quantitative mass spectrometry for the rapid adaptive optimisation of automated flow reactors, React Chem Eng, 2016, 1(1), 96–100.
[126] Parrott AJ, Bourne RA, Akien GR, Irvine DJ, Poliakoff M, Self-optimizing continuous reactions in supercritical carbon dioxide, Angew Chem Int Ed Engl, 2011, 50(16), 3788–3792.
[127] Caramelli D, Salley D, Henson A, Camarasa GA, Sharabi S, Keenan G, Cronin L, Networking chemical robots for reaction multitasking, Nat Commun, 2018, 9(1), 3406. 1–10.
[128] Sans V, Porwol L, Dragone V, Cronin L, A self optimizing synthetic organic reactor system using real-time in-line NMR spectroscopy, Chem Sci, 2015, 6(2), 1258–1264.
[129] Rubens M, Vrijsen JH, Laun J, Junkers T, Precise polymer synthesis by autonomous self-optimizing flow reactors, Angew Chem Int Ed, 2019, 58(10), 3183–3187.
[130] Rubens M, Van Herck J, Junkers T, Automated polymer synthesis platform for integrated conversion targeting based on inline benchtop NMR, ACS Macro Lett, 2019, 1437–1441.
[131] Haven JJ, Junkers T, Reactor automation in continuous flow polymerisation: on-demand delivery of precision polymer materials, Chem Oggi, 2018, 36(5), 42–44.

Genovéva Filipcsei, Zsolt Ötvös, Réka Angi, Balázs Buchholcz,
Ádám Bódis and Ferenc Darvas

5 Flow chemistry for nanotechnology

5.1 Introduction to nanotechnology

Flow chemistry was introduced and translated into practice at the turn of the twenty-first century to improve and promote the nanomaterial development and production of materials with unique characteristics.

i

Nanotechnology is a multidisciplinary field of material and engineering science that combines physics, chemistry, and biology with the field of engineering, at the nanoscale (about one to a few hundred nanometers). It also covers functional engineering at the atomic or molecular level. Due to the great diversity of the enabling technologies, such as physical, chemical, and biological processes realized at the nanometer scale, a generally accepted and widely used definition of nanotechnology is still absent.

Flow chemistry-based nanotechnology offers unique opportunities across industries, for example, development of novel materials for use in electronics and IT sectors, military and in-line security use, space exploration, advanced healthcare and personalized medicines, innovative cosmetic products, food technology, and improved environmental management. Tailored "nano" manufacturing could provide cheap and green technologies to develop and create novel materials with enhanced characteristics, thus resulting in sustainable twenty-first-century manufacturing processes.

This chapter provides a general introduction of nanotechnology, from theory to application, and introduces the role and advantages of flow chemistry in material production at the nanoscale.

5.2 Nanomaterials

5.2.1 Size, structure, and size-dependent properties

Nanoparticles can be considered as an intermediate state between macroscopic (bulk) phase and the atomic or molecular systems. Here, in general, we consider a material as nanomaterial when the particles are in the range of 1–100 nm.

The classification of nanomaterials is based on the number of dimensions of a material that are outside the nanoscale (>100 nm) range [1]. In **zero-dimensional** (0D) nanomaterials, all the dimensions are measured within the nanoscale (no dimension is

https://doi.org/10.1515/9783110693690-005

larger than 100 nm). In **one-dimensional** nanomaterials (1D), one dimension is outside the nanoscale. In **two-dimensional** nanomaterials (2D), two dimensions are outside the nanoscale. **Three-dimensional** nanomaterials (3D) are bulk materials possessing nanocrystalline structures or structures with the presence of features at the nanoscale.

Special features of nanoparticles include large surface-to-volume ratio, high percentage of atoms/molecules on the surface, unique light scattering properties and plasmon resonance of metal nanoparticles, confined energy states in the electronic band structure of semiconductor nanoparticles, unique chemical and physical properties, and same-sized scale as many biological structures.

The large surface area increases the proportion of atoms on the nanoparticle surface [2]. As a particle decreases in size, a much greater proportion of atoms are found at the surface as compared to that inside, which results in an increase of dominance of the atoms' behavior on the surface over the ones in the bulk, interior phase.

The increased surface-to-volume ratio leads to:

1) *Melting temperature depression:* Melting starts at the surface, where the amplitude vibration of atoms is more than that of the bulk and the transition temperature might depend on the size and shape of the materials [3].
2) *Mechanical properties:* The crystalline size reduction generally leads to improved hardness of the crystalline materials. Therefore, by micronization or nanoparticle formation, the mechanical strength of materials such as metals or ceramic materials can be improved [4].
3) *Solubility and dissolution rate:* Increased specific surface area results in enhanced interfacial (solid–liquid) effect, leading to increasing apparent solubility. This property highly relates to and determines the dissolution rate (e.g., active pharmaceutical or agrochemical ingredients).
4) *Reactivity:* Nanoparticles can be much more reactive than its bulk counterparts. Besides the high surface area, nanosize enables to tailor crystal lattice orientation and fine tune the adsorption energy of the surface [5]. This facet of fine tuning leads to reactive surfaces with atomic steps, ledges, kinks, and dangling bonds, which are generally used in the field of heterogeneous catalysis.

This high surface area has a key importance in reactivity, solubility, sintering performance, or in the field of catalysis electrodes or fuel cells.

Decreasing the particle size from the macroscopic regime to the nanoscale itself leads to a change in the ***optical properties*** of the metallic or semiconductor particles, resulting in the materialization of discrete, quantized energy levels, instead of the continuous valence and conduction bands [6]. This phenomenon leads to the "blue shift" in the emitted light in the case of semiconductor particles.

The size-dependent magnetic properties such as enhanced coercivity, enhanced magnetization, and superparamagnetism are also considerable observed phenomena [7].

A common challenge is the quantitative characterization of the surface of nanoparticles, like surface composition, termination, charge and function groups. These properties determine the fundamental properties of active nanomaterials. These properties have significant role in biomedical and agrochemical applications where the nanoparticles need to incorporate into the cells. [8].

5.2.2 Introduction to the diverse world of nanomaterials

5.2.2.1 Inorganic nanoparticles

Inorganic nanoparticles are the most represented nanostructures in the field of materials science. They can be heteroatomic (e.g., metals) or multicomponent (e.g., metal oxides) with various structures and morphologies. Structures made up of 40–50 atoms are called as clusters or cages.

5.2.2.1.1 Carbon structures

As mentioned in the introduction, fullerenes, carbon nanotubes, and graphene and their derivatives are the most popular groups of nanomaterials. The discovery of fullerenes and graphene was awarded the Nobel Prize in 1996 and 2010, respectively.

Carbon nanostructures have great thermal and electric conductivity. Covalent sp^2 bond between the C atoms provides excellent mechanical properties (Young's modulus: 1–5 TPa, tensile strength: ~100 GPa, elongation break: 15–20%) for carbon nanotubes, which makes them the only material suitable for the construction of the "Space elevator" [9].

Graphene is a single layer of graphite (2D nanostructure), which is made up of C_6 hexagonal subunits. Carbon nanotubes are rolled-up graphene sheets, which can be single- or multiwalled. Fullerene is a spherical (soccer ball-like) structure made from C_5 and C_6 units and generally consists of a few tens of carbon atoms.

Novel physical properties of graphene relate to its electronic band structure. It is typical of 2D materials that their electronic properties dramatically change with the number of layers. If this number is above 10, the electronic structure owns the characteristic of the bulk phase.

5.2.2.1.2 Metal nanoparticles

Metal nanoparticles have attracted considerable attention both in the academic and industrial field. These particles can be monometallic (mostly from noble metals like

Au, Ag, Pt, Pd), alloys (e.g., Ag–Au and Ni–Co) or composite structures with grain boundaries such as bimetallic Janus nanoparticles (Pt-Au). These nanostructures can be prepared from colloidal solution (e.g., sols) or by immobilization on a carrier to gain high dispersity.

5.2.2.1.3 Multielement nanoparticles

A large segment of inorganic nanotechnology deals with multielement materials such as metal oxides, nitrides, carbides, chalcogenides, oxyhalogenides, to name a few. These materials are generally classified by the structure (spinels, perovskites, titanates, etc.) or electronic properties (conductors, semiconductors, insulators).

5.2.2.2 Organic nanoparticles

In the past few years, there has been a growing interest in the use of nanoparticles prepared from biologically active small molecules such as active pharmaceutical or agrochemical ingredients, nutraceuticals or food supplements, and active ingredients of skin care products [10]. Nanoparticles of these biologically active molecules are characterized by increased solubility and dissolution rate, which can lead to an enhanced biological performance of the active ingredients.

Organic nanoparticles or nanobiomaterials can be classified based on their building blocks. They can be micelles, liposomes or polymersomes, lipid nanoparticles, polymeric nanoparticles, and dendrimers [11].

Micelles are spherical amphiphilic structures that have a hydrophobic core and a hydrophilic shell, with particle diameters from 5 to 100 nm range. The hydrophilic shell makes the micelle water soluble, while the hydrophobic core solubilizes the insoluble active ingredients.

Liposomes are made of phospholipids, lipids, or cholesterol. Phospholipids spontaneously self-assemble in water, forming spherical structures, in which the hydrophilic "head" faces toward the solvent and the hydrophobic "tails" form the lipid bilayer. Liposomes can have one or more lipid bilayers, but they all enclose an aqueous core, mimicking the morphology of cell membranes, and they can encapsulate both hydrophilic and hydrophobic drugs.

Polymersomes are composed of synthetic amphiphilic block copolymers. An amphiphilic block copolymer consists of two or more blocks of different polymers, linked together by covalent bonds; one of the blocks is a hydrophilic polymer and the other one can be any biocompatible polymer. Polymersomes possess higher stability, higher mechanical resistance, and reduced permeability, compared to liposomes.

Lipid nanoparticles (e.g., solid lipid nanoparticles (SLN)) have been designed to replace liposomes. SLNs are drug carriers in the submicron size range (50–500 nm) made of biocompatible and biodegradable lipids, which are solid at room and body temperature. Changing 30% of the lipid mass using liquid lipids can thus produce another kind of SLN; these particles are called nanostructured lipid carriers.

Polymer nanoparticles are either solid spheres or nanocapsules composed of biocompatible and biodegradable polymers or natural polymers. Nanogels are polymeric nanoparticles, where the polymers cross-link in a porous network that ensures high drug entrapment efficiency.

Dendrimers are nanosized particles prepared from highly branched, monodisperse macromolecules such as polyamidoamines, polypropyleneimines, and polyaryl ethers. The particles in the 10 nm size range have a compact spherical geometry in solution. Due to the tailor-made branches, well-defined molecular weight, and controlled surface functionality of the macromolecules, the dendrimers have great potential as drug delivery carriers [12].

5.2.2.3 Hybrid nanoparticles

Hybrid nanomaterials offer new opportunities for the development of novel nanostructures with improved or new synergetic properties. These materials are assembled nanosized structures, consisting of inorganic and organic materials with controllable particle size, surface functionality, high drug loading, entrapment of multiple therapeutic agents, tuneable drug release profile, and good serum stability [13].

Hybrid nanostructures have advantageous properties in the field of detection and biodiagnostic screening. For example, lipid-coated silica nanoparticles containing quantum dots can be used for direct visualization and quantification of the nanometer-sized structure [14].

Magnetic nanoparticles and other functional moieties such as fluorescence or radiogenic tags for multimodal imaging and gene delivery can be synergistically integrated into hybrid nanoparticles to provide more accurate information on in vitro and in vivo biological systems [15].

Hybrid nanoparticles can also be used in cancer therapies as targeted drug delivery systems to modify the biodistribution of the drugs and facilitate their accumulation in the tumors. For example, magnetic nanoparticles containing immobilized drug molecules can be locally administered into the tumor. The drug release from these nanoparticles can be controlled by applying an external magnetic field to the site of the action [16].

5.2.2.4 Composite nanoparticles

Compositions consisting of at least one nano-sized component are called nanocomposites. In these multiphase systems, nanoparticles are dispersed in a matrix to enhance its physical and chemical properties such as optical, dielectric, mechanistic, and heat resistance. Considering their matrix material, nanocomposites can be ceramic-, polymeric-, carbon- or metal-matrix-based materials. Nanocomposites offer exciting opportunities for pharmaceutical, medical, biotechnology, electronics, or energy sectors.

5.3 Principles of nanoparticle synthesis

Nanoparticles can be easily prepared, however, the particles immediately aggregate and form larger clusters. Therefore, stabilization of the nanoparticles plays an essential role in their preparation. Stabilization is also important to control the particle size, particle distribution, and the growth rate of particles.

The particle formation and growth can be explained by the classical nucleation theory [17]. The theory describes the nucleation process in terms of the change in Gibbs free energy of the system upon transfer of molecules from the liquid phase to a solid cluster, with the following equation:

$$\Delta G = -\frac{4}{V}\pi r^3 k_B T \ln(S) + 4\pi r^2 \gamma$$

where V is the molecular volume of the precipitated species, r is the radius of the nuclei, k_B is the Boltzmann constant, T is the absolute temperature, S is the saturation ratio, and γ is the surface free energy per surface area.

It is clearly seen that the particle formation is strongly dependent on the saturation ratio. Particles can be formed only if $S > 1$; the system is supersaturated and the free energy term is negative, favoring generation of solid particles and their growth.

The maximum value of ΔG corresponds to the nucleus with critical radius (r^*). A thermodynamically stable nucleus exists when the radius of the nucleus reaches r^*. Therefore, the slope of ΔG at the critical radius of nucleus will be zero [18]:

$$\frac{d\Delta G}{dr} = 0$$

In this situation, the critical radius (r^*) of the spherical nucleus can be obtained by

$$r^* = \frac{2V\gamma}{3k_B T \ln(S)}$$

The behavior of the formed solid particles in a supersaturation solution depends on their size. Particles with radius smaller than r^* will dissolve because this is the only way that leads to a reduction of the particle's free energy. Similarly, if $r > r^*$, particle growth will occur.

The stabilization of the nanoparticles can be either electrostatic or steric stabilization. In the case of electrostatic or steric stabilization, charged molecules are adsorbed at the surface of the particle, resulting in an electrical double layer to prevent the particle aggregation through electrostatic repulsion force (Coulombic repulsion force). In steric stabilization, nonionic surfactants or polymers are absorbed at the surface, producing strong steric repulsion between the nanoparticles [19].

The stability is by the balance of the various interaction forces such as van der Waals attraction, double-layer repulsion, and steric interaction. These interaction forces have been described at a fundamental level by Derjaguin, Landau, Verwey, and Overbeek theory. In this theory, the van der Waals attraction is combined with the double-layer repulsion and an energy–distance curve can be established to describe the conditions of stability/instability. The electrical forces increase exponentially as particles approach one another and the attractive forces increase as an inverse power of separation. As a consequence, these additive forces may be expressed as a potential energy versus separation curve. A positive resultant corresponds to an energy barrier and repulsion, while a negative resultant corresponds to attraction and hence, aggregation. It is generally considered that the basic theory and its subsequent modifications provide a sound basis for understanding stability of the nanoparticles [20].

According to specific application requirements, the properties of the nanoparticles composed of different building blocks need to be tailored and optimized. Thus, different preparation strategies of the nanoparticles have been explored and developed. In general, these strategies can be classified into two main categories: bottom-up and top-down. Bottom-up approach relies on the arrangement of smaller component at the molecular level into more complex assemblies, while top-down approach basically relies on mechanical attrition to render large components into nanosized substances. The combination of these two approaches is also applicable to nanoparticle production. It relies on microprecipitation, which is followed by particle size reduction.

Top-down approaches include technologies that are derived from conventional solid-state methods that are capable of creating materials smaller than 100 nm. In "top-down" techniques, the large objects are modified to result in smaller species. This approach includes various lithographic methods [21], condensation from a vapor using laser ablation [22], arc discharge [23], electrically heated generators, wire electrical explosion [24] or sputtering techniques [25], and mechanical methods such as milling [26] or sonochemical technique [27]. For the preparation of organic nanoparticles, for example, poorly water-soluble active pharmaceutical ingredients, milling techniques in various media are widely used. Using dry- or wet-milling process, different nanoparticulated active molecules are fabricated that have enhanced pharmacological and/or physicochemical properties such as increased bioavailability or solubility [28, 29]. However, the top-down techniques often involve more expensive equipment, higher energy consumption, and high chance of contamination, all of which have limited their wider applications.

Beside these techniques, biomimetic approach is also available to employ biological methods and systems to design nanomaterials. Nanobiotechnology is the use of biomolecules for applications in nanotechnology, including the use of viruses and lipid assemblies [30].

Though both "top-down" and "bottom-up" methods are suitable for nanoparticle synthesis, the "bottom-up" approach has become favored due to controllability of the process. The solution phase synthesis of metal, semiconductor, or organic nanoparticles can be carried out at elevated or room temperatures and in batch or continuous mode. Key challenges regarding nanoparticle synthesis are the ability to fine-tune the size of the nanoparticle and control its distribution. To meet these requirements, the precise control of reaction parameters and the careful selection of the ingredients are essential.

5.4 Flow chemistry–based nanoparticle synthesis in practice and their application

Both, top-down and bottom-up approaches, are applicable to produce nanosized particles. The properties of the particles are independent of whether they are synthetized using batch methodology or continuous technologies. However, flow technologies offer several advantages over batch technologies. Batch technologies frequently suffer from high costs associated with producing large quantities of uniform, nanosized particles. Flow chemistry–based approaches offer inexpensive, scalable, reproducible, and safe routes for the development and production of nanosized materials.

Nanosized particles sometimes show increased toxicological effect on humans, animals, and environment. Nanoparticle toxicology, an emergent field, aims to determine the hazard of nanoparticles and, therefore, their potential risk due to their increased use and the likelihood of exposure. Flow chemistry–based nanoparticle production approach offers a great tool to significantly reduce the toxic human exposure during the production process. This technology is a solution-based method. It uses reagent solutions or suspensions of micron-sized particles for the production of colloid solution, eliminating the potential exposure to the dust of ultrafine particles.

This section provides a brief overview of the principles of nanoparticle synthesis and introduces the bottom-line flow technologies.

5.4.1 Synthesis and application of organic nanoparticles

5.4.1.1 Synthesis of drug nanoparticles

Traditional organic nanoparticle fabrication approaches rely on particle size reduction using dry- or wet-milling [31] processes. The size reduction occurs by collision of particles with the surfaces of the equipment as well as with each other. The process itself consumes a lot of energy and is characterized by the wear of the milling media and with product contamination. Due to the heat-induced active form, conversion of a large number of active compounds cannot be nanoformulated with this approach, for example, salt or active compounds with low melting point cannot be milled.

Of late, there has been a growing interest in exploiting the benefits of the existing flow chemistry–based technologies to overcome the shortcomings and drawbacks of the conventional batch processes.

Homogenization is a process by which the particle size of a suspension or an emulsion can be reduced down to the nanometer range. This approach combines high shear, turbulence, impact, as well as cavitation in the homogenizer. The dispersion of an active compound containing the stabilizer is passed through a very narrow gap under high pressure [32]. The number of cycles, pressure, percentage of solid content, type and combination of surfactants, hardness of active compound, and temperature of the homogenization process has strong influence on the particle size.

High-pressure homogenizers can be microfluidizers and piston gap homogenizers. In piston gap homogenizers, the powder of the active substance is dispersed in an aqueous solution and passed through a narrow homogenization gap (25 mm) at a high streaming velocity under high pressure (500–1,500 bars). The particles are broken by cavitation, high-shear forces, and collision. Microfluidizers are jet stream homogenizers of two fluid streams. They collide frontally at a high velocity (up to 1,000 m/s) under pressures of up to 4,000 bar [33].

Bottom-up technologies [34] provide more control over particle size, shape, and morphology as compared to mechanical processes such as milling and homogenization. Precipitation is driven by a deviation from phase equilibrium conditions, where supersaturation driving forces are gradients in concentration or temperature. In general, the active compound is typically dissolved in a solvent and precipitated by the addition of an antisolvent or solvent evaporation. During the precipitation process, the solubility of the compound decreases drastically, resulting in supersaturation, which drives nucleation. Once nucleation occurs, the particles grow by condensation [35].

Flow chemistry–based bottom-up precipitation method was successfully used to produce nanoparticles of biologically active small organic molecules. For example, nanoparticles of a nonsteroidal inflammatory drug, Ibuprofen, were prepared in a microfluidic device. Ibuprofen sodium salt was dissolved in dextran (stabilizer) solution at 25 °C, and then the nanoparticles were continuously produced at atmospheric pressure by the precipitating effect of the added hydrochloric acid solution. The particle size of the produced colloid solution was monitored by online dynamic light scattering measurements [36]. The particle size of the synthesized nanoparticles could be controlled in a wide range by the flow rate and the amount of dextran. In Fig. 5.1, the effects of the antisolvent flow rate on the particle size and size distribution are shown. By changing the flow rates, the particle size varied from 10 to 18 nm.

The robustness of the flow process was also investigated. Experimental results showed that the Ibuprofen nanoparticles can be reproduced with a deviation of ±0.15 nm (Tab. 5.1.)

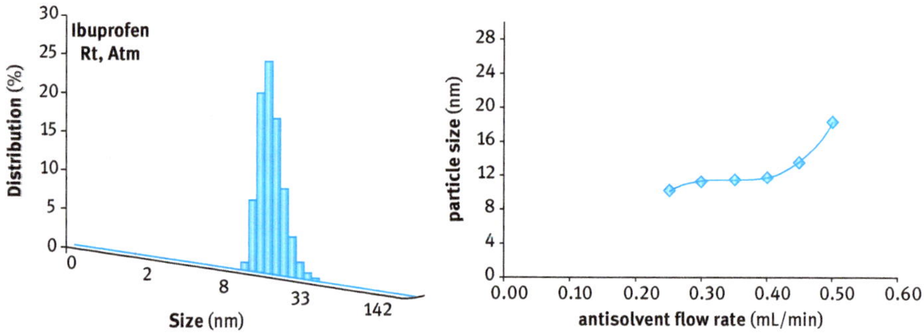

Fig. 5.1: Particle size (a) and size distribution (b) of Ibuprofen nanoparticles using different flow rates.

Tab. 5.1: Reproducibility of Ibuprofen nanoparticle production.

Transformation no.	1	2	3	4	5	6	7	**Average**	**Deviation**
Particle size (nm)	16.25	15.75	16.15	16.00	16.00	15.75	16.00	**15.99**	**0.15**

Precipitation processes can also use compressed or supercritical fluids as antisolvents.

Bottom-up processes are often easier to scale-up and require less particle handling than milling and homogenization operations, resulting in higher process yields and lower impurity risks, as well as simplified cleaning and sterilization procedures. Additionally, precipitation technologies may be operated as continuous or semi continuous processes, whereas milling and homogenization operations are usually batch processes.

Nanoparticles of biologically active small molecules can be prepared by using combination technologies such as spray drying or lyophilization and high-pressure homogenization. The spray-dried or lyophilized product is dispersed in a surfactant solution and passed through a homogenizer [37]. Microprecipitation process can also be combined with high-pressure homogenization with piston gap to produce nanoparticles. Treatment of a precipitated suspension with energy (e.g., high shear forces) avoids particle growth in precipitated suspensions (annealing process).

5.4.1.2 Synthesis of agrochemical nanoparticles

Nanotechnology is one of the breakthrough approaches in our toolbox and, currently, has an advancing and important role in agriculture. In 2019, IUPAC selected the field of nanopesticides as one of the 10 major emerging technologies that has the potential to solve the world's ever-growing feeding challenge [38].

Current formulation techniques of agrochemical active ingredients (AAIs) are limited to create (1) water-miscible formulations such as emulsifiable concentrate, wettable powder, liquid soluble concentrate, and soluble powder and (2) suspension concentrate, capsule suspensions, and water dispersible granules. But there is a growing need for novel formulations for sustainable crop production to decrease the effective amount of pesticide used to reduce the harmful environmental overload.

Nanonization of AAIs leads to similar advantages such as drug nanoformulations: increased apparent solubility (10–10,000 times), better dissolution rate, enhanced bioavailability, and lower environmental impact. This similarity is also shown in the synthetic methods. Nanoparticle synthesis used for the drug nanoparticle productions can also be used to prepare nano-sized AAIs.

A microfluidic approach for the preparation of AAI nanoparticles can be employed by mixing the water insoluble AAI dissolved in organic solvent with an aqueous phase in the presence of selected stabilizers to nanoprecipitate the AAI at 25 °C and under flow conditions. Different solid formulation techniques such as freeze- or spray-drying, and dialysis or solvent evaporation can be used to obtain solid powder for further processing. Both size and stability of the nanoparticles can be controlled by the choice of stabilizers, concentrations, and flow rates.

The effect of the combined flow rates (sum of the solvent and antisolvent phases) on the particle size of a nanoherbicide prepared in a microfluidic reactor is shown in Fig. 5.2.

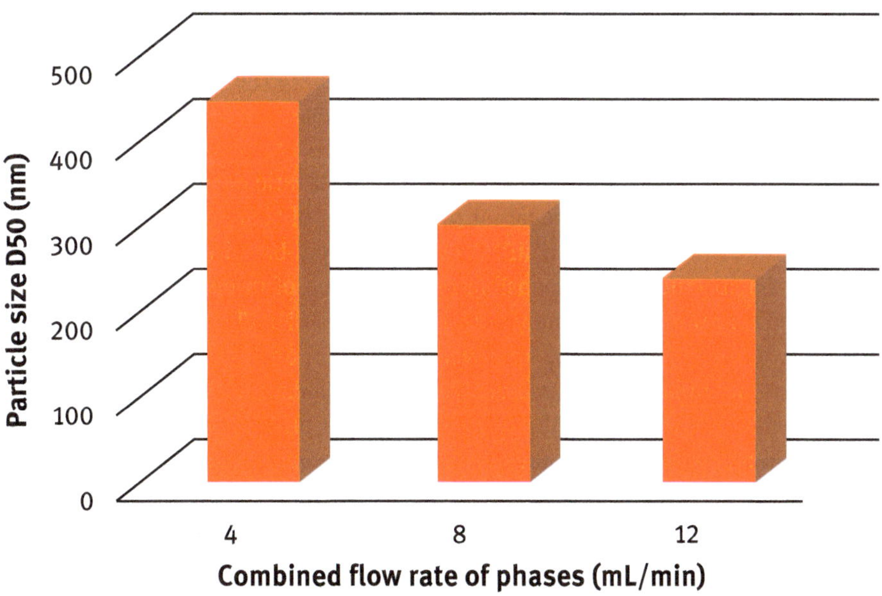

Fig. 5.2: Effect of the combined flow rates on the particle size of the AAI. Flow rates of the organic phase flow rate was set to 1; 2; 3 ml/min, while the aqueous phase flow rate was 3;6;9 ml/min, ml/min, respectively [39].

Continuous flow technology allows us to fine-tune the size and size distribution by controlling the parameters such as flow rate, residence time, mixer type, mixing time, and temperature. Moreover, there is no limit for the batch size and ensures high reproducibility.

As was shown earlier in this chapter, the size of the particles in connection with their material and biological properties can be finely tuned using microfluidic reactors. In contrast, batch technology can provide only limited reaction parameters to control the particle size and size distribution of the final product.

5.4.1.3 Application of organic nanoparticles

The application of nanomaterials in the pharmaceutical science and medical fields is rapidly growing and has drawn increasing attention, of late. They can be used as fluorescent biological labels, drug and gene delivery systems, and MRI contrast agents. Nanostructures can also be used in protein detection, DNA structure probing, tissue engineering, tumor destruction via heating (hyperthermia), and separation and purification of biological molecules and cells [39].

The first liposomal drug delivery system (Doxil®) was approved by the FDA in 1995. The first nanodrug consisting of albumin and the anticancer drug, paclitaxel, was approved in the United States in 2005 [40]. This formulation has proven extremely beneficial in ovarian and breast cancer and has a market size that is beginning to rival those of the most successful cancer drugs of any type. Since then, several nanocarrier-based drugs have been introduced to the market, such as drug nanocrystals [41]. Recently, three nanomedicines have been approved: Patisiran/ONPATTRO, VYXEOS, and NBTXR3/Hensify. VYXEOS is a combination chemotherapy nanoparticle that encapsulates a synergistic molar ratio of cytarabine to daunorubicin of 5:1 and received approval for the treatment of acute myeloid leukemia in 2017. Patisiran/ONPATTRO is an siRNA-delivering lipid-based nanoparticle for the silencing of a specific gene responsible for expression of transthyretin, which can cause hereditary transthyretin amyloidosis. Patisiran/ONPATTRO was approved in 2018 and was the first clinically approved example of an RNAi therapy-delivering nanoparticle administered intravenously. NBTXR3/Hensify is a 50 nm crystalline hafnium oxide nanoparticle with negatively charged phosphate coating. NBTXR3/Hensify enhances external radiotherapy via a physical mode of action that relies on hafnium's natural radioenhancing properties. NBTXR3/Hensify received CE Mark in 2019 for the treatment of locally advanced soft tissue sarcoma [42].

Nanotechnology has an emerging interest in the field of cosmetics and dermal products as it offers a more effective treatment for several skin diseases than the traditional therapy. Nanocarriers used for the targeted delivery of active medicament as well cosmetic ingredients have added the advantage of improved skin penetration and depot effect with sustained release drug action. Nanomaterials in cosmetics include but not limited to SLNs, lipid carriers, vitamin and gold loaded nanofibers facial mask, and so on [43].

Nanoparticles also have a great potential in food-related applications, such as packaging, quality monitoring, and safety control. Nanomaterials can improve certain properties of the food products by encapsulating food components (e.g., controlling the release of flavors) or increasing the bioavailability of nutritional components [44].

In packaging, nanoparticles are incorporated into traditional packaging materials (e.g., metal, glass, paper, and plastic) to slow down the leakage of gases through the packaging.

In the agrochemical industry, nanotechnology is an enabling tool to significantly improve leaf/cuticle penetration, killing capacity of a chemical (herbicide, insecticide or fungicide), reduce application rates of chemical per hectare (reduced environmental impact), and reduce worker exposure (improve safety by removing inflammable solvents) [45].

Novel nanopesticides are designed for sustained release applications, enabling the continuous and long-term crop treatment, or their controlled released forms, to allow the periodic treatment triggered by environmental stimuli. Smart delivery systems of herbicides could be applied seasonally in a controlled manner, resulting in greater production of crops and less injury to agricultural workers.

Decreasing the applied effective dose of the herbicides, the risk of environmental pollution can be significantly reduced. Moreover, the greener plant protection process cost is less when compared to the industrial standards process using traditional formulations. The global estimated applied amount of the traditional formulation of the tested herbicides is 150,000 tons per year for the 2015–2026 period [46]. Using a nanoformulated herbicide, this amount can be significantly reduced, at least by one-fifth.

Nanofertilizers can be used to reduce nitrogen loss caused by leaching, emissions, and incorporation in soil microorganisms. Nanomaterials allow the selective release of fertilizers linked to time or environmental condition. Slow-controlled-release fertilizers may also improve the soil by decreasing the toxic effects associated with overapplication of fertilizer.

Nanosensors can detect contaminants, pests, nutrient content, and plant stress due to drought, temperature, or pressure. They could also potentially help farmers increase efficiency by applying inputs only when necessary.

5.4.2 Synthesis and application of inorganic nanoparticles

5.4.2.1 Synthesis of inorganic nanoparticles

Synthesis of nanoparticles by chemical reduction of metal ions to their 0 oxidation states is a commonly used method, whose advantage is the ability to fabricate particles having different shapes such as nanorods, nanowires, and hollow nanoparticles. With the chemical reduction method, the form and size of the nanoparticles can be tuned by changing the reducing agent, the dispersing agent, the reaction time, and temperature. The process uses non-complicated equipment or instruments and can yield large quantities of nanoparticles at a low cost in a short time. Based on this method, a wide range of metallic and bimetallic nanoparticles was synthesized such as gold [47], silver [48] platinum [49], palladium [50], iron/gold [51], iron/copper [52], and cobalt/platinum [53].

Metallic nanoparticle fabrication by the decomposition of an organometallic complex is a widely used method, which is a suitable route for the preparation of Fe [54], Co [55], Pt [56], Au [57], FeMo [58], and so on.

For bimetallic nanoparticle preparation, the two above mentioned synthesis routes are often mixed and used, resulting in, for example, monodisperse iron–platinum (FePt) nanoparticles by the reduction of platinum acetylacetonate and decomposition of iron pentacarbonyl in the presence of oleic acid and oleyl amine stabilizers [59] or FePd nanoparticles [60].

Synthesis of gold [61] or semiconductor [62] nanoparticles in microfluidic reactors is also reported in literature, besides flow synthesis of other metallic nanoparticles such as silver [63], titania [64], silica [65], or ZnO [66].

Gold nanorods (GNRs) are prepared using a coaxial flow device coupled to a coiled aging reactor [67]. The inner phase consists of the gold source ($HAuCl_4$), and lysine (used instead of a surfactant as a colloidal stabilizer) is introduced to an outer phase of the reductant tetramethylammonium hydroxide (TMAOH). This reaction mixture is directly injected into an aging loop, where the gold nuclei are grown into GNRs. The small volumes of the microreactor allows fast screening of reactants and further optimization leads to high-quality materials in a shorter amount of time, as demonstrated by the short (16 s) residence time in the "aging" section of the reactor.

Others have discovered the utility of micro reactors for synthesizing magnetic cobalt ferrite nanoparticles. The composite particles are prepared via multistep synthesis in a continuous-flow device [68]. These particles have significant importance for applications entailing magnetic imaging and therapies.

The size dependency of platinum nanoparticles on the applied flow rate in a microfluidic reactor is demonstrated in Fig. 5.3. The Pt nanoparticles are synthesized by the reduction of $H_2PtCl_6{\cdot}6H_2O$ using methanol as the reducing agent at 150 °C and 50 bar. It is clearly seen that the increasing flow rate induces a reduced residence time, resulting in the formation of smaller particles.

The successful use of metal nanoparticles in catalysis is based on the high activity and selectivity of nanoparticle-containing homogeneous or heterogeneous catalysts. Supported Au nanoparticle catalysts are found to be effective in hydrogenation or oxidation of alkenes, aldehydes and alcohols, oxidative decomposition of dioxines, C–C coupling reactions, hydroamination of terminal alkynes, water gas shift reaction, and so on. Pt nanocatalysts are used in selective hydrogenations, oxidation reactions [69], catalytic decomposition of alcohols [70], NO reduction [71], fuel cell technology, and so on. Various C–C coupling, CO oxidation-, methane oxidation

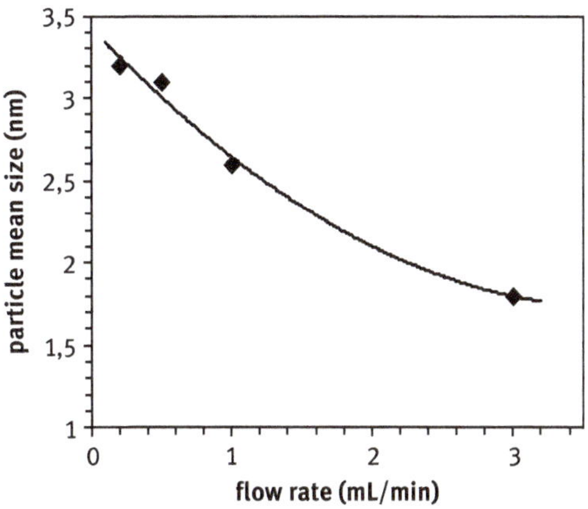

Fig. 5.3: Size dependency of platinum nanoparticles on the applied flow rate.

[72], and NO-reduction [73] reactions are carried out over catalysts containing Pd nanoparticles.

Flow chemistry–based methods can be effectively used for the continuous production of semiconductor nanoparticles. For example, CdSe nanoparticles are prepared by thermal decomposition of the precursor materials (selenium in trioctylphosphine solution, cadmium acetate and trioctylphosphine) using a microfluidic device. The precise reaction parameter control allows us to tailor both the size and the surface morphology of the particles. To avoid precipitation and blockages from the evaporation of the solvent, the applied pressure was chosen to be 70 bar in each case. Optical spectra of the colloid samples were recorded online using a flow-through cell lined with a xenon light source and a spectrophotometer. Increasing the temperature of the reaction from 180 to 295 °C and keeping the flow rates constant, the emission wavelength of the colloid solution containing the semiconductor nanoparticles shifted to higher values, indicating the formation of larger particles (Fig. 5.4).

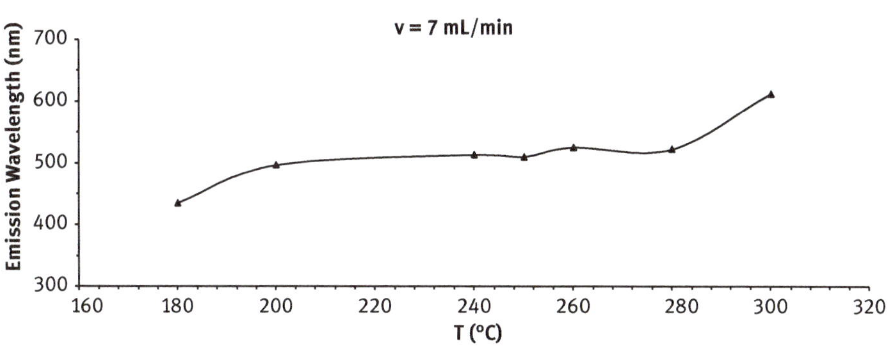

Fig. 5.4: Dependency of the optical properties on the synthesis temperature at a constant flow rate.

Varying the flow rate between 3 and 7 mL min^{-1} (decreasing the residence time from 27 to 12 s) resulted in a blueshift in the emission wavelength from 624 to 526 nm, indicating the formation of smaller particles at 280 °C (Fig. 5.5). Transmission electron microscopy measurements showed that the smallest particle size was 2–3 nm.

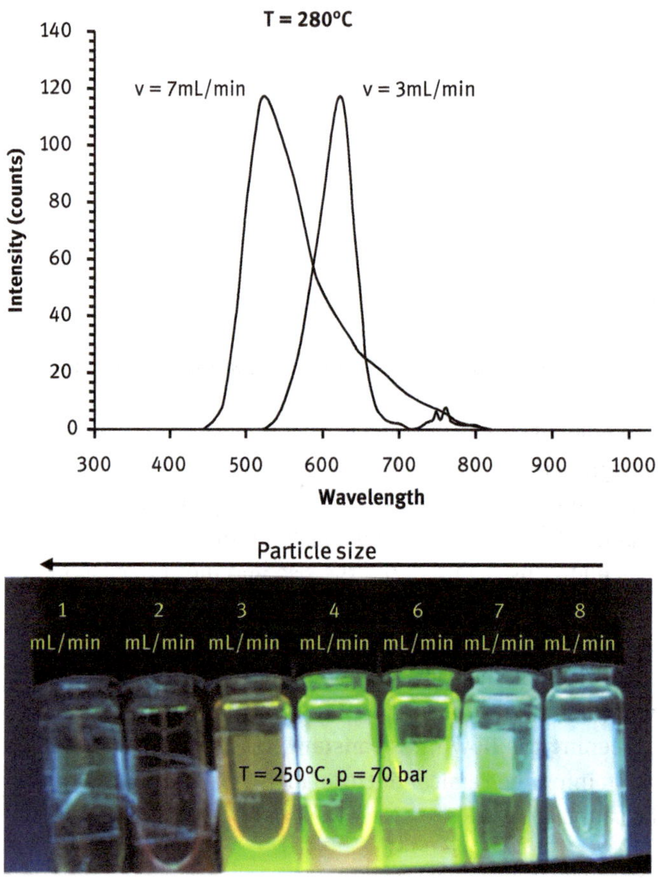

Fig. 5.5: Emission spectra recorded on samples prepared at constant temperature and different flow rates.

Use of this reactor setup and high-throughput methodology made it easy to monitor the effect of each parameter change on the nanocrystal structure.

Key factors affecting the properties of inorganic nanoparticles prepared by flow technology are summarized in Fig. 5.6.

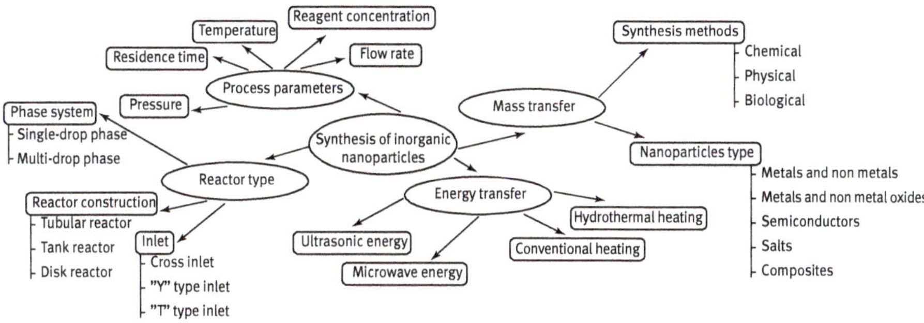

Fig. 5.6: Factors affecting the properties of flow synthesized inorganic nanoparticles.

5.4.2.2 Application of inorganic nanoparticles

Inorganic nanoparticles have versatile applications in various fields. In this section, we will focus on the most important ones.

5.4.2.2.1 Coatings

Nanoparticles started a revolution in the market of "smart coatings". Mimicking the lotus effect, fractal structured metal oxides have superhydrophobic character that can be used in the paint, glass, and ceramic industries. Self-cleaning photocatalytic surfaces have also been used in the paint industry for many years. Nanoparticles recently entered the industry of touch screens in order to replace the commonly used indium tin oxide units.

5.4.2.2.2 Sensors

Sensors are able to measure physical quantities and convert them to signals. They can be chemical-, bio-, plasmonic-, magnetic-, catalytic-, electrochemical-based, etc. Gold nanoparticles are the most popular components in the nano sensor industry. They can be functionalized easily to use as biosensors. The surface plasmon resonance effect (which depends on the particle size) can also be used in many fields.

5.4.2.2.3 Biomedical applications

A vast number of researches deal with inorganic nanoparticle-based cancer treatments. Nano-sized gold, silver, and mesoporous silica are well-studied candidates in this field. Magnetic resonance imaging also uses magnetic nanoparticles, such as mixed iron oxides.

5.4.2.2.4 Heterogeneous catalysis

Heterogeneous catalysts generally consist of metal-based active centers. In order to increase the number of active sites and reactive interfaces, active component is used to disperse using an appropriate support (metal oxide, carbon allotrope, zeolite, etc.). It can be used in thermal-, photo-, and electro-activated reactions.

(For further details on heterogeneous catalysis please see Volume 1, Chapter 10, Title: Catalysis in flow, and Chapter 11, Title: Gaseous reagents in flow chemistry)

Flow nanotechnology generally deals with the synthesis of nanoparticles under continuous flow conditions; however, nanoparticles can be used in the field of flow heterogeneous catalysis. Immobilized/supported nanoparticles can be loaded on fixed bed reactors (like CatCart® by ThalesNano), avoiding the leaching of nano-sized solids. Materials science has vital importance for the effective immobilization of nanoparticles. Mesoporous structures have a considerable evolution over the past few decades, where the metal organic frameworks became the state-of-the-art class of the catalyst supports besides the well-known mesoporous silicas.

A great example for flow nanocatalysis is the biomass valorization. Luque and coworkers investigated the catalytic transfer hydrogenation of furfural in isopropanol, over activated carbon-supported bimetallic (W and Ni) nanoparticles in batch and flow conditions. They found a vital metal leaching during the batch experiments, while the catalyst showed better stability and high (>90%) 2-methylfuran stability in flow experiments [74].

Platinum nanoparticles also result in great selectivity in aldehyde hydrogenation reactions by continuous flow. Uozumi and coworkers used amphiphilic polymer resin as support in a fixed bed flow reactor. With a 22 s residence time, they achieved ≤99% yield of benzylic or aliphatic alcohols. Moreover, in the case of benzaldehyde, they ran long-term reactions (8 days) and achieved a turnover of 997 [75].

A huge advantage in continuous flow heterogeneous catalysis is the possibility to form in situ supported nanoparticles and maintain reactions subsequently. Paun and coworkers reported about the continuous flow modification of Ru/Al_2O_3 catalyst bed with silver. $AgNO_3$ (1.5 mM, 0.3 mL min^{-1}) was introduced into the catalyst cartridge and reduced by H_2 at 10 bar, 80 °C for 100 min using a H-Cube® instrument by Thalesnano. The process obtained supported Ag nanoparticles with size of ca. 220 nm. The as-prepared bimetallic catalyst was successfully used in continuous flow hydrogenation of D-xylose to xylitol [76].

In situ sulfidation of the fixed nanocatalyst bed is also an interesting example for the combination of the catalyst modification and its subsequent application in chemical reactions. Moore and coworkers applied diphenyl sulfide to poisoning nano-Pd-supported carbon catalyst. The sulfidation led to high conversion in olefin hydrogenation reactions and enhanced selectivity in α/β-unsaturated carbonyl compounds [77].

5.4.3 Synthesis of composite nanoparticles

Ceramic matrix nanocomposites can be prepared by a conventional powder method, polymer precursor route, spray pyrolysis, vapor techniques, and chemical methods. The last includes the sol–gel process, colloidal and precipitation approaches, and template synthesis [78].

Most widely used techniques for the preparation of polymer–matrix nanocomposites are the intercalation of the polymer or pre-polymer from solution, in situ intercalative polymerization, melt intercalation, and template synthesis (sol–gel technology) [79].

Main synthesis processes for metallic-matrix nanocomposites are spray pyrolysis, liquid metal infiltration, rapid solidification, vapor techniques, electrodeposition, and chemical methods, which includes colloidal and sol–gel processes.

5.4.3.1 Application of composite nanoparticles

Ceramic–matrix nanocomposites can be used in a wide range of industrial applications, such as electronic (sensors, microelectromechanics systems, optical fibers), automotive (catalyst, engineering parts), chemical (high-performance catalyst), or military. Unique properties of the nanophased ceramic composites, such as enhanced resistance/hardness, improved toughness/durability, and decreased elasticity, evince increasing interest in biomedical applications [80].

Due to the more advanced development status, polymeric-matrix nanocomposites are in the forefront of applications. The advantageous properties such as increased strength, reduced flammability, and heat resistance provide the possibility of using polymeric-matrix composites in numerous industrial fields. Polymer-based nanocomposites may include semiconducting or metallic nanoparticles, which also widen their field of applications. Among others, the application opportunities are in packaging, barrier films, painted structural elements, wire, cable, optical devices, microelectronics, coatings, dental applications, catalyst support, parts of batteries of fuel cells, medical devices, and so on.

Metal–matrix and carbon–matrix nanocomposites have promising application opportunities in catalysis, coatings, photoelectrochemical applications, automotive structures, electronic packaging, machinery, tooling, optical devices, and so on.

5.4.3.1.1 The future of flow nanotechnology: an outlook

The unique material properties of nanosized particles, such as different conductivity, improved reactivity, and enhanced bioavailability, are attributed to the small size, surface structure, chemical composition, shape, and solubility. They make these nanostructures attractive for different applications from catalysis to medicaments.

Moreover, nanotechnology presents a unique opportunity to offer a more sustainable approach to protect public health and the environment.

Nowadays, nanotechnology is more frequently being referred to in connection with green chemistry, and green engineering and manufacturing. The principles of green chemistry can be applied to produce safer and more sustainable nanomaterials. They employ more efficient and sustainable nano manufacturing processes. Main elements of the future vision concerning green nanotechnology are nanostructured photovoltaics (organic, inorganic), artificial photosynthesis for fuel production, nanostructures for energy storage (batteries), solid state lighting, thermoelectrics, water treatment, medical nanorobots, and smart drug delivery systems [81].

Flow technologies in nanoparticle synthesis will certainly continue to grow in importance and in diversity of application due to the easiness of fine-tuning their process parameters, their inherently green nature, and the reduction of toxic exposure, relative to other nanoization technology approaches. One of the most promising signs indicating the bright future of flow nanotechnology is the most recent result of synthesizing mRNA-based anti-COVID vaccine by continuous flow lipid nanoformulation. Even more, the Pfizer vaccine of worldwide reputation is also packed in lipid nanoparticles, where the nanoformulation is an active factor contributing to its high druggability [82].

Further readings

- Guo Z, Tan L. Fundamentals and Applications of Nanomaterials. Artech House, 2009.
- Wilde G. Nanostructured Materials, Elsevier, Oxford, UK, Elsevier Science, 2008.
- Vajtai R. Springer Handbook of Nanomaterials, Springer, Berlin, Heidelberg, 2013.
- Vollath D. Nanomaterials: An Introduction to Synthesis, Properties and Applications, Weinheim, Germany, Wiley-VCH Verlag GmbH&Co. KGaA, 2008.
- Cao G. Nanostructures & Nanomaterials: Synthesis, Properties & Applications, 1st Edition. London, UK. Imperial College Press, 2004.
- Hiemenz PC. Principles of Colloid and Surface Chemistry, Third Edition, Revised and Expanded (Undergraduate Chemistry Series), NY, USA, CRC Press, 1997.
- Antonietti M. Colloid Chemistry I. Germany, Springer-Verlag Berlin Heidelberg, 2003.
- Ahmed W, Jacksom MJ. Emerging Nanotechnologies for Manufacturing, 1st Edition, Oxford, UK, Elsevier, 2009.
- Bhushan B. Springer Handbook of Nanotechnology, Springer-Verlag Berlin Heidelberg, 2010.
- Rotello V. Nanoparticles: Building Blocks for Nanotechnology, New York, USA, Springer Science and Business Media Inc., 2004.
- Feldheim DL, Foss CA Jr. Metal Nanoparticles: Synthesis, Characterization and Applications, New York, USA, Marcel Dekker Inc., 2002.
- Kumar CSSR. Semiconductor Nanomaterials, Weinheim, Germany, WILEY-VCH Verlag GmbH & Co. KGaA, 2010.

Study questions

5.1 What is the definition of nanomaterials?

5.2 What are the advantages of flow technology in preparing nanomaterials?

5.3 What are the theoretical principles adaptable to prepare nanoparticles?
5.4 What are the main technologies to prepare nanoparticles?
5.5 Give a few examples of metal nanoparticles.
5.6 Give a few examples of organic nanoparticles.

References

[1] Introduction to nanotechnology (https://www.nanowerk.com/nanotechnology/introduction/introduction_to_nanotechnology_1.php)
[2] Sun X, Zhang Y, Chen G, Gai Z, Application of nanoparticles in enhanced oil recovery: A Critical review of recent progress, Energies, 2017, 10, 345.
[3] Olson EA, Efremov MY, Zhang M, Zhang Z, Allen LH. Size-dependent melting of Bi nanoparticle, J Appl Phys, 2005, 97, 034304.
[4] Wu Q, Miao WS, Gao HJ, Hui D, Mechanical properties of nanomaterials: A review, Nanotechnol Rev, 2020, 9, 259–273.
[5] Ahmadi M, Mistry H, Roldan CB, 2016. Tailoring the catalytic properties of metal nanoparticles via support interactions, J Phys Chem Lett, 2016, 7, 3519–3533.
[6] Baskoutas S, Terzis AF, Size-dependent band gap of colloidal quantum dots, J Appl Phys, 2006, 99, 013708.
[7] Gubin SP, Magnetic nanoparticles, John Wiley & Sons, 2009.
[8] Grassian VH, When size really matters: Size-dependent properties and surface chemistry of metal and metal oxide nanoparticles in gas and liquid phase environments, J Phys Chem C, 2008, 112, 18303–18313.
[9] Bradley CE, Design and deployment of space elevator, Acta Astronaut, 2000, 47, 735–744.
[10] Professor Steve Rannard. Opportunities for organic nanoparticles; Speciality Chemicals Magazine; June 2008 (https://www.researchgate.net/publication/265147982_Opportunities_for_organic_nanoparticles).
[11] Martinelli C, Pucci C, Ciofani G, Nanostructured carriers as innovative tools for cancer diagnosis and therapy, APL Bioeng, 2019, 3, 011502.
[12] Christensen JB, Boas U, Heegaard PMH, Peng L, Dendrimers in Medicine and Biotechnology: New Molecular Tools, 1st, Royal Society of Chemistry, 2006.
[13] Sailor MJ, Park JH, Hybrid Nanoparticles for detection and treatment of cancer, Adv Mater, 2012, 24, 3779–3802.
[14] Van Schooneveld MM, Gloter A, Stephan O, Zagonel LF, Koole R, Meijerink A, Milder JMM, De Groot FMF, Imaging and quantifying the morphology of an organic–in-organic nanoparticle at the sub-nanometre level, Nature Nanotechnol, 2010, 5, 538–544.
[15] Cheong J, Lee JH, Synergistically Integrated nanoparticles as multimodal probes for nanobiotechnology, Accounts Chem Res, 2008, 41, 1630–1640.
[16] Yang HW, Hua MY, Liu HL, Huang CY, Wei KC, Potential of magnetic nanoparticles for targeted drug delivery, Nanotechnol Sci Appl, 2012, 5, 73–86.
[17] Becker R, Doring W, Kinetic treatment of germ formation in supersaturated vapour, Ann Phys-Berlin, 1935, 24, 719.
[18] Mersmann A, Crystallization Technology Handbook, Basel, Taylor & Francis Group, 2001.
[19] Anand M, Bell PW, Fan X, Enick RM, Roberts CB, Synthesis and steric stabilization of silver nanoparticles in neat carbon dioxide solvent using fluorine-free compounds, J Phys Chem B, 2006, 110, 14693–14701.

[20] Tharwat FT, Colloid Stability: The Role of Surface Forces, Part I, Weinheim, WILEY-VCH Verlag GmbH & Co. KGaA, 2007.

[21] Langford RM, Focused ion beam nanofabrication: A comparison with conventional processing techniques, J Nanosci Nanotechnol, 2006, 6, 661–668.

[22] Semaltianos NG, Nanoparticles by Laser Ablation, Crit Rev Solid State, 2010, 35, 105–124.

[23] Xie SY, Ma ZJ, Wang CF, Lin SC, Jiang ZY, Huang RB, Zheng LS, Preparation and self-assembly of copper nanoparticles via discharge of copper rod electrodes in a surfactant solution: A combination of physical and chemical processes, J Solid State Chem, 2004, 177, 3743–3747.

[24] Kotov YA, Electric explosion of wires as a method for preparation of nanopowders, J Nanopart Res, 2003, 5, 539–550.

[25] Chung BX, Liu CP, The synthesis of Cobalt nanoparticles by DC magnetron sputtering and the effects of electron charging, Materi Lett, 2004, 58, 1437–1440.

[26] Chakka VM, Altuncevahir B, Jin ZQ, Li Y, Liu JP, Magnetic nanoparticles produced by surfactant-assisted ball milling, J Appl Phys, 2006, 99, 08E912.

[27] Okitsu K, Ashokkumar M, Grieser F, Sonochemical synthesis of gold nanoparticles: Effects of ultrasound frequency, J Phys Chem B, 2005, 109, 20673–20675.

[28] Liu P, Rong X, Laru J, Van Veen B, Kiesvaara J, Hirvonen J, Laaksonen T, Peltonen L, Nanosuspensions of poorly soluble drugs: Preparation and development by wet milling, Int J Pharmaceut, 2011, 411, 215–222.

[29] Ghosh I, Schenck D, Bose S, Ruegger C, Optimization of formulation and process parameters for the production of nanosuspension by wet media milling technique: Effect of Vitamin E TPGS and nanocrystal particle size on oral absorption, Eur J Pharm Sci, 2012, 47, 718–728.

[30] Niemeyer CM, Mirkin CA, Nanobiotechnology: Concept, Applications and perspectives, Weinheim, Wiley-VCH Verlag GmbH&Co KGaA, 2004.

[31] Morales JO, Watts AB, McConville JT, Mechanical particle-size reduction techniques, In: Williams RO, Ab W, Da M, Formulating Poorly Water Soluble Drugs, Springer, 2012, 133–163.

[32] Sawant SV, Kadam JV, Jadhav KR, Sankpal SV, Drug nanocrystals: Novel technique for delivery of poorly soluble drugs, Int J of Sci Innov Discov, 2011, 1, 1–15.

[33] Bruno RP, McIlwrick R, Microfluidizer processor technology for high performance particle size reduction, mixing and dispersion, Eur J Pharm Biopharm, 1999, 56, 29–36.

[34] Rowe JM, Johnston KP, Precipitation technologies for nanoparticle production, In: Williams RO, Ab W, Da M, Formulating Poorly Water Soluble Drugs, Springer, 2012, 501–553.

[35] Weber M, Thies M, Understanding the RESS process, In: Sun YP, Dekker M, New York, 2002, 387–437.

[36] PatentWO2015121836A1. 2015.

[37] Shegokar R, Muller RH, Nanocrystals: Industrially feasible multifunctional formulation technology for poorly soluble actives, Int J of Pharmaceut, 2010, 399, 129–139.

[38] International Union of Pure and Applied Chemistry, Top ten emerging technologies in chemistry. (April 1, 2019, at https://iupac.org/what-we-do/top-ten/)

[39] Salata OV, Applications of nanoparticles in biology and medicine, J of Nanobiotechnol, 2004, 2, 1–6.

[40] FDA Approval for Paclitaxel Albumin-stabilized Nanoparticle Formulation. National Cancer Institute at the National Institutes of Health, 2013. (accessed September 6, 2013 at https://www.cancer.gov/about-cancer/treatment/drugs/nanoparticlepaclitaxel)

[41] Bamrungsap S, Zhao Z, Chen T, Wang L, Li C, Fu T, Tan W, Nanotechnology in Therapeutics: A Focus on Nanoparticles as a Drug Delivery System, Nanomedicine, 2012, 7, 1253–1271.

[42] Anselmo AC, Mitragotri S, 2019. Nanoparticles in the clinic: An update, Bioeng Transl Med, 2019, 4, e10143.

[43] Bangale MS, Mitkare SS, Gattani SG, Sakarkar DM, Recent nanotechnological aspects in cosmetics and dermatological preparations, Int J Pharm Pharm Sci, 2012, 4, 88–97.
[44] Ducan TV, Applications of nanotechnology in food packaging and food safety: Barrier materials, antimicrobials and sensors, J Colloid Interf Sci, 2011, 363, 1–24.
[45] Gruère G, Narrod C, Abbott L Agriculture, food, and water nanotechnologies for the poor: Opportunities and constraints 2011. (Accesed September 5, 2013, at http://www.ifpri.org/sites/default/files/publications/bp019.pdf)
[46] Maggi F, Tang FH, La Cecilia D, McBratney A, PEST-CHEMGRIDS, global gridded maps of the top 20 crop-specific pesticide application rates from 2015 to 2025, Sci Data, 2019, 6, 1–20.
[47] Tapan KS, Murphy CJ, Room temperature, high-yield synthesis of multiple shapes of gold nanoparticles in aqueous solution, J Am Chem Soc, 2004, 126, 8648–8649.
[48] Panáček A, Kvitek L, Prucek R, Kolář M, Večeřová R, Pizúrová N, Zbořil R, Silver colloid nanoparticles: Synthesis, characterization, and their antibacterial activity, J Physl Chem B, 2006, 110, 16248–16253.
[49] Chen J, Herricks T, Xia Y, Polyol synthesis of platinum nanostructures: Control of morphology through the manipulation of reduction kinetics, Angew Chem, 2005, 117, 2645–2648.
[50] Cookson J, The preparation of palladium nanoparticles: Controlled particle sizes are key to producing more effective and efficient materials, Plat Met Rev, 2012, 56, 83–98.
[51] Lin J, Zhou W, Kumbhar A, Wiemann J, Fang J, Carpenter EE, O'connor CJ, Gold-coated iron (Fe@Au) nanoparticles: Synthesis, characterization, and magnetic field-induced self-assembly, J Solid State Chem, 2001, 159, 26–31.
[52] Tanori J, Duxin N, Petit C, Lisiecki I, Veillet P, Pileni MP, Synthesis of nanosize metallic and alloyed particles in ordered phases, Colloid Polym Sci, 1995, 273, 886–892.
[53] Zhang X, Chan KW, Microemulsion synthesis and electrocatalytic properties of platinum–cobalt nanoparticles, J of Mater Chem, 2002, 12, 1203–1206.
[54] Park SJ, Kim S, Lee S, Khim ZG, Char K, Hyeon T, Synthesis and magnetic studies of uniform iron nanorods and nanospheres, J Am Chem Soc, 2000, 122, 8581–8582.
[55] Puntes VF, Zanchet D, Erdonmez CK, Alivisatos AP, Synthesis of hcp-Co Nanodisks, J Am Chem Soc, 2002, 124, 12874–12880.
[56] Wang C, Daimon H, Onodera T, Koda T, Sun S, A general approach to the size- and shape-controlled synthesis of platinum nanoparticles and their catalytic reduction of oxygen, J Am Chem Soc, 2007, 129, 6974–6975.
[57] Zhang J, Gao Y, Alvarez-Puebla RA, Buriak JM, Fenniri H, Synthesis and SERS properties of nanocrystalline gold octahedra generated from thermal decomposition of $HAuCl_4$ in block copolymers, Adv Mater, 2006, 18, 3233–3237.
[58] Li Y, Liu J, Wang Y, Wang Z, Preparation of monodispersed Fe-Mo nanoparticles as the catalyst for CVD synthesis of carbon nanotubes, Chem Mater, 2001, 13, 1008–1014.
[59] Sun S, Murray CB, Weller D, Folks L, Moser A, Monodisperse FePt nanoparticles and ferromagnetic FePt nanocrystal superlattices, Science, 2000, 287, 1989–1992.
[60] Chen M, Nikles DE, Synthesis of spherical FePd and CoPt nanoparticles, J Appl Phys, 2002, 91, 8477–8479.
[61] Nishikawa H, Morita T, Sugiyama J, Kimura S, Formation of gold nanoparticles in microreactor composed of helical peptide assembly in water, J Colloid Interf Sci, 2004, 280, 506–510.
[62] Krishnadasan S, Tovilla J, Vilar R, deMello AJ, deMello JC, On-line analysis of CdSe nanoparticle formation in a continuous flow chip-based microreactor, J Mater Chem, 2004, 14, 2655–2660.
[63] Lin XZ, Terepka AD, Yang H, Synthesis of silver nanoparticles in a continuous flow tubular microreactor, Nano Lett, 2004, 4, 2227–2232.

[64] Takagi M, Maki T, Miyahara M, Mae K, Production of titania nanoparticles by using a new microreactor assembled with same axle dual pipe, Chem Eng J, 2004, 101, 269–276.
[65] Khan SA, Günther A, Schmidt MA, Jensen KF, Microfluidic synthesis of colloidal silica, Langmuir, 2004, 20, 8604–8611.
[66] He S, Liu Y, Uehara M, Maeda H, Continuous micro flow synthesis of ZnO nanorods with UV emissions, Mater Sci Eng B, 2007, 137, 295–298.
[67] Sebastián V, Lee SK, Zhou C, Kraus MF, Fujimoto JG, Jensen KF, One-step continuous synthesis of biocompatible gold nanorods for optical coherence tomography, Chem Commun, 2012, 48, 6654–6656.
[68] Abou-Hassan A, Neveu S, Dupuis V, Cabuil V, Synthesis of cobalt ferrite nanoparticles in continuous-flow microreactors, RSC Adv, 2012, 2, 11263–11266.
[69] Kim SC, Shim WG, Lee MS, Jung SC, Park YK, Preparation of platinum nanoparticle and its catalytic activity for toluene oxidation, J Nanosci Nanotechnol, 2011, 11, 7347–7352.
[70] Mostafa S, Croy JR, Heinrich H, Cuenya BR, Catalytic decomposition of alcohols over size-selected Pt nanoparticles supported on ZrO_2: A study of activity, selectivity, and stability, Appl Catal A, 2009, 366, 353–362.
[71] Miyazaki A, Balint I, Nakano Y, Morphology control of platinum nanoparticles and their catalytic properties, J Nanopart Res, 2003, 5, 69–80.
[72] Yuranov I, Moeckli P, Suvorova E, Buffat P, Kiwi-Minsker L, Pd/SiO_2 catalysts: Synthesis of Pd nanoparticles with the controlled size in mesoporous silicas, J Mol Catal A, 2003, 192, 239–251.
[73] Castegnaro MV, Kilian AS, Baibich IM, Alves MCM, Morai J, On the reactivity of carbon supported Pd nanoparticles during NO reduction: Unraveling a metal–support redox interaction, Langmuir, 2013, 29, 7125–7133.
[74] Wang Y, Prinsen P, Triantafyllidis KS, Karakoulia SA, Yepez A, Len C, Luque R, Batch versus continuous flow performance of supported mono-and bimetallic nickel catalysts for catalytic transfer hydrogenation of furfural in isopropanol, ChemCatChem, 2018, 10, 3459–3468.
[75] Osako T, Torii K, Hirata S, Uozumi Y, Chemoselective continuous-flow hydrogenation of aldehydes catalyzed by platinum nanoparticles dispersed in an amphiphilic resin, ACS Catal, 2017, 7, 7371–7377.
[76] Paun C, Lewin E, Sá J, Continuous-flow hydrogenation of d-xylose with bimetallic ruthenium catalysts on micrometric alumina, Synth Cat Open, 2017, 2.
[77] Moore JC, Howie RA, Bourne SL, Jenkins GN, Licence P, Poliakoff M, George MW, In situ sulfidation of pd/c: a straightforward method for chemoselective conjugate reduction by continuous hydrogenation, ACS Sustainable Chem Eng, 2019, 7, 16814–16819.
[78] Din SH, Nano-composites and their applications: a review, Charact Appl Nanomat, 2019, 2, 1–9.
[79] Omanović-Mikličanin E, Badnjević A, Kazlagić A, Hajlovac M, Nanocomposites: A brief review, Health Tech, 2020, 10, 51–59.
[80] Camargo PHC, Satyanarayana KG, Wypych F, Nanocomposites: synthesis, structure, properties and newapplication opportunities, Mater Res, 2009, 12, 1–39.
[81] OECD, Nanotechnology for Green Innovation, OECD (2013), "Nanotechnology for Green Innovation", OECD Science, Technology and Industry Policy Papers, No. 5, OECD Publishing.
[82] Editorial. Nanomedicine and the COVID-19 vaccines, Nat Nanotechnol, 2020, 15, 963.

Francesco Ferlin, Nam Nghiep Tran, Aikaterini Anastasopoulou, Marc Escribà Gelonch, Daniela Lanari, Federica Valentini, Volker Hessel and Luigi Vaccaro

6 From green chemistry principles to sustainable flow chemistry

Objective of this chapter

This book chapter intends to give an overview on the quantitative assessment of the sustainability features of the processes conducted in flow conditions. Whereas possible, the key parameters and the related mathematical calculations that are relevant for the evaluation of the sustainability of a processes conducted in flow and the advantages compared to batch protocols will be discussed. The chapter starts with the definition of green chemistry principles and metrics emphasizing their utility as tool for the greenness assessment at both academic and industrial level. A selection of representative flow chemistry processes is discussed paying major attention to those using safer conditions compared to batch and/or employing waste- or biomass-derived reaction media and heterogeneous catalysts aiming at the minimization of waste associated and at implementing circularity into chemical processes design.

6.1 Quantitative sustainability assessment and outlook to flow chemistry

6.1.1 Green metrics for use in flow chemistry

In recent years, the consequences of environmental damages caused by human activities urged researchers worldwide to reinvent and modify conventional chemical processes in order to make them more efficient and more environmentally friendly. Waste reduction, the substitution of hazardous substances, and energy saving are subjects of great attention from researchers and are also fundamental topics that are part of several government strategies for sustainable development. In 1987, the World Commission on Environment and Development introduced the "Brundtland Report" where for the first time the concept of "sustainable development" was mentioned [1]. Since then, a plethora of green chemistry strategies have been developed and introduced into real cases of industrial production, for example, green synthesis of ibuprofen proposed by BHC Company in the early 1990s [2].

https://doi.org/10.1515/9783110693690-006

Green chemistry represents the area where chemical products and processes are developed to generate less hazardous substances, reduce pollution and waste, encourage the use of renewable feedstocks, lead to safer designs, optimize materials and energy usage, and eventually minimize production costs [3–6]. This concept places environmental issues as an integral part of the initial process design. Anastas and Warner introduced the 12 principles of green chemistry, with the intention of proposing a standardized approach for the development of a green process [7]. A truly green process should satisfy all 12 principles but it is also obvious that such an ideal scenario can be achieved with great difficulty in the real world. On the other hand, to evaluate and classify the eco-compatibility of a given process, a series of green metrics have been developed and introduced to quantify the actual level of sustainability achieved [8]. These metrics allow to measure the possible improvement of a designed process and a quantified comparison with the processes utilized previously [9]. In the last four decades as shown in Figure 6.1, numerous scientific works have been dedicated to the metrics of green chemistry. To quantify the greenness of a process, the measurement of a large number of different parameters has been proposed [7, 9–13]. To find a path in this intricate labyrinth it is therefore important to underline that a measurement of the efficiency of which can actually be useful and must be based on parameters that can be simple, clearly defined, measurable, and above all objective [14].

The main aim of this section is to summarize the most popular and simplest metrics adopted in the literature to develop novel green synthetic tools.

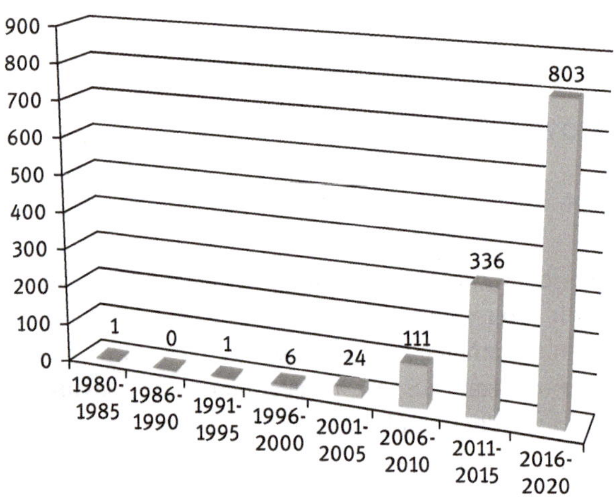

Figure 6.1: Publications with keyword "green chemistry metrics" for each period of 5 years from 1980 to 2018 (obtained using Google Scholar citation data 10/10/2020).

6.1.1.1 Green chemistry principles

As mentioned above, the 12 principles of green chemistry introduced by Anastas and Warner have been considered as the fundamental guideline to approach the design of a green process and to develop an efficient chemical process [7]. The principles are presented in Table 6.1.

Table 6.1: The 12 principles of green chemistry.

No.	Principle	Definition
1	Prevention	Preventing the formation of waste is better than treating or cleaning waste at the end of the process
2	Atom economy	Maximizing the inclusion in the final products of all components used in the reaction
3	Less hazardous chemical syntheses	Utilizing and producing less toxic and hazardous substances and any materials
4	Designing safer chemicals	Minimizing the toxicity of the products while maintaining their desired function
5	Safer solvents and auxiliaries	Minimizing the use of auxiliary substances (e.g., solvents and separation agents) whenever possible or use them safely.
6	Design for energy efficiency	Minimizing energy consumption based on environmental and economic analysis
7	Use of renewable feedstock	Encouraging the use of renewable resources wherever technically and economically feasible
8	Reduce derivatives	Minimizing or preventing unnecessary generation of derivatives due to extra reagents required which might lead to more waste generation.
9	Catalysis	Catalytic reagents (as selective as possible) are preferred to those stoichiometric
10	Design for degradation	Products should be designed to break down into harmless degradation products at the end of their lives
11	Real-time analysis for pollution prevention	Enabling real-time, in-process surveillance, and control before the creation of hazardous substances, analytical methodologies need to be further developed
12	Inherently safer chemistry for accident prevention	Minimizing accident potential, including discharge, explosion, and fire, substances used in a chemical process should be selected

It is expected that the application of these green chemistry principles will lead to the definition of the possibly ideal chemical production in terms of low environmental impacts, economic feasibility, safety, and sustainability [15]. As shown in Figure 6.1, it is also noteworthy that more than 800 publications have published worldwide only in the period 2016–2020 in the effort to apply the green chemistry principles at laboratory, pilot, and industrial scale [15–19]. Interestingly, the discussion of green chemistry principles has also been used as a key tool for high schools and universities to train next generation chemists who are aware of the crucial role greener manufacturing plays in shaping a better future for our planet [18, 20].

6.1.2 Green chemistry metrics

The 12 principles of green chemistry are useful to provide an idea of how a green chemistry process should be designed, yet they cannot be utilized to assess quantitatively the greenness of a process and most importantly to make a precise comparison between different processes. In fact, it is crucial to implement the principles with a set of measurable parameters with metrics that allows to calculate in a numerical manner how a certain process fulfill the principles. Green metrics become in this manner the fundamental tool steering the researchers toward the most efficient route to improve and make a process more efficient and sustainable [21–25]. Apart from this, green chemistry metrics play also the important role to raise awareness among chemists pushing the development of eco-friendly chemical syntheses, greener product design, and pollution prevention at the early stage of the production process [14]. Generally, green metrics have been developed and classified based on the mass and energy balances, environmental impacts, social and economic aspects, and process safety [23, 26, 27].

In this section we report the most popular metrics and their use for assessing the sustainability of representative flow processes and, when it is possible, in comparison with the known batch protocols.

6.1.2.1 Basic and simple green metrics

One of the most fundamental and simplest scales for the greenness evaluation of a process consists in the so-called mass-based metrics that take into account the quantities of all the materials used during the entire process. The mass metrics measure mass yields, excess, losses, and waste. The process material may be the reactants, solvents, water, catalysts, and other auxiliaries that may not directly appear in the global reaction scheme. We will therefore consider as a general reaction:

$$aA + bB \rightarrow pP + qQ$$

where A and B are reagents, P is the main product, and Q is a byproduct while a, b, p, and q are the stoichiometric coefficients. The most popular metrics are listed below:

Reaction yield (Y): this is one of the most common metric used by the chemists to quantify the amount of reactant being converted to the desired product (Equation (6.1)). The value of Y is between 0 and 1 [3]:

$$Y = \frac{\text{Real mass of product [kg]}}{\text{Theoretical mass of product [kg]}} \tag{6.1}$$

Atom economy (AE): AE (equation (6.2)) relates the reaction efficiency to the degree of integration of atoms of the reagents in the desired product (maximum number of atoms of reactants appearing in the products) [28]. This is one of the most commonly used measures of process efficiency [20]:

$$\text{AE} = \frac{\text{MW}_{\text{product}}\left[\frac{\text{g}}{\text{mol}}\right]}{\text{MW}_{\text{reagents}}\left[\frac{\text{g}}{\text{mol}}\right]} \tag{6.2}$$

where MW is the molecular weight (g mol^{-1}).

For a generic multistage reaction used for producing R:

$$\text{A} + \text{B} \rightarrow \text{P} + \text{X}$$

$$\text{P} + \text{C} \rightarrow \text{Q} + \text{Y}$$

$$\text{Q} + \text{D} \rightarrow \text{R} + \text{Z}$$

The AE is calculated as follows:

$$\text{AE} = \frac{\text{MW}_\text{R}}{\text{MW}_\text{A} + \text{MW}_\text{B} + \text{MW}_\text{C} + \text{MW}_\text{D}} \tag{6.3}$$

Carbon economy (CE): this metric quantifies the number of carbon atoms in the reagents integrated into the main product (Equation (6.4)). Based on this, a reaction that does not contain carbon atoms in any by-products will have a CE of 1 and an AE less than 1. Indeed, the AE reports the loss of atoms while CE reflects the loss of carbons [29]:

$$\text{CE} = \frac{\text{Number of carbon in products}}{\text{Number of carbon in reactants}} \times 100\% \tag{6.4}$$

Reaction mass efficiency (RME): this was proposed to measure the effect of solvents, catalysts, and other reactants on the greenness of a process. RME (Equation (6.5)) is calculated based on the ratio between the actual mass of the products over the mass of the reactants:

$$\text{RME} = \frac{\text{Actual mass of products}}{\text{Mass of reactants}} \times 100\% \tag{6.5}$$

Effective mass efficiency (EME): this is defined as the percentage of the mass of the desired product relative to the mass of all non-benign reagents (by-products, reagents, or solvents that have environmental risk associated with them) used in its synthesis (Equation (6.6)). The reagents here may include any used reactant, solvent, or catalyst [30]:

$$\mathrm{EME} = \frac{\text{Actual mass of desired products}}{\text{Mass of nonbenign reagents}} \times 100\% \tag{6.6}$$

Stoichiometric factor (SF): this takes into account the use of excess reagents and enables a comparison between an experimental process and a similar process in which all reactants should be stoichiometrically used [31]. SF is measured by the following equation:

$$\mathrm{SF} = 1 + \frac{\sum \text{Mass of excess reagents} \quad [\text{kg}]}{\sum \text{Mass of stoichiometric reagents}\,[\text{kg}]} \tag{6.7}$$

Material recovery parameter (MRP): it considers all the possible recycled materials (excess reagents, solvents, and catalysts) [4]. MRP is calculated by the following equation:

$$\mathrm{MRP} = 1 + \frac{\text{Mass of all materials recovered} \quad [\text{kg}]}{\sum \text{Mass of stoichiometric reagents}\,[\text{kg}]} \tag{6.8}$$

Process mass intensity (PMI): it was first developed to evaluate the effectiveness of a process by considering masses of all reactants (and water) used to produce a unit mass of the desired product (Equation (6.9)) [32]. However, while we believe that it is always correct to include water in the calculation, in some works, water was excluded from the total mass of materials in the PMI calculation [23]:

$$\mathrm{PMI} = \frac{\text{Mass of all materials}\,[\text{kg}]}{\text{Mass of products} \quad [\text{kg}]} \tag{6.9}$$

Solvent intensity (SI) and water intensity (WI): in a synthetic process, solvent and water may constitute the main contribution to waste production and economic loss. This is important for those industrial sectors where solvent and water are utilized in large amounts, as for the synthesis of active pharmaceutical ingredients (APIs). Importantly, the shortage of clean water all over the world (e.g., Africa) encourages the use of WI as a green chemistry parameter. Thus, SI (Equation (6.10)) and WI (Equation (6.11)) are developed to quantify the solvent and water loss to prevent additional waste generation [4, 23, 33]:

$$\mathrm{SI} = \frac{\text{Mass of all solvent used excluding water}\,[\text{kg}]}{\text{Mass of products} \quad [\text{kg}]} \tag{6.10}$$

$$\mathrm{WI} = \frac{\text{Mass of all water used}\,[\text{kg}]}{\text{Mass of products} \quad [\text{kg}]} \tag{6.11}$$

Environmental factor (E-factor): it was proposed by Sheldon in 1992 and it is based on the first of the 12 principles of green chemistry. According to the definition of E-factor, any compounds in the product mixture (excluding the ones that can be recycled) apart from the main product are considered waste (Equation (6.12)). E-factor will be zero in the ideal case where no waste is generated during a process [34]:

$$\text{E}-\text{factor} = \frac{\text{Mass of waste}\ [\text{kg}]}{\text{Mass of products}\ [\text{kg}]} \tag{6.12}$$

Simple environmental factor (sEF) and complete environmental factor (cEF): they were proposed by Roschangar et al. to assess different stages of a process. The sEF (Equation (6.13)) is applied at the early planning stage without considering water and solvents while the cEF (Equation (6.14)) takes into account solvents, water, and all process products and is widely applied in a complete waste stream assessment [33]:

$$\text{sEF} = \frac{\sum \text{Mass}_{\text{raw materials}} + \sum \text{Mass}_{\text{reagents}} - \text{Mass of product(kg)}}{\text{Mass of product(kg)}} \tag{6.13}$$

$$\text{cEF} = \frac{\sum \text{Mass}_{\text{raw materials}} + \sum \text{Mass}_{\text{reagents}} + \sum \text{Mass}_{\text{solvents}} + \text{Mass}_{\text{water}} - \text{Mass of product(kg)}}{\text{Mass of product (kg)}} \tag{6.14}$$

Synthesis ideality (Si): it was proposed by Gaich and Baran in 2010 to focus on the impact of the reaction steps [35]. This parameter follows the sixth principle of green chemistry, targeting at reducing chemical transformations required and the comparison of different routes for achieving the final complex structure [35]. The Si is given by the following equation:

$$\text{Si} = \frac{\text{Number of reactions} + \text{Number of strategic redox reactions}}{\text{Number of reactions}} \tag{6.15}$$

Construction reactions are described as chemical modifications forming bonds of carbon–carbon or carbon–heteroatom while strategic redox reactions are a sort of building reactions that immediately set the right functions found in the main product and include asymmetric reductions or oxidations.

Renewable intensity (RI): The use of renewable materials is one of the principles of green chemistry, and RI (Equation (6.16)) is a straightforward metric to evaluate how renewable a reaction is:

$$\text{RI} = \frac{\text{Mass of all renewably derived materials used}\ [\text{kg}]}{\text{Mass of products}\ [\text{kg}]} \tag{6.16}$$

6.2 Flow chemistry and green metrics

In this section, we have described the relevant connection points between flow chemistry and the principles and metrics of green chemistry. At this aim we have also selected and described some examples chosen among those reporting the use of (i) safer solvents, (ii) recoverable heterogeneous catalysts, and (iii) metrics to calculate and quantify the waste produced. We have also tried to cover different areas of interest reporting the use of flow chemistry in the synthesis and/or use of green solvents deriving from waste valorization or biomass in the context of the circular green economy. Other examples include also the use of recoverable heterogeneous catalysts that represent an effective tool to achieve continuous chemical production while minimizing the waste associated. Particular attention has been also directed toward those protocols that are also accompanied by data that allow the calculation of green metrics associated with the flow application, allowing a quantified evaluation of the environmental impact. Where possible, and when these data are available, an useful comparison among batch and flow protocols has been presented.

At the end of the chapter a question-and-answer section is provided.

6.2.1 Application of green metrics to flow chemistry

Flow chemistry is a useful tool for the definition of modern and innovative synthetic strategies which therefore aim at the closely related principles of green chemistry [36–39]. Indeed, the employment of flow reactors allows to effectively improve crucial reaction parameters such as reaction time, mixing and heat transfer, scale-up, and energy efficiency.

The precise control of these parameters may also leads to additional advantages and therefore they can influence the general efficiency of the whole process in terms of: waste production, safety and hazard prevention, overall operational cost of the process.

Furthermore, a great, but difficult to quantify, advantage of flow technologies is the intrinsic possibility of accessing new reactivity and process outcomes due to the wide range of new reaction conditions that can be adopted compared to classic batch reactors.

It is worthy to emphasize that flow reactors may not only allow to develop a modern and efficient version of known process, but may constitute a key tool for inventing and defining brand-new synthetic strategies (opening novel process scenarios) based on photo-, homogeneous, and heterogeneous catalysis, multistep and telescoped protocols, and more [40–43].

In the following two examples reported, it will be shown two different types of flow reactors and conditions and the associated results in terms of improved sustainability supported by the green metrics calculations.

In the first example, the palladium-catalyzed conjugate reduction of olefins to α,β-unsaturated carbonyl compounds in the presence of hydrogen as a reductive agent is presented [44]. This process takes advantage of the use of diphenyl sulfide to produce in situ a reactive and chemoselective hydrogenation catalyst allowing the use of environmentally benign isopropyl acetate as reaction medium. The optimization was directly performed in flow conditions using an H-Cube® hydrogenation reactor (Figure 6.2) and it was aimed at the identification of the proper amount of sulfide necessary to achieve high catalytic efficiency while controlling the possible poisoning of the palladium catalytic centers. Optimal conditions (>95% conversion to the desired product) consisted in the employment of a 1.25 mM concentration of sulfide at 100 °C with a flow rate of 0.5 mL min^{-1} and these proved to be suitable for a variety of chalcones including natural occurring substrates.

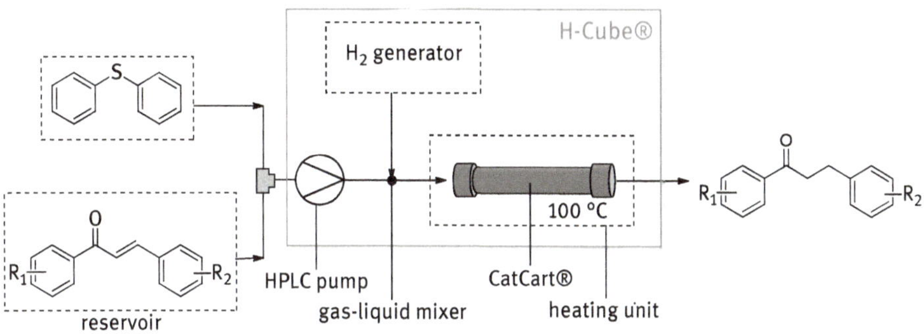

Figure 6.2: Chemoselective hydrogenation of α,β-unsaturated carbonyl.

To assess the greenness of the developed methodology, the authors have used a metric tool named Green Motion™. This tool was developed by the flavor and fragrance company, Mane, with the goal of evaluating health, safety, and environmental impacts of their manufacturing processes and it is based on a 0–100 scale [45]. The Green Motion™ calculation process is very simple yet effective, and it deals with the 12 principles of the green chemistry in a penalty point system. The lower the impact to the environment is, the higher the rating is (100 value is ideal). Relevant criteria adopted in the calculation refer to different concepts:

(1) Raw material: origin and process naturalness
(2) Solvent category
(3) GHS pictogram: hazard and toxicity of reagents and products
(4) Reaction parameters: mass yield, number of steps and solvents, CE, and overall process time
(5) Conditions: heating/cooling, vacuum, pressure
(6) E-factor: waste

Based on this, the overall sustainability of the palladium-catalyzed conjugate reduction in the H-Cube® system can be summarized as follows:
(1) High selectivity, therefore reduced waste: low E-factor of 1.34 (Principle 1)
(2) Hydrogenation with H_2: high AE (Principle 2)
(3) Low toxicity of reagents (Principle 3)
(4) Catalytic quantity of the auxiliary: diphenyl sulfides (Principle 5)
(5) Operation at atmospheric pressure, and despite heating, no cooling is required (Principle 6)
(6) Use of a renewable solvent: isopropyl acetate (Principle 7)
(7) Catalytic reagents (Principle 9)
(8) No storage of H_2: reduced chance of accidents (Principle 12)

A second example that we wish to report deals with a detailed comparison of the batch and flow procedures available for the azidation of α,β-unsaturated carbonyls for which different green metrics have been calculated and taken into account: for example, material efficiency, environmental impact, and safety-hazard impact [46]. The flow protocol has been optimized for different α,β-unsaturated carbonyls using trimethylsilyl azide ($TMSN_3$) as N_3-source in solvent-free conditions. The mixture was cyclically pumped through the reactor for the desired time, air-solvent valve was open at the end of the reaction to collect the product with a final washing (Figure 6.3). The results obtained with this setup are compared below with other procedures reported in the literature. It has been proven that the flow protocols feature a greener profile compared to the corresponding batch procedures.

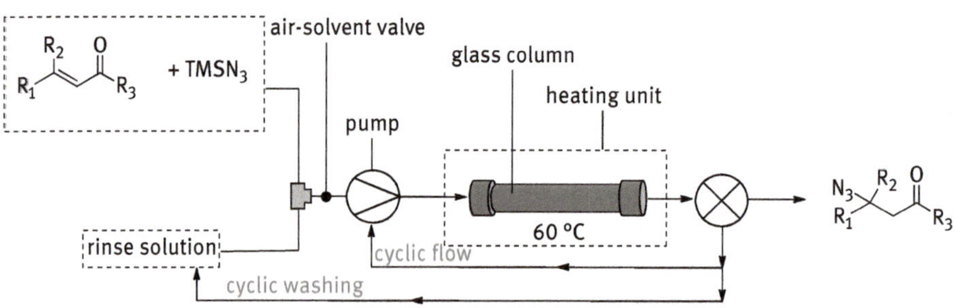

Figure 6.3: General flow scheme for β-azidation of α,β-unsaturated carbonyl compounds.

As mentioned above, in this analysis three groups of green metrics were examined.
(1) Material efficiency metrics: reaction yield, AE, global RME, E-factor, PMI
(2) Environmental impact metrics (eight parameters): acidification–basification potential, ozone depletion potential, smog formation potential, global warming potential, inhalation toxicity potential, ingestion toxicity potential, bioconcentration potential, and abiotic resource depletion potential

(3) Safety-hazard metrics: corrosive gas potential, corrosive liquid/solid potential, flammability potential, oxygen balance potential, hydrogen gas generation potential, explosive vapor potential, explosive strength potential, impact sensitivity potential, occupational exposure limit potential, skin dose potential, and risk phrase potential

While the selected protocols and the use of heterogeneous catalysts proved to be effective for a variety of substrates, the specific evaluation of these metrics was applied to a specific substrate employed in all the protocols. The main differences among the available procedures consist in the azidation reagents and the auxiliary materials used such as catalysts, reaction media, work-up, and purification materials. It is therefore noteworthy that in azidation the most toxic and critical aspects are related to the formation of hydrazoic acid or a metal azide species. $TMSN_3$ is a sort of "protected" safer source of azido ion and its toxicity is anyway related to the possible release of hydrazoic acid.

To make the results of this study as easily accessible and comparative as possible, some assumptions have been made:
(1) the volumes of solvents used for the work-up extraction were set equal to the volume of the reaction medium;
(2) the mass of drying agent is set as 2 g for a 20 mL or less volume of reaction medium;
(3) the mass of drying agent is 10 g for a volume of reaction medium between 100 and 500 mL.

Combining the output for the evaluation of each metrics, radial polygons can be constructed and the vector magnitude ratio (VMR), obtained from the ratio of the vector magnitude for the parameters considered versus those for the ideal green case, was used to measure the greenness associated with each procedure.

The assessment showed below depends on the specific class of substrates as follows:

(1) Linear α,β-unsaturated ketones

This example allows the direct comparison of the flow procedures [46] with previously developed batch [47, 48] protocols. Indeed, the very similar reaction conditions adopted for accessing of β-azidated products lead to the same AE value. This aspect is advantageous for the specific evaluation of the differences in terms of environmental impact and safety-hazard metrics as shown in Table 6.2.

The latest developed flow procedure [46] features the maximum VMR associated with the lowest PMI value obtained. Indeed, this protocol does not require the use of dry conditions and allows to eliminate drying agent. In addition, only the stoichiometric amount of water needed for the reaction has been considered in

Table 6.2: Metrics comparison in β-azidation of linear α,β-unsaturated carbonyl compounds.

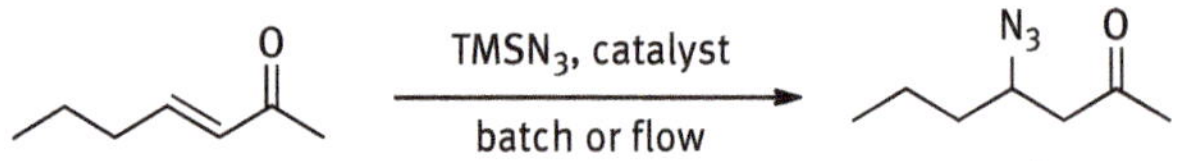

Reference	Yield (%)	AE (%)	PMI	BI	EI	VMR	SHZI
[47] (batch)	87	63.3	7.05	0.8243	26.05	0.6291	53.1
[48] (batch)	91	63.3	12.07	0.8040	26.05	0.6055	53.1
[48] (flow)	93	63.3	4.73	0.8199	36.38	0.6728	53.6
[46] (flow)	95	63.3	1.92	0.6243	20.95	0.7848	29.1

the E-factor calculation. Anyway, the benign index (BI) associated with this procedure has the lowest value compared to the others. This apparently contradictory feature is ascribable to the calculation methodology for this parameter which is based on mass weighted impact potential. So even if the flow protocol produces an extremely low amount of waste, the mass percentage (not the actual amount) of the toxic trimethylsilanol (a reaction byproduct) produced by the reaction is higher than previously reported batch protocols [47, 48] (63% and 5%, respectively).

(2) Cyclic α,β-unsaturated ketones

The selection of the azido ion source for the compared procedures mainly influences the AE and SHZI values. Indeed, as shown in Table 6.3 during a process design, not only the environmental metrics need to be considered but also a careful plan of the safety-hazard metrics must be taken into account.

Table 6.3: Metrics comparison in β-azidation of cyclic α,β-unsaturated carbonyl compounds.

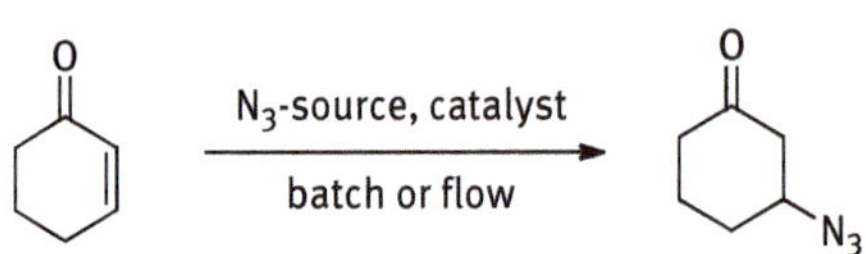

Reference	Yield (%)	AE	PMI	BI	EI	VMR	SHZI
[49] (batch)	90	51.3	59	0.9890	558.0	0.5987	207.4
[50] (batch)	94	62.9	19.8	0.9267	5025.5	0.6239	1173.5
[51] (batch)	95	62.9	25.2	0.9429	4913.4	0.6277	1389.1
[46] (flow)	98	60.7	3.2	0.9761	290.1	0.7572	89.8

The highest AE values have been obtained by using sodium azide [50, 51] instead of $TMSN_3$ [46, 49]. On the other hand, the choice of sodium azide dramatically influenced the SHZI as this reactant is highly toxic. The safety-hazard impact is strictly related to (i) the use of dichloromethane for the work-up procedure [49]; (ii) exposition limit values of NaN_3 [50, 51]; (iii) amount of trimethylsilanol generated as byproduct [46].

Although the BI values for all the selected processes are high, the latest flow protocol [46] developed showed the highest VMR, thanks to the low PMI value.

(3) Acrylic acid

Also in this example the reaction conditions of the reported examples differ for the N_3-source. Batch protocols [52, 53] used NaN_3 in acetic acid, while flow protocols [46] employed $TMSN_3$ with different fluoride-based bifunctional heterogeneous catalysts. The choice of $TMSN_3$ resulted in a slightly lower AE if compared to the protocol using NaN_3, however, the flow protocols feature values largely higher for VMR and lower for PMI. These values derived not only from the high yields obtained, but also from the optimized solvent-free conditions which led to a significant waste minimization. Indeed, in the protocol from reference [53] 500 mL of solvent were used in the work-up procedure for only 3 g of acrylic acid affording in a PMI value of two orders of magnitude higher than the flow protocol [46].

Table 6.4: Metrics comparison in β-azidation of acrylic acid.

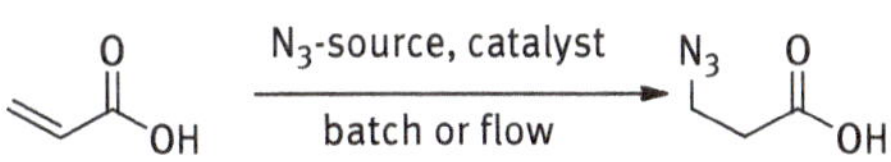

Reference	Yield (%)	AE	PMI	BI	EI	VMR	SHZI
[52] (batch)	24	58.4	39	0.9287	7916.0	0.5349	1315.2
[53] (batch)	50	58.4	296	0.9910	7797.9	0.5279	1320.6
[46] (flow)[a]	99	56.1	3.6	0.9955	3270.8	0.7519	216.0
[46] (flow)[b]	80	56.1	4.5	0.9541	3270.8	0.6978	216.0

[a]PS-DABCOF as catalyst; [b]Amberlyst-F as catalyst.

EI values are extremely high for all the proposed protocols, this observation is strictly related to the toxicity of acrylic acid used as reagent. This toxicity has a particular weight also in the determination of safety-hazard impact. Indeed, the higher SHZI values evidenced in the batch procedures [52, 53] are due to both unreacted acrylic acid and to the exposition limits to NaN_3 in safety-hazard impact of waste as well as in safety-hazard impact of input materials (Table 6.4).

(4) Mesityl oxides

As above commented, the use of NaN_3 [52, 54] as N_3 source in comparison with $TMSN_3$ [46] allows to obtain better AE values also for the β-azidation of mesityl oxide. On the other hand, when hydrazoic acid has been implemented in batch, as described in reference [55], as shown in Table 6.5 the AE still decreases due to the necessity for using triethylaluminum. In addition, this last described protocol has the highest PMI due to the reaction solvent used in the reaction and for work-up procedure. It is worth mentioning that the PMI for references [52, 54] have been intentionally under-estimated due to the lack of detailed experimental work-up procedure.

Table 6.5: Metrics comparison in β-azidation of mesityl oxide.

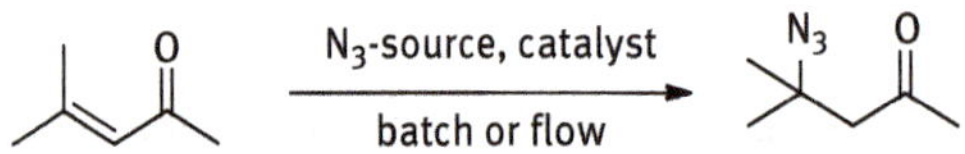

Reference	Yield (%)	AE	PMI	BI	EI	VMR	SHZI
[52] (batch)	38	63.2	30.3	0.9493	4721.5	0.5619	1173.2
[54] (batch)	66	63.2	137	0.9896	4633.9	0.5657	1400
[55] (batch)	88	45.6	238	0.9975	4074.8	0.5804	227.1
[46] (flow)	90	61	3.5	0.8921	76.3	0.7220	73.9

The flow procedure [46], even in this example, has the higher VMR associated with the lowest PMI and EI values. This protocol presents the lowest SHZI, however, the BI is lower in this case if compared with the other batch protocols. This behavior is related with the percentage in mass of the hazardous waste: trimethylsilanol waste from flow procedure is 26% of the total waste mass, while in the other processes unreacted NaN_3 represents only 5% and 0.9%, respectively, for [52] and [54] and hydrazoic acid is the 0.2% waste in protocol described by reference [55].

By the analysis of all these data collected, it is possible to effectively measure the improvements in terms of sustainability achieved with the latest flow protocol developed [46] in comparison to the previously reported batch or flow protocols. These advances are mainly ascribable to the increase in material use efficiency and therefore in the minimization of waste generation and PMI values. The adoption of flow conditions allowed the use of the minimal amounts of reagents and solvents with considerable improvements in efficient recycle and reuse of the catalyst. This comparison study shows the usefulness in terms of sustainability derived from the adoption of flow conditions. It is worthy to note that, in general, the parameters analysis suffers from the lack of data and therefore it needs to be adequately interpreted by using the appropriate uncertainty factors. Importantly, good evaluation of environmental impact metrics is not always strictly connected with good values of safety-hazard impact metrics and a critical evaluation of the greenness of the process is always required also considering possible practical applications.

6.2.2 Biomass-derived and/or waste-derived alternatives to classic solvents

Green chemistry also means the development of processes that reduce or, ideally, eliminate the generation of hazardous substances or intermediates (Anastas and Warner, 1998). Green chemistry principles 3 and 7 specifically emphasize the importance of the use of less toxic solvents and chemicals, and preferably derived from renewable resources. Accordingly, and also considering that solvents constitute the largest contribution to the waste associated with a synthetic process, many efforts have been made for the classification and listing of the so-called *preferable solvents* [56–58]. In the scenario of sustainability, the life cycle assessment [59] should be used to implement the evaluation of environmental impact of chemical production.

In this arena, a growing interest of numerous researchers has allowed several new alternative solvents to emerge as possible alternatives to the toxic and fossil derived classic ones. In this context, we like to focus the attention on two main interesting classes of solvents that are emerging with promising safety and toxicity properties but also fulfilling some of the green chemistry principles:
(1) Solvents derived from renewable sources
(2) Solvents derived from the valorization of a chemical waste

Both categories are object of extensive studies and the latter, very recently, has started to gain significant attention also thanks to major chemical companies that are converting their side production into valuable new chemicals. In this realm, it is worthy to mention the case of cyclopentyl methyl ether (CPME) produced as a green solvent by Zeon corporation by valorizing cyclopentene, a waste of their petrol fraction manipulation [60], and also Rhodiasolv© Polarclean, produced by Solvay from glutaronitrile: a waste derived from nylon 6,6 production [61]. To date, no flow process has been reported to produce these chemicals.

6.2.3 Biomass-derived solvent production in flow

A different scenario is otherwise presented for the biomass-derived solvents that have received much attention in the last decades with several examples in the literature demonstrating their utility and good properties as media for synthetic processes (Figure 6.4) [62–65]. More importantly, the interest for these chemicals is also corroborated by several protocols developed to implement their productions from different types of biomasses. Aiming at the production of fundamental chemicals,

lignocellulosic biomass is certainly the most interesting and abundant renewable feedstock which can be divided into three major components:
(1) Lignin: complex biopolymer constituted by a mixture of phenolic monomers. Major valorization of lignin leads to the production of bio-oil whose subsequent transformations allow to obtain value-added chemicals.
(2) Cellulose: glucose-based polysaccharide that after treatment may lead to the formation of platform molecules which can be efficiently converted in fuel, chemicals, and also valuable solvents.
(3) Hemicellulose: branched biopolymer constituted by five- and six-carbon sugars that, as for cellulose, can be treated to produces interesting solvents besides fuel, and chemicals. (Figure 6.9).

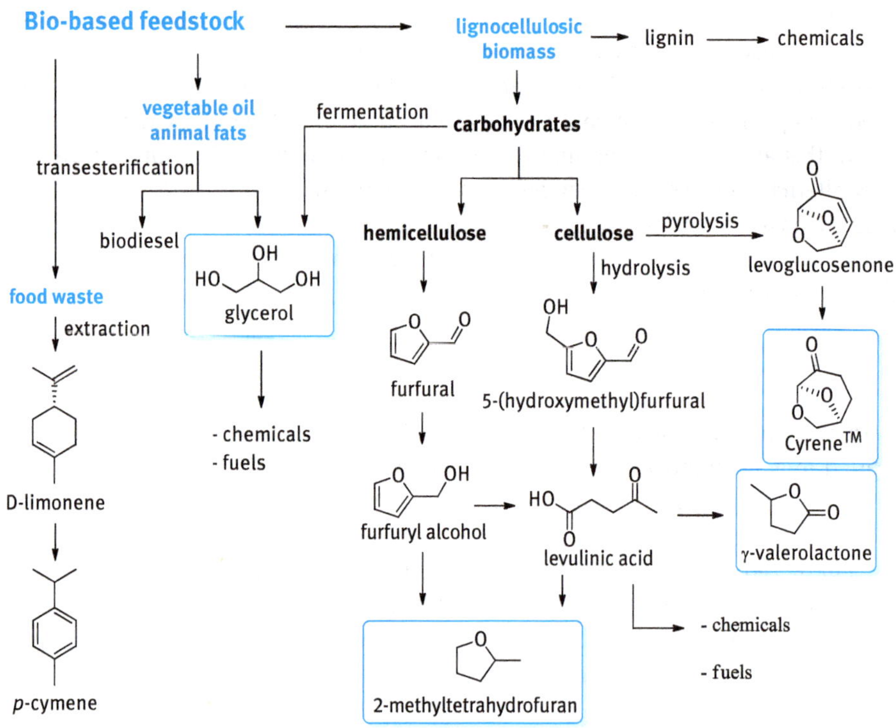

Figure 6.4: Biomass-derived solvents.

Among the platform chemicals derived from biomass valorization, levulinic acid (LA) is of central interest also considering the production of biomass-derived reaction media [66]. LA can be obtained by treatment of both hemicellulose and cellulose and after transformation can be converted into γ-valerolactone (GVL) and 2-methyltetrahydrofuran (2Me-THF). These latter are both largely employed as media [67–70] and effective protocols in flow for their production have also been defined. These processes required the use of molecular hydrogen or liquid H-source such as formic acid (FA) [71–74]. FA in turn is produced by hydrolysis of 5-(hydroxymethyl) furfural (HMF) accompanying the formation of LA in equimolar amounts, and it is widely used for biomass-upgrading [75]. Among the biomass-derived solvents dihydrolevoglucosenone (Cyrene), derived from selective reduction of levoglucosenone [76–78], is another platform chemical obtained from cellulose pyrolysis, and it has been used both in organic synthesis [79] and in material synthesis [80, 81]. However, to date no flow protocol has been developed to produce or use this solvent. The synthesis of glycerol instead is strictly related to its isolation as side-product from the production of biodiesel [82, 83].

6.2.3.1 Levulinic acid (LA)

This first example here presented consists in a simple application for the preparation of LA directly from D-fructose using a continuous-flow protocol (Figure 6.5) [84]. An important and useful detail to highlight is that, to ensure a continuous operation of the flow reactor, a filter section was installed to block the carbon particles formed during the reaction. The procedure is very simple as an aqueous solution of D-fructose in a 1:2 mixture of 2 M HCl aq. and methanol is pumped at a 0.13 mL/min rate into the Vapourtech flow reactor heated to 140 °C. After 80 min of operation, the reaction mixture was diluted with water and basified to extract the side product hydroxymethyl furfural. Then the aqueous phase was acidified to pH = 1 and LA extracted in 72% yield.

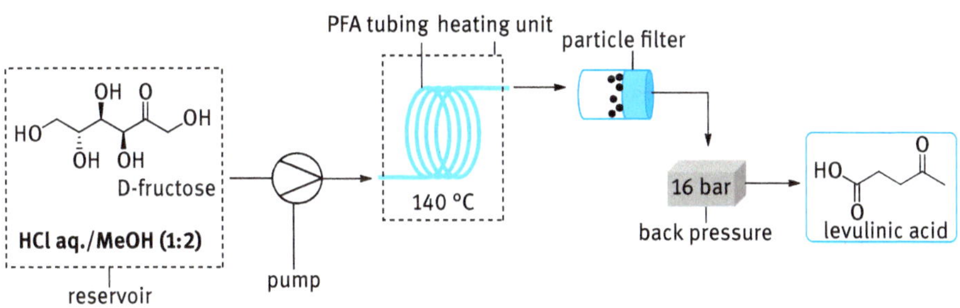

Figure 6.5: Continuous flow production of LA from D-fructose.

6.2.3.2 GVL

Starting from LA, it is possible to obtain a variety of products and arguably one of the most interesting is GVL that can be prepared in flow *via* hydrogenative cyclization of LA (Figure 6.6) using a commercial flow apparatus (H-Cube® and H-CubePro®, developed by ThalesNano Inc.) [85]. In this commercially available reactor system, hydrogen is generated in situ by electrolysis of water and it is continuously mixed with the substrate allowing to avoid the expensive and hazardous issues related to hydrogen production and storage. The reaction mixture was continuously pumped through a cartridge (CatCart®) packed with the catalyst of choice. Advantages related to the use of this reactor are mainly related to the effective alternative for transport, storage, and manipulation of gaseous hydrogen: (i) avoid use of high-pressure cylinder; (ii) minimize fire and explosion hazard; (iii) simplicity of operation; and (iv) high throughput of the reaction.

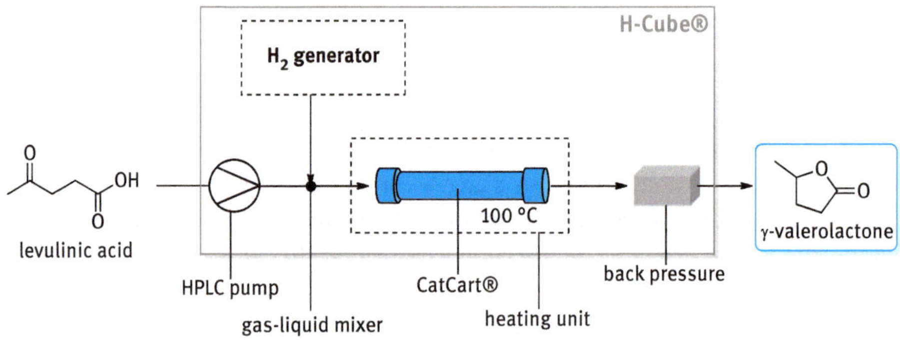

Figure 6.6: Flow synthesis of GVL from LA in H-Cube® flow apparatus.

With this flow setup, a 0.1 M aqueous solution of LA was directly passed into the reactor at a flow rate of 1 mL/min using 5% Ru/C or RANEY Ni CatCart® under 100 bar of H_2 at 100 °C. The conversion to GVL reached 82.9% with 5% Ru/C as catalyst, and it was further increased to 98.5% by adding Bu-DPPDS as ligand. The safe manipulation of hydrogen pressure needed to achieve good conversion to GVL allowed for a fast reductive cyclization with a productivity of 0.844 $mol_{GVL} \times g_{METAL} \times h^{-1}$ for 3 h by using Ru/C CatCart®.

6.2.3.3 2-Methyl-tetrahydrofuran (2-Me-THF)

Alternatively, LA can be a precursor for 2-Me-THF, another well-known green solvent proposed as a biomass-derived alternative to THF. In the following example, a bio-based scenario has recreated in flow conditions (Figure 6.7) [86]. In nature, cellulose is hydrolyzed to glucose and 5-hydroxymethylfurfural intermediate which give LA and FA in 1:1 ratio via dehydration. This FA cogenerated in the dehydration process could then be utilized for the sustainable and selective production of 2-Me-THF. Additional environmentally benign feature of the protocol developed relies in the utilization of different heterogeneous catalysts (Cu based vs. Pd based). An initial optimization of the reaction condition was performed under microwave irradiation in batch and then a flow chemistry equipment (X-Cube, Thalesnano Inc.) was utilized to translate the optimized microwave batch results. The benefits of flow conditions in comparison with batch protocols include: an improved control of reaction parameters; quick and efficient reagent mixing; shorter times of reactions (similar to microwave batch reactions); enhanced heat and mass transfer which can significantly promote catalyzed processes.

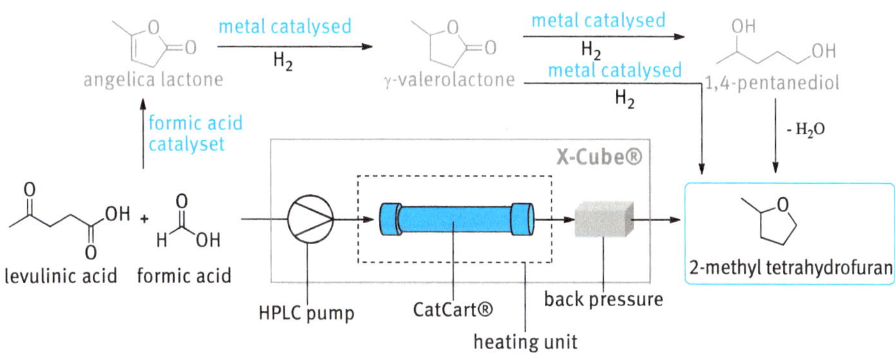

Figure 6.7: Flow synthesis of 2-Me-THF from LA and FA.

The deactivation and leaching of the catalysts were investigated directly in flow conditions showing a major tendency for the Cu-based catalyst to leach metal species in comparison to the more stable commercial Pd/C. On the other hand, selectivity to 2-Me-THF was better with copper-based catalyst. Indeed, when noble metals (Pd or Ru) are employed, different reaction pathways take place giving 4-hydroxyvaleric acid and pentanoic acid as major side-product. With the flow condition presented, LA can be converted to 2-Me-THF with a selectivity varying from 28% to 85%, depending on the catalyst used, in 1 min of residence time and with a flow rate of 0.1 mL min^{-1}.

6.2.4 Flow protocols combining biomass-derived solvents and heterogeneous catalysis

The utilization of biomass-derived chemicals as reaction media in flow reactors is an effective strategy to prove and expand their general utility in a more sustainable chemical production [87]. In this scenario the definition of different conditions and processes that can effectively promote the shift from oil- to bio-based chemistry plays a crucial role.

As described in previous chapters and mentioned above in the flow synthesis of bioderived solvents, another powerful tool to allow sustainable flow protocols is the use of stable recoverable and reusable heterogeneous catalytic systems [88, 89]. In these section examples regarding the use of biomass- or waste-derived solvents as medium in flow synthetic protocols in combination with heterogeneous catalysts will be presented. A major focus will be given to the comprehension of the different approaches adopted to improve of the sustainability of the process moving from batch to flow protocols.

The Fujiwara–Moritani reaction also known as the "oxidative Heck" reaction is a representative process based on a Pd-catalyzed C–H activation technology. If compared to its cross-coupling equivalent strategy, namely the Mizoroki–Heck reaction, the Fujiwara–Moritani represents a more atom-economical approach being a cross-dehydrogenative coupling [90]. Pd(II) is generally used as the catalyst in combination with an oxidant responsible for controlling and preventing Pd (0) formation while continuously regenerating the actual active catalytic species. This reaction also requires the addition of an acid which coordinates the metal favoring the key C–H activation step. Here is showed a variant of the Fujiwara–Moritani reaction in which the ortho-functionalization of acetanilides has been realized using commercially available Pd/C as cheap and reusable heterogeneous catalyst and benzoquinone as the stoichiometric oxidant (Figure 6.8) [91]. The generally employed toxic polar aprotic solvents, for example, DMF or DMSO, were efficiently replaced by GVL that is comparable in terms of polarity, but it features a very green profile and being a lactone, a less coordinating ability compared to amides. Palladium leaching has been measured during the optimization in batch to identify the best reaction conditions and obtaining a very low value of 4 ppm compared to other procedures [91]. The batch optimization study has allowed to decide the proper flow strategy and to build the adequate reactor by considering (i) the percentage of palladium lost in the optimized reaction conditions and (ii) therefore predict the actual quantitative durability of the catalyst in flow conditions.

Although leaching is a requirement for this reaction to occur with Pd(II) as the active catalytic species, Pd(0) formed at the end of the catalytic cycle is hypothesized to be redeposited on the support at the end of the reaction following a kind of *release and catch* mechanism. Under batch conditions, the catalyst can be reused for five runs with little decrease in its efficiency but performing the process in the continuous-flow packed bed reactor, the reaction productivity has been significantly increased up to 4 g h^{-1}. The possibility to continuously process several hundred mmol of substrate with no decrease in efficiency is largely due to the adoption of the flow reactor but also to the use of GVL selected as the reaction medium. Indeed, compared to other commonly used solvents (DMF in AcOH above all) the more limited coordination ability of GVL results in a perfect balance to achieve high reactivity and minimal palladium leaching.

Another instructive example on the use of biomass-derived solvent to affect the reaction efficiency is shown in the development of asymmetric nitro-aldol reaction in 2-Me-THF (Figure 6.9) [92]. Also known as the Henry reaction, this is, in principle,

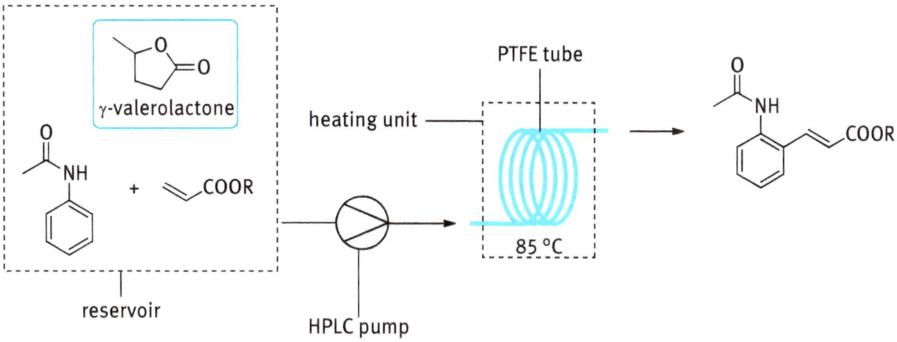

Figure 6.8: Fujiwara–Moritani reaction in GVL under flow conditions.

a preferable one for sustainable transformation due to the wide availability of substrates and its 100% AE.

The reaction generally proceeds under proton transfer conditions by the actions of a catalyst installing consecutive hydroxyl and nitro groups on stereogenic carbons. The addition of the proper chiral catalyst allows therefore to drive the enantio- and diastero-selectivity of the final condensation product.

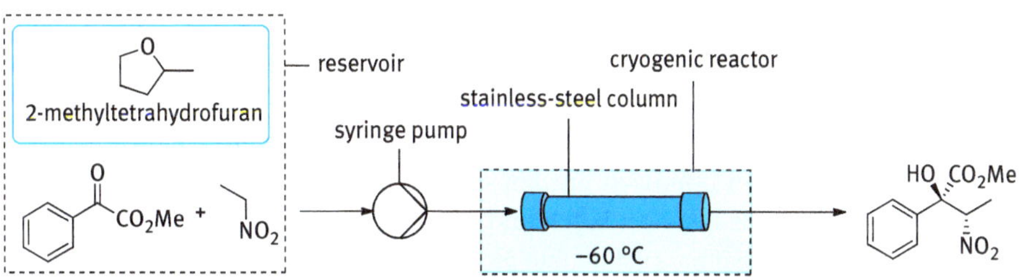

Figure 6.9: Nitro-aldol reaction in 2-Me-THF under flow conditions.

In this example, a careful evaluation of the differences between THF and 2-Me-THF has been reported by using an immobilized Nd/Na heterobimetallic chiral catalyst. The catalytic system furnishes the *anti*-products with high stereoselectivity. A relevant solvent effect has been observed between THF and 2-MeTHF as the latter allowed to enhance the reaction rate affording better stereoselectivities and complete conversions to products with a low loading of catalyst (1 mol%). These differences in the reaction rates between THF and 2-Me-THF are ascribable to the accessibility of the Nd cation that as a Lewis acid is responsible to bind the α-keto esters for the C–C bond construction. On the other hand, the sterically less demanding THF might be more prone to occupy the coordination sphere, thereby making the reaction slower. The heterogeneous catalyst showed compatibility with flow conditions, and the enantioenriched products obtained allowed the synthesis of APIs (efinaconazole and albaconazole). The heterogeneity of the Nd/Na catalyst was implemented in a continuous-flow apparatus. To this end, the catalyst has been further modified by immobilization onto multiwalled carbon nanotubes. The substrate mixture in 2-Me-THF has been allowed to pass through a stainless-steel column charged with the catalyst affording the desired product in 92% yield after 92 h without loss in stereoselectivity compared with batch conditions. Importantly, the cooling volume has been reduced, decreasing the energy input for cryogenic conditions (–60 °C), and evaporation of the eluents allowed to isolate the crude product without any additional work-up procedure resulting in an operationally simple protocol. In this example, the combination of flow technology with the use of heterogeneous catalyst in biomass-derived 2-Me-THF clearly leads to several interesting advantages: (i) 2-Me-THF enhanced the efficiency and selectivity of the reaction in comparison with THF; (ii) under flow conditions, the stereoselectivity of the process is preserved; (iii) energy consumption for the necessary cryogenic conditions is minimized, thanks to flow reactor technology.

6.2.5 Waste minimization

The quantification of the benefits raised by the adoption of flow conditions, in combination with heterogeneous and recoverable catalytic systems, can be calculated using appropriate metrics as shown in the previous examples. Whereas the overall sustainable assessment that belongs to the usage of biomass- and/or waste-derived solvents is not immediate and needs to be carefully evaluated with a more Life Cycle Assessment (LCA)-based approach. In this section, selected examples are shown to focus on the differences between batch and flow conditions highlighting the strategies followed to achieve more sustainable processes.

6.2.5.1 Use of biomass-derived GVL in C–C bond formation via Heck–Mizoroki coupling with heterogeneous catalyst

This first example [93] focuses on the design of an immobilized heterogeneous catalytic system the reaction conditions optimization, the most adequate work-up procedures and the consequent flow reactor settings to access a generally sustainable protocol. The important factors that have to be considered in the design of an immobilized metal-based heterogeneous catalyst aiming at a waste minimize process are the high accessibility to the catalytic sites, the stability and *inertia* of the heterogeneous support or the matrix, and the consequent most appropriate immobilization strategy. The immobilization of palladium salts onto a cationic imidazolium moiety (tag) supported on a polystyrene matrix is a consolidated approach and it has been implemented in this case for the preparation of a specific polymeric ionic tag (POLITAG) catalytic system. The catalyst was designed considering that ionic moieties possess the ability to scavenge metal species, allowing their immobilization.

The styrene/vinylbenzylchloride/1,4-bis(4-vinylphenoxy)benzene copolymer used in this case as support exhibits thermal and mechanical stability, and more importantly, compatibility with various polar solvents. The ionic ligand selected for the palladium coordination is a pincer-type moiety that showed the formation of complex with enhanced stability compared to similar monoionic ligands. The initial optimization in batch conditions aimed at the obvious identification of the most efficient reaction conditions: (i) the use of a biomass-derived safe solvent as GVL, (ii) the quantification of the leached palladium species which directly influences the reusability and durability of the catalyst and the purity of the isolated product, and (iii) the optimization of the conditions for the reusability of the catalyst.

The choice of a heterogeneous base was in this case functional to allow its removal (and reuse) along with the catalyst. This strategy, in fact, led to a very simple work-up procedure at the end of the reaction that, after the simple removing of the reaction medium, allowed to isolate the products by recrystallization. The batch protocol features E-factor values ranging from 21 to 75 depending on the effective yield of the isolated pure product.

The flow protocol developed from the batch results was realized by charging a stainless-steel column reactor with a mixture of catalyst and polymer-supported base (PS-TEA) both dispersed over glass beads (∅ = 1 mm) (Figure 6.10).

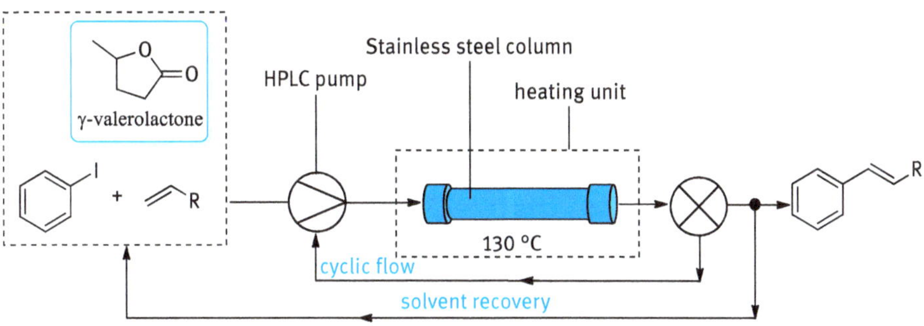

Figure 6.10: Heck–Mizoroki reaction in GVL under flow conditions.

During the flow procedure a pre-mixed solution of the starting materials was charged in a reservoir. The reactor connected to a HPLC pump was installed into a thermostated box at 130 °C and the reaction mixture was continuously pumped at a flow rate of 1.0 mL min^{-1} until complete conversion of the reactants into product was achieved (2 h). Next, a solution of TEA (1.5 equiv. in 2 mL of GVL) was cyclically pumped through the base–catalyst column (flow rate: 1 mL min^{-1}; 30 min) to regenerate PS-TEA. By employing the same flow reactor, different substrates were quantitatively allowed to react under the conditions previously described.

The adoption of the above-mentioned flow protocol allowed to achieve improvements in terms of efficiency and sustainability by (i) reducing leaching of palladium species, (ii) increasing stability of the catalyst, iii) significantly minimizing the E-factor reaching values ranging from 2.4 to 5 (vs. 21–75 of the batch procedures).

6.2.5.2 Use of cyclopentyl methyl ether in the multistep flow synthesis of benzoxazoles

CPME is a clear example of an alternative solvent [94] derived from industrial waste, and it is produced by Zeon Corporation by reacting cyclopentene waste with methanol in a 100% atom economical process. This solvent features several intriguing properties that make it very interesting for the definition of safe sustainable processes. Representatively, lower peroxide generation ability compared to other commons ethereal solvent, stability in acidic/basic conditions relatively high boiling point (106 °C), good toxicity profile.

These features, and mainly the low tendency to generate peroxide, make CPME a safer candidate especially for the development of oxidative reaction. In this example it is reported the development of a continuous-flow protocol for the telescoped synthesis of benzoxazole via oxidative process (Figure 6.11) [95]. The synthetic procedure is composed by three steps: benzylic alcohol oxidation to benzaldehyde; imine formation from benzhaldeyhe and 2-aminophenols; and oxidative cyclization to furnish benzoxazoles.

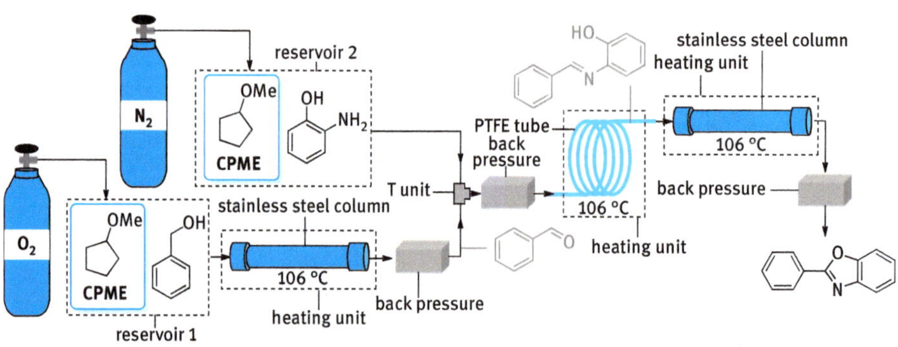

Figure 6.11: Multistep synthesis of benzoxazoles in CPME under pump-free flow conditions.

The first and the third steps require oxidative conditions and therefore, after an optimization in batch of the reaction conditions, two different manganese-based octahedral molecular sieves (OMS-2) catalysts have been selected to access the desired reactivity in combination with a molecular oxygen stream which acts as both oxidant and carrier of the reaction mixture.

At this stage, the flow system has been set up as depicted in Figure 6.10 and further optimization have been carried out to identify the proper flow rate also by the use of back pressure regulator. The flow reactor showed good ability in the telescoped synthesis of benzoxazoles that could be completed in approximately 1 h of reaction time. Importantly, by adopting these flow conditions, the output of the flow furnishes a very simple mixture in which only product and unreacted benzaldehyde are present. After evaporation of the solvent CPME and final co-evaporation of benzaldehyde with small amount of ethanol, pure products have been obtained. The features that improve the sustainability of this flow procedure in respect to other literature protocols are therefore, side-product minimization, easy work-up via recrystallization; E-factor minimization (46.4 in batch vs. 1.7 in flow).

6.2.6 Flow-assisted sustainable synthesis of drugs and intermediates

Flow chemistry approach has recently received the attention of the chemical industry, especially in the synthesis of APIs and their intermediates [96]. Flow technologies indeed represent a chance to create more efficient strategies to the target compounds compared to common batch processes and leading to a more sustainable chemical production. Green chemistry and economic principles become even more evidently linked when the large volumes of chemicals and the related waste disposal issues that chemical industries need to be handled and considered. In this context, minimization of waste and energy consumption in addition to safety improvements becomes the real goal of modern chemistry for which the development of innovative technologies as flow chemistry becomes inevitably crucial.

6.2.6.1 Use of 2-Me-THF to reduce the waste associated with the synthesis of drug (Diazepam)

A well-representative example of the importance of flow chemistry for the waste minimized production of complex highly valued chemicals is given by the synthesis of the psychotropic drug Diazepam (Figure 6.12) [97]. The authors' strategy is based on two steps and focus on (i) solvent minimization; (ii) maximization of the synthetic efficiency; (iii) minimization of by-product formation. 2-Me-THF has been used to eliminate the need for an additional solvent at the work-up extraction stage affording in a sustainable production, thanks to its peculiar properties: increased solubility of the starting materials; low density (0.85 g mL^{-1}) that minimizes the weight per kilogram of waste; the inertia of this solvent at high temperatures; and nonmiscibility with water, enabling an *in-line* aqueous extraction.

Diazepam production was achieved via an initial synthetic step between benzophenone (in 2-Me-THF) and neat chloroacetyl chloride at 90 °C that led to the quantitative formation of the corresponding amide. In the second step the resulting amide was combined with an ammonium hydroxide stream to give Diazepam in 55% yield. It is noteworthy that this flow procedure allowed to reach an E-factor value of 9, while previous reports by the same authors showed an E-factor value of 36.86.

6.2.6.2 Flow synthesis of paroxetine intermediate with a heterogeneous organocatalyst

In this example, a continuous-flow procedure for the asymmetric synthesis of chiral phenylpyridine derivative was developed funding further application in the total synthesis of paroxetine drug [98]. A two-step optimization process has been performed as follows:

First step: Asymmetric conjugate addition of 4-fluoro-cinnamaldehyde and methyl malonate over a polystyrene-supported prolinol derivative was used as a heterogeneous organocatalyst. The optimal conditions (93% conversion and 97% ee) were found by adopting SolFC, using a molar ratio between cinnamaldehyde and malonate 1:2, and with a residence time of 20 min (70 mL min^{-1}) at 60 °C. These conditions allowed to obtain 6.26 g of the chiral aldehyde by simply evaporating the unreacted reagents with a very good overall 84% yield and a productivity of 2.47 g h^{-1}. Importantly, this process generates an extremely low amount of waste with a calculated E-factor value of 0.7.

Second step: The first step was telescoped with a reductive amination/lactamization/ester reduction sequence to furnish the desired chiral phenylpyridine (Figure 6.13). Taking advantages of a heterogeneous catalytic hydrogenation, the use of neat BH_3DMS to reduce the ester, and the use of highly concentrated condition using 2-Me-THF as a sole medium, which avoid solvents swap procedure, this consecutive flow procedure led to a productivity of phenylpyridine derivative of 2.97 g h^{-1} which resulted in 4.95 g (83% yield) after 100 min of continuous operation.

In this example, the adoption of flow conditions allowed to conduct an asymmetric synthesis under neat or highly concentrated conditions; achieve a productivity of multigram per hours; and reduce the waste generation with a cumulative E-factor values for all transformations of 6.

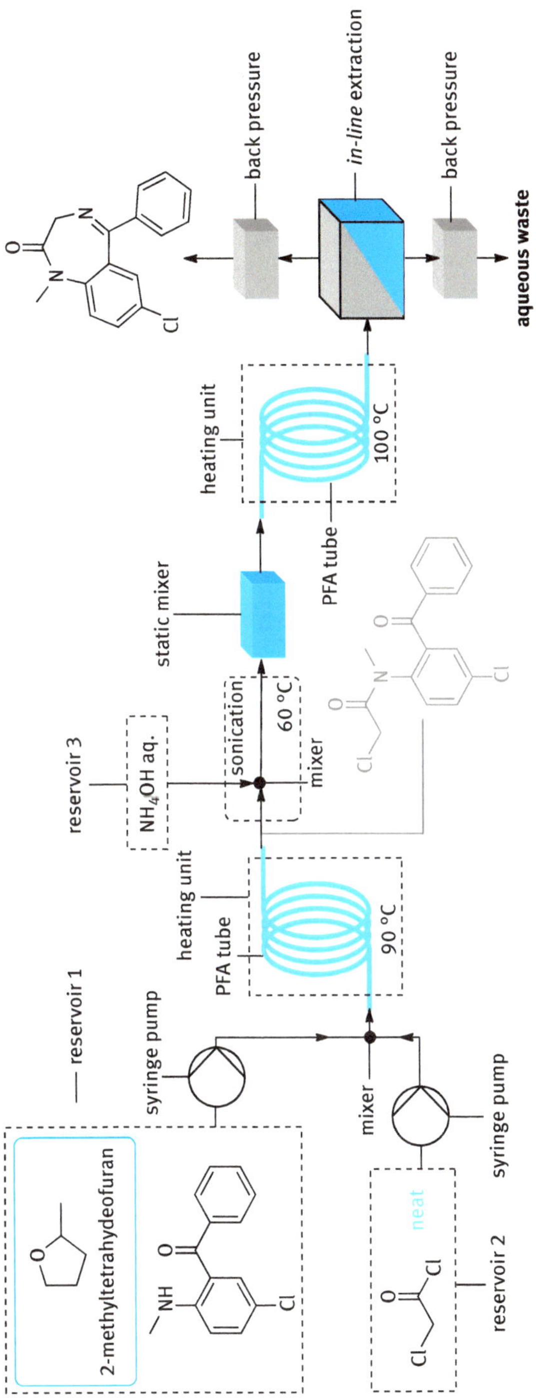

Figure 6.12: Multistep synthesis of diazepam in 2-Me-THF under flow conditions.

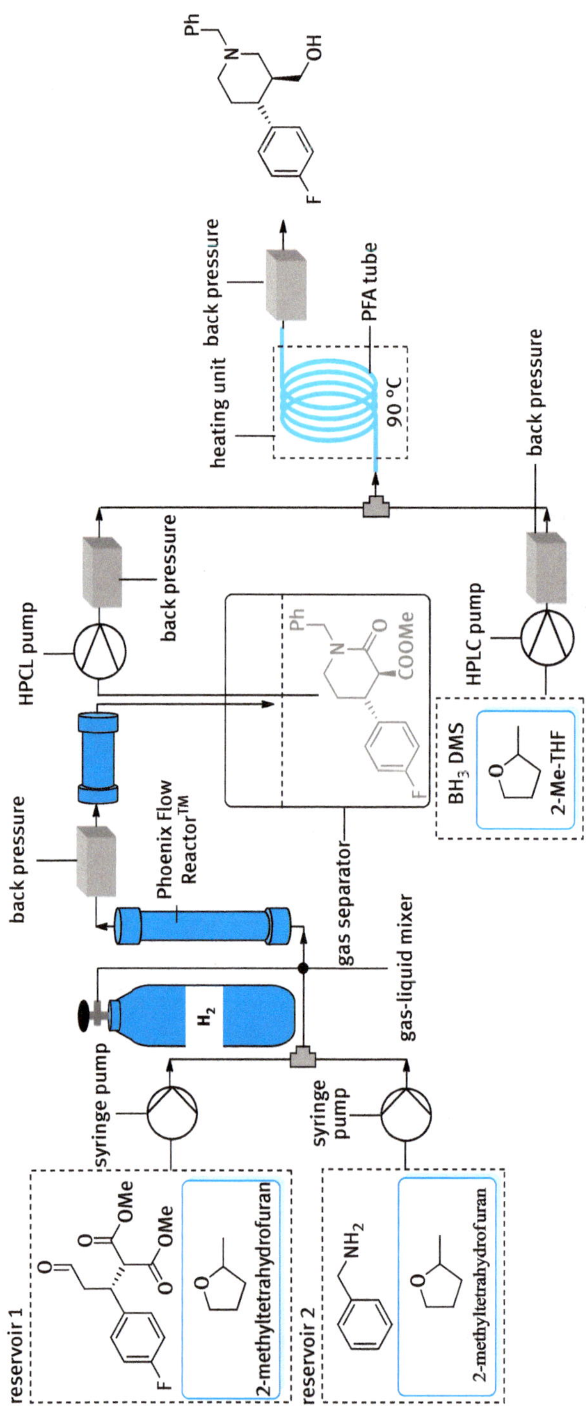

Figure 6.13: Multistep flow synthesis of chiral intermediate phenylpiperidine of (−)-paroxetine.

6.2.7 Critical evaluation to assess the greenness of synthetic procedures

As in this chapter the advantages of the synergic use of flow technology, heterogeneous catalysis, and biomass-derived solvents have been reported, it is worthy to highlight their role as fundamental pillars for the definition of sustainable strategies involving waste minimization and/or renewable feedstock valorization. It is also worthy to note that the comparison between batch and flow procedures is very useful to achieve a real quantification of the benign environmental impact accessible and also justify the change of technology.

Critical considerations on the sustainability of a process should always be made also on the impact of starting materials' production, manipulation, and their safety and toxicity profiles.

Study questions

?

6.1 Why bio- and waste-derived solvents are a preferable choice?
6.2 How it is possible to minimize hazards in transport and storage of hydrogen?
6.3 Which is the role of the solvent when it is used in combination with heterogeneous catalysts?
6.4 How the choice of the solvent is reflected in a waste minimization?
6.5 Choosing among the different examples illustrated, calculate the percentage of the solvent utilization in the waste and its impact in the E-factor.
6.6 Choosing among the examples in which E-factor or other metrics were not calculated, try to propose a waste-minimized strategy.

References

[1] Keeble BR, The Brundtland report: 'Our common future', J Medicine War, 1988, 4, 17–25.
[2] Murphy MA, Early industrial roots of green chemistry and the history of the BHC Ibuprofen process invention and its quality connection, Found Chem, 2018, 20, 121–165.
[3] Albini A, Protti S, Paradigms in Green Chemistry and Technology, Springer, 2016.
[4] Andraos J, Sayed M, On the use of" green" metrics in the undergraduate organic chemistry lecture and lab to assess the mass efficiency of organic reactions, J Chem Ed, 2007, 84, 1004.
[5] Amelio A et al., Guidelines based on life cycle assessment for solvent selection during the process design and evaluation of treatment alternatives, Green Chem, 2014, 16, 3045–3063.
[6] Lima-Ramos J, Tufvesson P, Woodley JM, Application of environmental and economic metrics to guide the development of biocatalytic processes, Green Process Synth, 2014, 3, 195–213.
[7] Gałuszka A, Migaszewski Z, Namieśnik J, The 12 principles of green analytical chemistry and the SIGNIFICANCE mnemonic of green analytical practices, Trends Anal Chem, 2013, 50, 78–84.
[8] Sheldon RA, Metrics of green chemistry and sustainability: past, present, and future, ACS Sustain Chem Eng, 2018, 6, 32–48.

[9] Weinberg L, Industrial Environmental Performance Metrics: Challenges and Opportunities. National Academy of Engineering, Washington, DC, National Academy Press, 1999, 252. $42.95 paperback. Environmental Practice, 2000, 2, 319–321.
[10] Tang SL, Smith RL, Poliakoff M, Principles of green chemistry: productively, Green Chem, 2005, 7, 761–762.
[11] Tang SY et al., The 24 principles of green engineering and green chemistry:"IMPROVEMENTS PRODUCTIVELY", Green Chem, 2008, 10, 268–269.
[12] Gilbertson LM et al., Designing nanomaterials to maximize performance and minimize undesirable implications guided by the Principles of Green Chemistry, Chem Soc Rev, 2015, 44, 5758–5777.
[13] Ivanković A et al., Review of 12 principles of green chemistry in practice, Int J Sustain Green Energy, 2017, 6, 39–48.
[14] Constable DJ, Curzons AD, Cunningham VL, Metrics to 'green'chemistry – which are the best?, Green Chem, 2002, 4, 521–527.
[15] Outili N, Meniai AH, Green Chemistry Metrics for Environmental Friendly Processes: Application to Biodiesel Production Using Cooking Oil, in Sustainable Green Chemical Processes and their Allied Applications, 2020, Springer, 63–95.
[16] Whiteker GT, Applications of the 12 principles of green chemistry in the crop protection industry, Org Proc Res Dev, 2019, 23, 2109–2121.
[17] Louaer M, Outili N, Reaction mechanism choice using green chemistry principles, Algerian J Eng Res, 2017, 16–20.
[18] Grieger K, Leontyev A, Promoting student awareness of green chemistry principles via student-generated presentation videos, J Chem, Ed., 2020, 97, 2657–2663.
[19] Koel M, Kaljurand M, Application of the principles of green chemistry in analytical chemistry, Pure Appl Chem, 2006, 78, 1993–2002.
[20] Lam CH et al., Teaching atom economy and E-factor concepts through a green laboratory experiment: aerobic oxidative cleavage of meso-hydrobenzoin to benzaldehyde using a heterogeneous catalyst, J Chem Ed, 2019, 96, 761–765.
[21] Andraos J, The Algebra of Organic Synthesis: Green Metrics, Design Strategy, Route Selection, And Optimization, CRC Press, 2016.
[22] Sheldon RA, The E factor 25 years on: the rise of green chemistry and sustainability, Green Chem, 2017, 19, 18–43.
[23] Jiménez-González C, Constable DJ, Ponder CS, Evaluating the "Greenness" of chemical processes and products in the pharmaceutical industry-a green metrics primer, Chem Soc Rev, 2012, 41, 1485–1498.
[24] Sheldon RA, Fundamentals of green chemistry: efficiency in reaction design, Chem Soc Rev, 2012, 41, 1437–1451.
[25] Tabone MD et al., Sustainability metrics: life cycle assessment and green design in polymers, Environ Sci Technol, 2010, 44, 8264–8269.
[26] Tobiszewski M et al., Green chemistry metrics with special reference to green analytical chemistry, Molecules, 2015, 20, 10928–10946.
[27] Dunn PJ, The importance of green chemistry in process research and development, Chem Soc Rev, 2012, 41, 1452–1461.
[28] Trost BM, The atom economy–a search for synthetic efficiency, Science, 1991, 254, 1471–1477.
[29] Curzons AD et al., So you think your process is green, how do you know?. Using principles of sustainability to determine what is green–a corporate perspective, Green Chemi, 2001, 3, 1–6.
[30] Boodhoo K, Harvey A, Process Intensification Technologies For Green Chemistry: Engineering Solutions For Sustainable Chemical Processing, John Wiley & Sons, 2013.

[31] Vanden Eynde JJ, How efficient is my (medicinal) chemistry?, Pharmaceuticals, 2016, 9, 26.
[32] Dicks AP, Hent A, The E Factor and Process Mass Intensity, in Green Chemistry Metrics, 2015, Springer, 45–67.
[33] Roschangar F, Sheldon R, Senanayake C, Overcoming barriers to green chemistry in the pharmaceutical industry–the Green Aspiration Level™ concept, Green Chemi, 2015, 17, 752–768.
[34] Augé J, Scherrmann M-C, Chimie Verte: Concepts et Applications, EDP sciences, 2017.
[35] Gaich T, Baran PS, Aiming for the ideal synthesis, J Org Chem, 2010, 75, 4657–4673.
[36] Santoro S, Ferlin F, Ackermann L, Vaccaro L, C–H functionalization reactions under flow conditions, Chem Soc Rev, 2019, 48, 2767–2782.
[37] Wiles C, Watts P, Continuous process technology: a tool for sustainable production, Green Chem, 2014, 16, 55–62.
[38] Vaccaro L, Lanari D, Marrocchi A, Strappaveccia G, Flow approaches towards sustainability, Green Chem, 2014, 16, 3680–3704.
[39] Plutschack MB, Pieber B, Gilmore K, Seeberger PH, The Hitchhiker's guide to flow chemistry, Chem Rev, 2017, 117, 11796–11893.
[40] Maljuric S, Jud W, Kappe CO, Cantillo D, Translating batch electrochemistry to single-pass continuous flow conditions: an organic chemist's guide, J Flow Chem, 2020, 10, 181–190.
[41] Govaerts S, Nyuchev A, Noel T, Pushing the boundaries of C–H bond functionalization chemistry using flow technology, J Flow Chem, 2020, 10, 13–71.
[42] Brzozowski M, O'Brien M, Ley SV, Polyzos A, Flow chemistry: intelligent processing of gas–liquid transformations using a tube-in-tube reactor acc, Chem Res, 2015, 48, 349–362.
[43] Britton J, Raston CL, Multi-step continuous-flow synthesis, Chem Soc Rev, 2017, 46, 1250–1271.
[44] Moore JC, Howie RA, Bourne SL, Jenkins GN, Licence P, Poliakoff M, George MW, In situ sulfidation of Pd/C: a straightforward method for chemoselective conjugate reduction by continuous hydrogenation, ACS Sustainable Chem Eng, 2019, 7, 16814–16819.
[45] Phan TVT, Gallardo C, Mane J, GREEN MOTION: a new and easy to use green chemistry metric from laboratories to industry, Green Chem, 2015, 17, 2846–2852.
[46] Andraos J, Ballerini E, Vaccaro L, A comparative approach to the most sustainable protocol for the β-azidation of α,β-unsaturated ketones and acids, Green Chem, 2015, 17, 913–925.
[47] Castrica L, Fringuelli F, Gregoli L, Pizzo F, Vaccaro L, Amberlite IRA900N3 as a new catalyst for the azidation of α,β-Unsaturated ketones under solvent-free conditions, J Org Chem, 2006, 71, 9536–9539.
[48] Angelini T, Lanari D, Maggi R, Pizzo F, Sartori G, Vaccaro L, Preparation and Use of Polystyryl-DABCOF2: an efficient recoverable and reusable catalyst for β-Azidation of α,β-unsaturated ketones in water, Adv Synth Catal, 2012, 354, 908–916.
[49] Guerin DJ, Horstmann TE, Miller SJ, Amine-catalyzed addition of azide ion to α,β-unsaturated carbonyl compounds, Org Lett, 1999, 1, 1107–1109.
[50] Xu LW, Xia CG, Li JW, Zhou SL, Efficient Lewis base-catalyzed conjugate addition of azide ion to cyclic enones in water, Synlett, 2003, 2246–2248.
[51] Xu LW, Li L, Xia CG, Zhou SL, Li JW, The first ionic liquids promoted conjugate addition of azide ion to α,β-unsaturated carbonyl compounds, Tetrahedron Lett, 2004, 45, 1219–1221.
[52] Boyer JH, Addition of hydrazoic acid to conjugated systems, J Am Chem Soc, 1951, 73, 5248–5252.
[53] Burlison JA, Blag B S J. synthesis and evaluation of coumermycin A1 analogues that Inhibit the Hsp90 protein folding machinery, Org Lett, 2006, 8, 4855–4858.
[54] Davies AJ, Donald ASR, Marks RE, The acid-catalysed decomposition of some β-azido-carbonyl compounds, J Chem Soc C, 1967, 2109–2112.

[55] Chung BY, Park YS, Cho IS, Hyun BC, Conjugate addition of hydrogen azide to the α,β-unsaturated carbonyl compounds: new azidoalumination reaction with diethylaluminum azide, Bull Korean Chem Soc, 1988, 9, 269.
[56] Alder CM, Hayler JD, Henderson RK, Redman AM, Shukla L, Shuster LE, Sneddon HF, Updating and further expanding GSK's solvent sustainability guide, Green Chem, 2016, 18, 3879–3890.
[57] Prat D, Pardigon O, Flemming HW, Letestu S, Ducandas V, Isnard P, Guntrum E, Senac T, Ruisseau S, Cruciani P, Hosek P, Sanofi's solvent selection guide: a step toward more sustainable processes, Org Process Res Dev, 2013, 17, 1517–1525.
[58] U.S. Food and Drug Administration, Q3C – Tables and ListGuidance for Industry (Revision 3), 2017, https://www.fda.gov/regulatory-information/search-fda-guidance-documents/q3c-tables-and-list-rev-3.
[59] Anastas PT, Lankey RL, Life cycle assessment and green chemistry: the yin and yang of industrial ecology, Green Chem, 2000, 2, 289–295.
[60] Sakamoto S, Contribution of Cyclopentyl Methyl Ether (CPME) to green chemistry, Chim Oggi – Chem Today, 2013, 31, 24–27.
[61] Vidal T. Sustainable solvents, products and process innovations. RSC Symposium. Available online at: http://www.rscspecialitychemicals.org.uk/docs/rsc-symposium/Sustainable-Solvents-Products-and-ProcessInnovations_Thierry-Vidal_-RSC-Symposium-2012.pdf (Accessed September 2017).
[62] Santoro S, Ferlin F, Luciani L, Ackermann L, Vaccaro L, Biomass-derived solvents as effective media for cross-coupling reactions and C–H functionalization processes, Green Chem, 2017, 19, 1601–1612.
[63] Sheldon RA, The greening of solvents: towards sustainable organic synthesis, Curr Opin Green Sustainable Chem, 2019, 18, 13–19.
[64] Clarke CJ, Tu WC, Levers O, Brohl A, Hallet JP, Green and Sustainable Solvents in Chemical Processes, Chem Rev, 2018, 118, 747–800.
[65] Gandeepan P, Kaplaneris N, Santoro S, Vaccaro L, Ackermann L, Biomass-derived solvents for sustainable transition metal-catalyzed C–H activation, ACS Sustainable Chem Eng, 2019, 7, 8023–8040.
[66] Farrán A, Cai C, Sandoval M, Xu Y, Liu J, Hernáiz MJ, Linhardt RJ, Green solvents in carbohydrate chemistry: from raw materials to fine chemicals, Chem Rev, 2015, 115(14), 6811–6853.
[67] Gao F, Bai R, Ferlin F, Vaccaro L, Li M, Gu Y, Replacement strategies for non-green dipolar aprotic solvents, Green Chem, 2020. Doi: 10.1039/D0GC02149K.
[68] Valentini F, Mahmoudi H, Bivona LA, Piermatti O, Bagherzadeh M, Fusaro L, Aprile C, Marrocchi A, Vaccaro L, Polymer-supported Bis-1,2,4-triazolium Ionic tag framework for an efficient Pd(0) catalytic system in biomass derived γ-valerolactone, ACS Sustainable Chem Eng, 2019, 7(7), 6939–6946.
[69] Marosvolgy-Haskj D, Lengyel B, Tukacs JM, Kollar L, Mika LT, Application of g-valerolactone as an alternative biomass-based medium for aminocarbonylation reactions, ChemPlusChem, 2016, 81, 1224–1229.
[70] Pace V, Hoyos P, Castoldi L, Dominguez de Maria P, Alcantara AR, 2-Methyltetrahydrofuran (2-MeTHF): a biomass-derivedsolvent with broad application in organic chemistry, ChemSusChem, 2012, 5, 1369–1379.
[71] Tukacs JM, Sylvester A, Kmecz I, Jones RV, Ovari M, Mika LT, Continuous flow hydrogenation of methyl and ethyl levulinate: an alternative route to γ-valerolactone production, R Soc Open Sci, 2019, 6, 182233.

[72] Yu Z, Lu X, Xiong J, Li X, Bai H, Ji N, Heterogeneous catalytic hydrogenation of levulinic acid to γ-valerolactone with formic acid as internal hydrogen source, ChemSusChem, 2020, 13, 2916–2930.
[73] Ye L, Han Y, Feng J, Lu X, A review about GVL production from lignocellulose: focusing on the full components utilization, Ind Crops Prod, 2020, 144, 112031.
[74] Liu P, Sun L, Jia X, Zhang C, Zhang W, Song Y, Wang H, Li C, Efficient one-pot conversion of furfural into 2-methyltetrahydrofuran using non-precious metal catalysts, Mol Catal, 2020, 490, 110951.
[75] Valentini F, Kozell V, Petrucci C, Marrocchi A, Gu Y, Gelman D, Vaccaro L, Formic acid, a biomass-derived source of energy and hydrogen for biomass upgrading, Energy Environ Sci, 2019, 12, 2646–2664.
[76] Mouterde LMM, Allais F, Stewart JD, Enzymatic reduction of levoglucosenone by an alkene reductase (OYE 2.6): a sustainable metal and dihydrogen-free access to the bio-based solvent Cyrene®, Green Chem, 2018, 20, 5528–5532.
[77] Camp JE, Bio-available solvent cyrene: synthesis, derivatization, and applications, ChemSusChem, 2018, 11, 3048–3055.
[78] Mazarío J, Romero MP, Concepción P, Chávez-Sifontes M, Spanevello RA, Comba MB, Suárez AG, Domine ME, Tuning zirconia-supported metal catalysts for selective one-step hydrogenation of levoglucosenone, Green Chem, 2019, 21, 4769–4785.
[79] Sherwood J, De bruyn M, Constantinou A, Moity L, McElroy CR, Farmer TJ, Duncan T, Raverty W, Hunt AJ, Clark JH, Dihydrolevoglucosenone (Cyrene) as a bio-based alternative for dipolar aprotic solvents, Chem Commun, 2014, 50, 9650–9652.
[80] Salavagione HJ, Sherwood J, De bruyn M, Budarin VL, Ellis GJ, Clark JH, Shuttleworth P S. Identification of high performance solvents for the sustainable processing of graphene, Green Chem, 2017, 19, 2550–2560.
[81] Zhang J, White GB, Ryan MD, Hunt AJ, Katz MJ, Dihydrolevoglucosenone (Cyrene) as a green alternative to N,N-Dimethylformamide (DMF) in MOF synthesis, ACS Sustainable Chem Eng, 2016, 4, 7186–7192.
[82] Jindapon W, Ruengyoo S, Kuchonthara P, Ngamcharussrivichai C, Vitidsant T, Continuous production of fatty acid methyl esters and high-purity glycerol over a dolomite-derived extrudate catalyst in a countercurrent-flow trickle-bed reactor, Renewable Energy, 2020, 157, 626–636.
[83] Gerardi R, Debecker DB, Estager J, Luis P, Monbaliu J-C M, Continuous flow upgrading of selected C2–C6 platform chemicals derived from biomass, Chem Rev, 2020, 120, 7219–7347.
[84] Brasholz M, von Kanel K, Hornung CH, Saubern Tsanaktsidis J, Highly efficient dehydration of carbohydrates to 5-(chloromethyl)furfural (CMF), 5-(hydroxymethyl)furfural (HMF) and levulinic acid by biphasic continuous flow processing, Green Chem, 2011, 13, 1114–1117.
[85] Tukacs JM, Jones RV, Darvas F, Dibo´ G, Lezsak G, Mika LT, Synthesis of γ-valerolactone using a continuous-flow reactor, RSC Adv, 2013, 3, 16283–16287.
[86] Bermudez JM, Menéndez JA, Romero AA, Serrano E, Garcia-Martinez J, Luque R, Continuous flow nanocatalysis: reaction pathways in the conversion of levulinic acid to valuable chemicals, Green Chem, 2013, 15, 2786–2792.
[87] Gérardy R, Emmanuel N, Toupy T, Kassin V-E, Tshibalonza NN, Schmitz M, Monbaliu J-CM, Continuous flow organic chemistry: successes and pitfalls atthe interface with current societal challenges, Eur J Org Chem, 2018, 2301–2351.
[88] Vaccaro L, Curini M, Ferlin F, Lanari D, Marrocchi A, Piermatti O, Trombettoni V, Definition of green synthetic tools based on safer reaction media, heterogeneous catalysis, and flow technology, Pure Appl Chem, 2018, 90.

[89] Ferlin F, Navarro PML, Gu Y, Lanari D, Vaccaro L, Waste minimized synthesis of pharmaceutically active compounds via heterogeneous manganese catalysed C–H oxidation in flow, Green Chem, 2020, 22, 397–403.
[90] Kitamura T, Fujiwara Y, From C–H to C–C Bonds: Cross-Dehydrogenative-Coupling, In: ed, Li C-J, The Royal Society of Chemistry, Cambridge, UK, 2014, 33–54.
[91] Ferlin F, Santoro S, Ackermann L, Vaccaro L, Heterogeneous C–H alkenylations in continuous-flow: oxidative palladium-catalysis in a biomass-derived reaction medium, Green Chem, 2017, 19, 2510–2514.
[92] Karasawa T, Oriez R, Kumagai N, Shibasaki M, Anti-selective catalytic asymmetric nitroaldol reaction of α-keto esters: intriguing solvent effect, flow reaction, and synthesis of active pharmaceutical ingredients, J Am Chem Soc, 2018, 140, 12290–12295.
[93] Mahmoudi H, Valentini F, Ferlin F, Bivona LA, Anastasiou I, Fusaro L, Aprile C, Marrocchi A, Vaccaro L, Tailored polymeric cationic tag-anionic Pd(II) complex as catalyst for the low-leaching Heck-Mizoroki coupling in flow and in biomass-derived GVL, Green Chem, 2019, 21, 355–360.
[94] Azzena U, Carraro M, Pisano L, Monticelli S, Bartolotta R, Pace V, Cyclopentyl methyl ether: anelectiveecofriendly etherealsolvent in classical and modern organic chemistry, ChemSusChem, 2019, 12, 40–70.
[95] Ferlin F, van der Hulst MK, Santoro S, Lanari D, Vaccaro L, Continuous flow/waste-minimized synthesis of benzoxazoles catalysed by heterogeneous manganese systems, Green Chem, 2019, 21, 5298–5305.
[96] Ferlin F, Lanari D, Vaccaro L, Sustainable flow approaches to active pharmaceutical ingredients, Green Chem, 2020, 22, 5937–5955.
[97] Bédard AC, Longstreet AR, Britton J, Wang Y, Moriguchi H, Hicklin RW, Green WH, Jamison TF, Minimizing E-factor in the continuous-flow synthesis of diazepam and Atropine, Bioorg Med Chem, 2017, 25, 6233–6241.
[98] Otvos SB, Pericàs MA, Kappe CO, Multigram-scale flow synthesis of the chiral key intermediate of (–)-paroxetine enabled by solvent-free heterogeneous organocatalysis, Chem Sci, 2019, 10, 11141–11146.

Antonio M. Rodríguez, Iván Torres-Moya, Angel Díaz-Ortiz, Antonio de la Hoz and Jesús Alcázar

7 Flow chemistry in fine chemical production

Flow chemistry technology has been extensively researched in the pharma and life science industry over the last two decades. In recent years, flow chemistry has also expanded to other industrial fields, such as agrochemicals and fragrance development. It is clear that flow chemistry can provide significant advantages over more commonly used methods and it has the potential to revolutionize certain aspects of research and development (R&D) in industries mentioned above. These new methods need to comply with environmental regulations and pollution prevention to avoid climate change and environmental damage, both factors that are now of critical importance.

7.1 Introduction

This section focuses on how flow chemistry is playing a pivotal role in helping the pharmaceutical, agrochemical, and fragrance industries to improve productivity, reduce waste, and improve quality and control in both R&D and manufacturing.

One of the most attractive alternatives to conventional methods for the discovery of new products is continuous-flow chemistry, which has recently emerged as a novel chemical tool to help chemists to combine efficiency and sustainability. Flow chemistry has been known for almost a century and it has been used in industrial settings like petroleum or agrochemical refineries, where continuous processing is the key to process intensification. The main goals are to reduce costs and the size of process equipment, to improve quality, to reduce energy consumption, and to minimize the number of solvents employed and the waste, thus following the 12 principles of green chemistry [1, 2].

The improvements associated with this emerging technology in comparison to traditional batch methods are due to its various advantages. In this respect, it is worth highlighting the following:

- Better control of the reaction conditions in comparison with batch processes, including quick and efficient mixing of reactants, short residence times with precise control and efficient heat and mass transfer with low operating volumes. As a result, yield and selectivity are considerably improved in such processes.
- High temperature and pressure, which are not commonly employed in organic chemistry, can be applied, and this contributes to the acceleration of reaction rates and reduction of reaction times.

https://doi.org/10.1515/9783110693690-007

- Reduction of the amount of solvent. This can be achieved by the use of high concentrations or a single solvent throughout the whole process. Reactions without solvents are also possible with neat or molten reagents, a situation that provides higher reactivity. A worrying problem in flow chemistry, namely, clogging, can be prevented by a careful choice of the reagents and solvents, the use of wide-bore channels or tubing, agitation, sonication, and control of the fluid velocity. As a consequence, emissions are lower than in batch production [3].
- Flow chemistry is a safer, faster, and cleaner technology than batch processes and it also has a reduced footprint. Hazards can be minimized because of the short residence times and better temperature control. This is achieved due to the small volume of the reactor. As a consequence, the amounts of hazardous reagents are minimal, thus reducing considerably the possibility of explosions or accidents.

The most important advantages that flow chemistry contributes are summarized in Fig. 7.1.

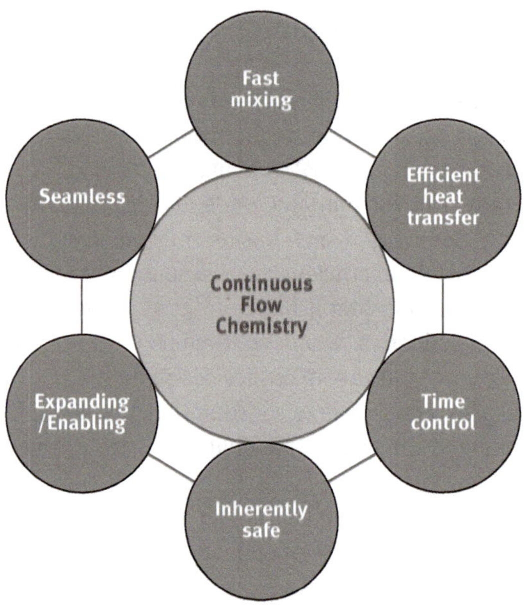

Fig. 7.1: The most important advantages offered by continuous-flow chemistry.

In flow chemistry, it is possible to combine incompatible reactions by the application of "telescoped flow chemistry." Multistep flow chemical sequences can be optimized using the smallest possible number of reactions, workup procedures, and equipment to minimize the complexity of the system and to improve reproducibility [4]. Examples of these processes will be discussed in later sections.

Another significant advantage of flow chemistry is that purification can also be performed in flow. Purification can be achieved by continuous crystallization, semi-batch recrystallization, neutralization, catch-and-release chromatography, and continuous-operated simulated moving bed chromatography [5]. The isolation can be carried out by in-line filtration of solid co-products, liquid–liquid phase separation, gas–liquid phase separation, or through the use of solid phase-supported scavengers.

Photo- and electrochemical reactors are also feasible because of the small size of continuous-flow reactors. Interest in photochemical reactions in recent years has increased enormously since the implementation of flow chemistry. The efficiency of these reactions has been improved under continuous flow when compared to batch conditions because of the homogeneous light irradiation, which is feasible through the use of polymer tube reactors (Perfluoroalkoxy alkanes (PFA), Fluorinated ethylene propylene (FEP)). Under batch conditions, the Lambert–Beer law limits the efficiency of photochemical reactions, a drawback that can be avoided on using flow chemistry [3, 6].

Finally, a green and sustainable process needs to be scalable, to industrial scale if possible, without extra environmental impact and without the need to optimize the reaction conditions again at every scale. A continuous-flow process can help to achieve these requirements in terms of solvent, concentration, reagents, product quality, energy consumption, equipment cost, waste generation, and safety, among other factors [7–9].

7.2 Advantages of flow technology in chemical production

At its inception, flow chemistry was principally used for the development of pharmaceutical derivatives. However, in the last decade the expansion of flow chemistry has had a major impact on the design of different fine chemicals in the agrochemical and fragrance industries. Flow chemistry is an emerging and novel system because it has numerous advantages in comparison to batch reactors, as outlined above. With flow chemistry the control of reaction parameters may improve the selectivity of some reactions, thus allowing access to intermediates and products in a lower number of steps and, very importantly, avoiding the use of protecting or directing groups, which in turn makes the synthetic process simpler. Furthermore, it is worth highlighting the possibility of using high temperatures and pressures as this allows access to chemical space that is difficult to achieve in other ways [10–12]. Apart from these relevant benefits, there are other advantages related to sustainability and these will be described below.

7.2.1 Cleaner chemistry

One of the most troubling problems in flow chemistry is the use of metals as catalysts in cross-coupling reactions due to their impact on the synthesis of natural products, agrochemical products, fragrances, and other biologically active products [13]. In the last decade, the use of catalysts formed by transition metals has been identified as an environmental limitation; hence, it is necessary to find different strategies to reduce the amount of metals in these kinds of processes.

The first alternative is the use of supported metal catalysts, which are generally contained in a column through which the reagents flow. The products are collected and the end of the column is free of metal catalyst and the corresponding ligands. This is an enormous advantage that allows the chemist to recycle and reuse the catalyst for future reactions while reducing its presence in the final product and increasing its turnover number (TON).

As an example, Fukuyama and coworkers produced on a 100 g scale a key intermediate for a matrix metalloproteinase inhibitor using a novel approach through supported and homogeneous catalysis. The method involved the reaction in a biphasic solution, with the catalyst and the reactants dissolved in immiscible solvents (Fig. 7.2). In this way, the catalyst could be separated from the crude product by an in-line separation device and fed back into the reaction system.

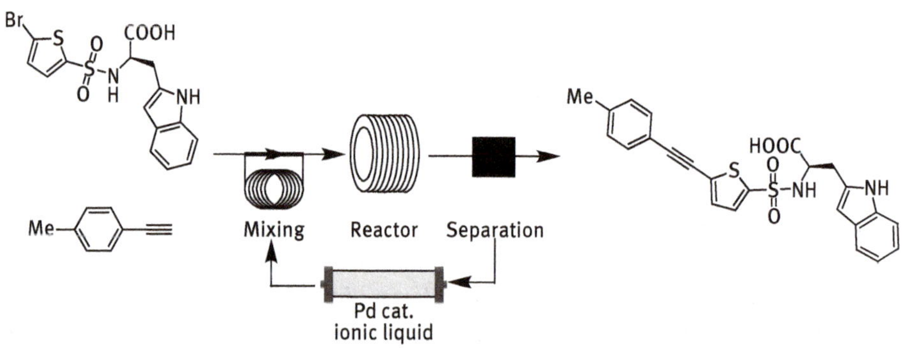

Fig. 7.2: Ionic liquid approach to matrix metalloproteinase inhibitor.

Another alternative to reduce metal waste and recover the metal catalyst is to carry out the reaction in solution but with the catalyst and the reactants dissolved in an immiscible solvent. Hence, in the two-phase system the reaction takes place at the interphase and the catalyst is separated from the crude reaction mixture in an online separation device. For example, the use of ionic liquids as media for transition metal NHC (*N*-heterocyclic carbene) complexes has been described and developed as a continuous catalyst-recycling system by connecting the outcome with a microextraction system.

This procedure has been widely employed for Heck reactions [14], Sonogashira reactions [15], and carbonylative Sonogashira cross-coupling reactions [16].

7.2.2 Enhanced synthesis

One of the essential advantages of flow chemistry in terms of industry in fields such as drug design, agrochemical products, or fragrances is enhanced synthesis. Reactions that were carried out under specific conditions have been improved by the use of flow chemistry, with increases in yields, regioselectivity, and the use of milder reaction conditions. One of the most critical functionalities in the synthesis of drugs, proteins, related peptides [17], and agrochemicals such as pesticides is the amide group [18]. For this reason, there are numerous different procedures to synthesize amides. Due to the importance of the amide group in drug design, the American Chemical Society Green Chemistry Institute Pharmaceutical Roundtable has focused its attention on research into amide formation [19].

In an effort to gain further knowledge about amide groups and to find more efficient and cheaper protocols, novel flow procedures for the direct aminolysis of esters, which are stable and abundant in nature, have been described in the last few years. The most representative example concerns the Bodroux protocol [20].

7.2.3 New reactivity patterns

Flow chemistry opens the door to new reactivity patterns or methodologies that were difficult to achieve under batch conditions or with conventional techniques. However, flow chemistry allows the performance of fast reactions in order to achieve the desired products with high selectivity. These possibilities led to the so-called flash chemistry approach [21]. This concept can be applied in different fields as follows:
- Highly exothermic reactions that are challenging to control in batch reactors
- Reactions where an intermediate quickly decomposes in conventional reactors
- Reactions where undesired products are formed in subsequent reactions in conventional reactors
- Reactions whose final products decompose readily in conventional reactors
- Polymer synthesis

In flash chemistry, reactions can be performed in a time range from milliseconds to seconds in reactors as shown in Fig. 7.3.

It should also be pointed out that flash chemistry can also contribute to green, sustainable chemistry in the following ways:
- Little or no use of auxiliary substances: Flash chemistry avoids the use of additional substances that slow down reactions to obtain better control.

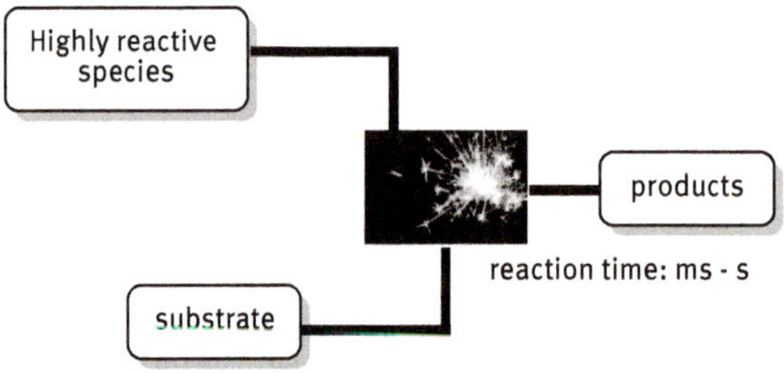

Fig. 7.3: Concept of flash chemistry in a microreactor.

- Energy-saving: Conventionally, cooling is often used to attain good control of fast reactions. The use of microreactors enables reactions to be conducted at higher temperatures, thus minimizing the energy required to control these reactions.
- High selectivity: Better control due to inherent properties of microreactors leads to high selectivity for the products. Therefore, the amount of waste is reduced considerably.
- On-demand and on-site synthesis: Synthesis based on speedy reactions using microreactors enables on-demand and on-site synthesis. This leads to a lower energy requirement for transportation.

7.2.4 Improved safety

Different compounds such as nitro derivatives, azides, or fluorinating agents are known because of their toxicity or hazardous nature. However, with flow chemistry the increase in safety is apparent. These hazardous materials are consumed during the process and there is no need to stockpile them. A good example is provided by diazo and diazonium derivatives. These compounds are of great interest in medicinal chemistry but they are incredibly toxic, highly energetic, and explosive when they are heated or exposed to concentrated acids [22].

Diazomethane is considered to be one of the most dangerous and hazardous reagents. Among the dangers associated with diazomethane, it is worth highlighting its sensitivity to shock and heat, because of which its preparation should be carried out in specialized glass equipment. Kappe and coworkers prepared this dangerous reagent in the inner tube or a tube-in-tube reactor from diazald and potassium hydroxide. The gas-permeable membrane of this tube allowed diazomethane to move to the other tube in which the reagent circulated. Methylation of alkenes and the Arndt–Eistert transformation of acyl chlorides (Fig. 7.4) were performed with this methodology [23].

The applicability of this methodology in medicinal chemistry was corroborated by the development of a continuous-flow process for the multistep synthesis of α-haloketones starting from *N*-protected amino acids. The α-halo ketones obtained in

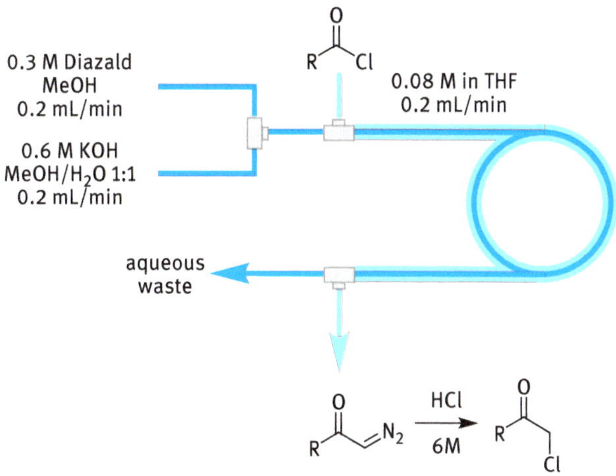

Fig. 7.4: Continuous flow of Arndt–Eistert reaction using a tube-in-tube reactor.

this process are chiral building blocks for the synthesis of HIV protease inhibitors such as darunavir and atazanavir (Fig. 7.5) [24].

This methodology reduces the storage, transport, and handling requirements of diazomethane and produces α-halo ketone building blocks in excellent yields in a multistep system without racemization.

In 2019, Kappe and coworkers described a safe and scalable methodology for the Wolff–Kishner reduction using continuous flow [25]. The Wolff–Kishner reaction involves the use of hydrazine in the presence of a strong base to convert aldehydes and ketones into the reduced methylene moiety. In this reaction, the use of low-cost hydrazine as a reducing agent in combination with caustic base provides an atom-efficient and environmentally friendly procedure for the deoxygenation of aldehydes and ketones to alkanes. The reaction requires challenging reaction conditions of temperature and pressure (200 °C and 50 bar). The use of hydrazine hydrate at production scale faces some problems regarding its high toxicity, potential carcinogenic properties, and highly corrosive nature. In particular, the accumulation of hydrazine can lead to dangerous explosions.

The use of corrosion-resistant silicon carbide allows the Wolff–Kishner reaction to be performed at a production scale in industry by continuous flow with significant advantages such as improved handling, lower excess requirements, and especially an increase in the safety of hydrazine reagents through the elimination of gaseous headspace. The safety of the operator is also improved since the exposure risk is minimized. The use of flow chemistry is shown in Fig. 7.6 and involves two steps: (a) a single-feed protocol for substrates that are not prone to fast hydrazone formation and (b) a two-feed protocol that allows preformation of the hydrazone before the addition of base.

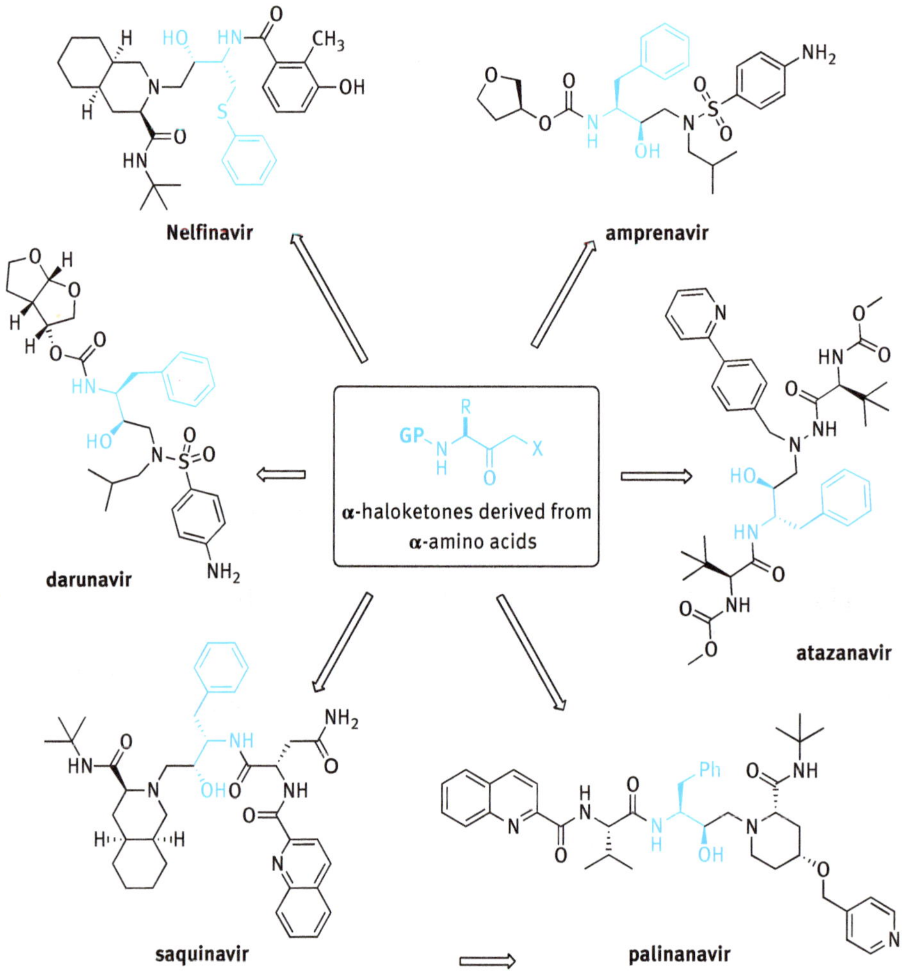

Fig. 7.5: HIV protease inhibitors from α-halo ketones.

It is undoubted that very recent advances in continuous-flow technologies have opened up new opportunities for the use of hydrazine as a versatile reagent with improved safety in comparison with batch processes. As mentioned above, continuous-flow chemistry has been postulated as a new tool for both synthetic and process chemistry. All of the advantages of flow chemistry are aligned with several of the 12 green chemistry principles [26, 27]. In the following sections, we will discuss how flow organometallic chemistry has proven its worth within the pharmaceutical sector in different applications, such as organometallic chemistry, multiple compound library preparations, other technologies applicable to drug discovery and their application in the fabrication of agrochemical products of fragrances.

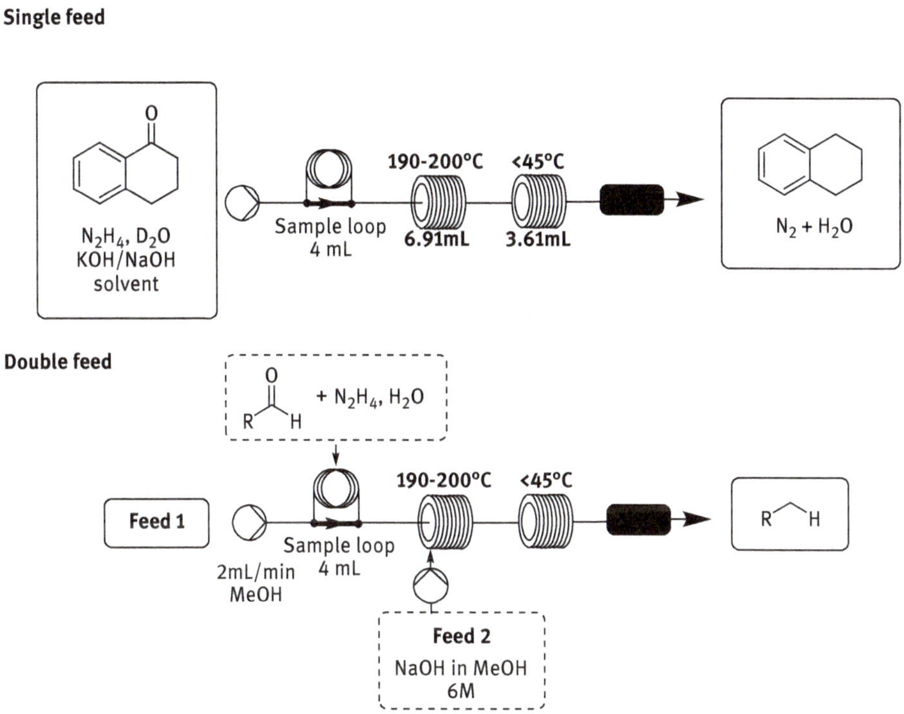

Fig. 7.6: (a) Single-feed and (b) double addition flow protocols.

(For related topics please see Volume 1, Chapter 2, Title: Principles of controlling reactions in flow chemistry, *Chapter 8, Title:* Mitigation of chemical hazards under continuous flow conditions, *and Chapter 11, Title:* Gaseous reagents *in flow chemistry)*

7.3 Flow chemistry in drug discovery

Organometallic chemistry has been an essential part of the pharmaceutical industry since the nineteenth century. Organometallics have been relevant compounds as intermediates or catalysts in the synthesis of biologically active molecules [28, 29], but they have also been used as medicines in their own right. Organometallic chemistry in continuous flow brings to organic synthesis the opportunity to perform reactions that are otherwise impossible in conventional laboratories and industrial equipment, thus significantly expanding the toolbox of synthesis [30]. In this section, we will discuss how flow organometallic chemistry has proven its value within the pharmaceutical sector in different applications such as organometallic chemistry, multiple compound library preparations, and other technologies applicable to drug discovery.

7.3.1 Heterogeneous organometallic catalysis

Flow technology implies a significant change to the practical necessities in the chain of carrying out heterogeneous organometallic synthesis, especially with the issue of handling solids, which presents a substantial challenge for the industrial adoption of this technology [31]. One of the main applications of flow chemistry in the area of organometallic chemistry is the use of packed-bed solid catalysts [32]. This approach allows chemists to recycle and reuse the catalyst for different reactions, reduce the leaching of the solid support and increase its TON. Among other metals, gold, iron, and zirconium packed beds have been developed for cycloisomerization and reduction reactions of different groups, such as nitro, aldehyde, and ketone groups [33–35]. However, palladium complexes are the most widely used organometallic species in continuous heterogeneous flow chemistry.

Due to the extensive use of palladium in reactions such as Suzuki, Heck or Negishi, numerous scientists have explored a variety of alternatives to immobilize palladium metal [36–40]. Among them, silica-supported palladium catalysts have proven to be among the most efficient [41, 42]. Janssen researchers and coworkers from Castilla-La Mancha University (UCLM) optimized the conditions to obtain a broad range of biaryl compounds in good to excellent yields with a low leaching catalyst [43]. In a follow-up article, the same researchers in collaboration with the Catholic University of Leuven (KULeuven) replaced boronic acids with organozinc reagents to obtain a large variety of alkyl–aryl Negishi coupling products [38]. Both reactions were performed under mild conditions and short residence times with good to excellent conversions. Products were isolated directly after aqueous workup or with minimal purification, thus increasing the sustainability of the process (Fig. 7.7).

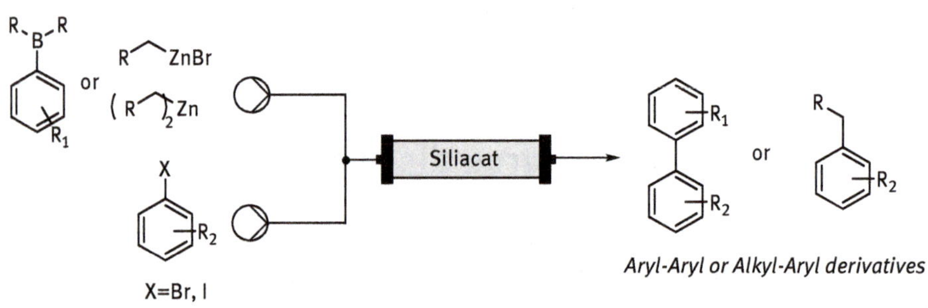

Fig. 7.7: Suzuki and Negishi coupling reactions using a silica-supported palladium catalyst.

A more elaborate approach was developed by Organ and coworkers. They supported Pd–PEPPSI–IPent on silica, as a precatalyst, and achieved higher reaction efficiency with hindered and electronically deactivated coupling partners in Negishi cross-coupling reactions [44]. Secondary alkyl zinc reagents were coupled to arenes and heteroarenes in a 3 min residence time at room temperature with good to excellent yields.

Very recently, Williams, Kappe, and coworker described the challenging *C*-glycosylation of a pyrrolotriazinamine via metal–halogen exchange to prepare remdesivir, the first approved treatment for COVID-19 (SARS-CoV-2) [45]. The authors transferred the multistep batch synthesis to a flow setup that avoids solid formation and permits stable processing. Detailed optimization of stoichiometries provided an improvement upon batch conditions with a total residence time of <1 min and a throughput of 51.8 mmol h^{-1} of the intermediate (Fig. 7.8).

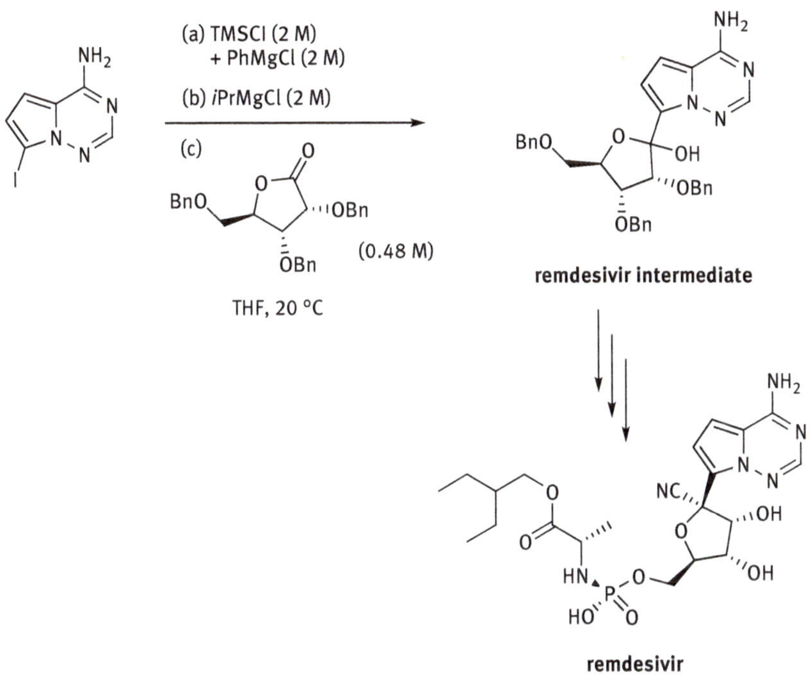

Fig. 7.8: Continuous-flow synthesis of the intermediate of remdesivir.

In the pharmaceutical industry it is common to encounter APIs that are large molecules with low solubilities and even suspensions in batch and semibatch operations. These reactions are significantly more challenging under continuous-flow conditions due to the increased potential for process failure by reactor, pump, or backpressure regulator blockages. For this reason, the application of continuous-flow equipment for the final stages of API synthesis may not be the best choice in all situations. Nevertheless, several approaches to continuous processing in the final stages of API syntheses have been proposed. A recent example of continuous crystallization makes use of a specific type of continuous-flow reactor, namely the oscillatory baffled reactor, which enables plug flow-type residence time distribution with long mean residence times [45]. The continuous mixing in the baffled reactor zones generates flow eddies and these prevent sedimentation of particles within the reactor.

In the realm of heterogeneous catalysis, there have been different approaches for the use of gases in flow reactors [46], with the use of supported catalysts and hydrogen standing as the prime example. There are multiple examples of continuous-flow hydrogenations in which homogeneous catalysts are more stable and easier to recycle than heterogeneous catalysts [47]. However, it is clear that continuous-flow chemistry has led to improvements with regard to reduced reaction times and increased selectivities, as well as new catalyst types and immobilization methods in pharmaceutical discoveries. One clear example is the synthesis of linezolid (Zyvox), a new class of antibiotic developed by Pfizer in early 2000. Gardiner and coworkers reported a novel continuous-flow hydrogenation reactor containing a series of 3D-printed catalytic static mixers (CSMs), which were coated with Pd catalysts by electroplating, for the synthesis of the critical intermediate for linezolid (Fig. 7.9) [48]. This system reduced the manufacturing cost of the reactor, eliminated the need for catalyst filtration, and allowed an easy scale up to produce 425 g of the product per day (with 99% conversion and 99% purity). Furthermore, a single set of 12 Pd-coated CSM inserts could be used almost continuously for hydrogenation reactions for over a year without significant loss of catalytic activity.

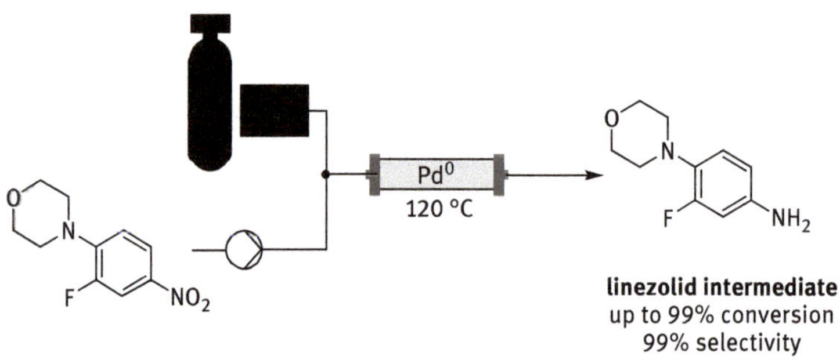

Fig. 7.9: Flow synthesis of the intermediate of linezolid using 3D-printed catalytic static mixers. MFC, mass flow controller.

More recently, Smyth and Manyar developed a continuous-flow packed-bed catalytic reactor for the hydrogenation of aromatic nitrobenzoic acids in water. These hydrogenations are green, more efficient, less consumptive, and safer than the conventional reduction process [49].

7.3.2 Homogeneous organometallic catalysis

The advantages of heterogeneous catalysis are numerous, but multiple research groups have proposed that metal leaching of the organometallic species may be responsible for catalysis [50]. Kappe and coworkers confirmed that this phenomenon limits the application of packed bed reactors [51]. In this way, homogeneous catalysis still offers advantages over supported catalysts, such as better control over the catalyst loading, as metal leaching does not apply in all cases. Therefore, a number of inflow cross-couplings that use homogeneous catalysts have been reported [52].

One of the advantages of continuous homogeneous organometallic chemistry is the wide range of reagents at our disposal, including those that would be considered difficult to handle or unstable and dangerous intermediates. The inherently safer working practice under flow conditions arises for several reasons, including lower reaction volumes, better temperature control, and the ability to accommodate higher pressures without risk. Therefore, unstable, highly reactive, or toxic intermediates can be generated in situ from benign, readily available and cheap precursors in a closed, pressurized system and converted directly into more stable, nonhazardous intermediates or products. For these reasons, it is very common in the literature to see reactions with azides [53], diazomethane [54], hydrogenations [55], carbonylations [46], fluorinations [56, 57], or the use of Grignard agents and others [58].

Similar to the family of the Grignard reagents, organolithium species are among the most reactive organometallic species that are commonly used in organic synthesis. However, the high reactivity of these organometallic species limits their applications, as cryogenic temperatures are usually required to avoid speedy and highly exothermic reactions, especially in the case of large-scale industrial production. Therefore, the use of flow microreactors to perform the preparation and use of such highly reactive organometallic species in a controlled way may be a viable alternative. With this in mind, Yoshida defined the concept of 'flash chemistry' to carry out extremely fast metalations using microreactors [59]. Other examples of such reagents can be seen in the literature and these include magnesiation reactions [60, 61] and zincation [62, 63].

Good examples of the multiple uses of different unconventional reagents include the flow trapping of carbon dioxide in the formation of a carboxylate group in the reaction with Grignard reagents. Grignard reagents were passed through a conventional tube-in-tube reactor to deliver the carbon dioxide (Fig. 7.10) [64]. The optimum conditions were determined to be 4 bar of carbon dioxide, which enabled near quantitative conversions even at moderate flow rates. The authors obtained a set of 10 different carboxylic acids in good to excellent yields (75–100%). They also developed an efficient catch-and-release protocol to facilitate in-line purification of the carboxylic acid using a cartridge containing a polymer-supported ammonium hydroxide species. Similarly, Rutjes and coworkers reported the formation of benzoic acid through the hydroxycarbonylation of phenylmagnesium bromide in flow, with a throughput of 0.52 g h^{-1} achieved using a commercial (FlowStart Evo, Future Chemistry) pumping system [65].

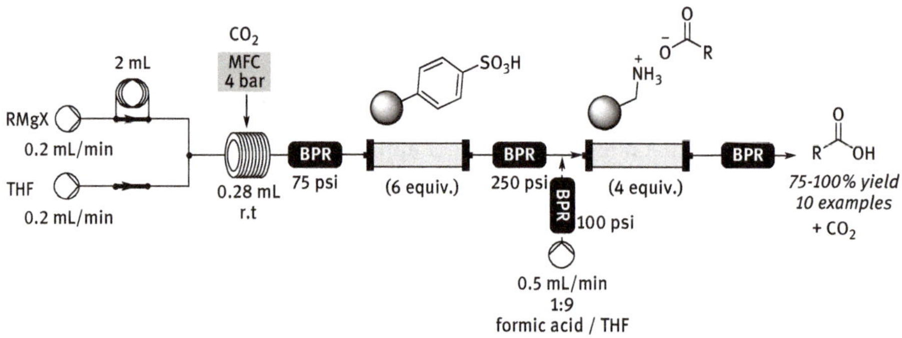

Fig. 7.10: Flow carboxylation of Grignard substrates using carbon dioxide.

In recent examples carbon monoxide has been used for the carbonylation of aryl halides and pseudohalides. This work was carried out in a collaboration between KU Leuven, Janssen, and Universidad de Castilla-La Mancha [66] and phenyl formates were used as CO precursors in coupling reactions in a safe way [67].

7.3.3 Multistep and telescoped flow synthesis

As discussed in previous sections, continuous-flow chemistry has a large box of tools to perform reactions under diverse and flexible conditions that enable demanding chemical challenges to be met [68]. One specific set of applications, especially useful in the fine chemical industry in recent years, involves multistep reactions under continuous-flow and reaction telescoping. When more than one transformation occurs the procedure can be denoted as a multistep continuous-flow reaction. Indeed, this is the preferred approach when developing a strategy to obtain an API in a single flow sequence. However, unlike reaction telescoping, multiple transformations take place without intermediate purifications and continuous-flow systems allow the possibility of in-line purification and reagent introduction at different points in the continuous-flow sequence.

An example of the preparation of an API under these conditions is the synthesis of the antiseizure medicine rufinamide. The time required for this synthesis was reduced to 11 min by using three reactor coils to mediate sequential transformations (Fig. 7.11) [69]. Once again, copper tubing catalyzed the azide–alkyne cycloaddition to generate the 1,2,3-triazole ring core of rufinamide. This process is notable as the first API synthesis to use reactor tubing in such a way, with the reaction sequence affording 217 mg h^{-1} of rufinamide in 98% yield.

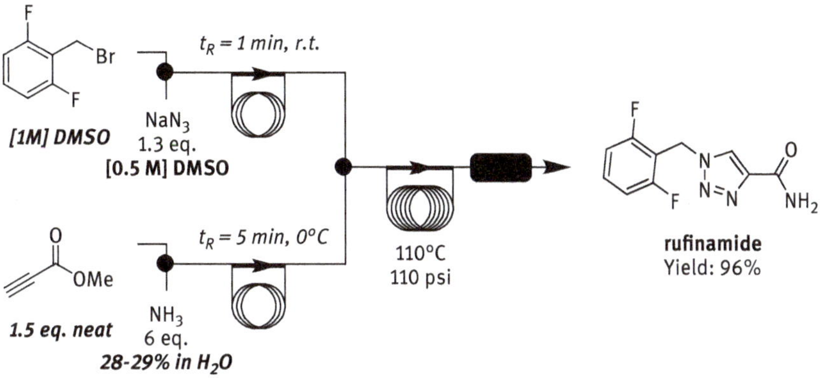

Fig. 7.11: Continuous-flow synthesis of rufinamide.

Another excellent example of a multistep synthesis with in-line purification steps is the continuous-flow synthesis in a three-step route to efavirenz, which is an essential medicine for the treatment of HIV [70]. Previous routes to efavirenz required up to five steps and this demonstrates that continuous flow improved both the efficiency of the synthesis and the viability of scale-up. In this continuous-flow system, a novel Ullmann cyclization produced efavirenz in 45% yield with a residence time of less than two hours. A silica-scavenging column was employed to remove by-products from the continuous-flow stream to provide a method for in-line purification but – as with other columns – this had a finite lifetime.

A prime example of telescoped synthesis was reported by Ley and coworkers, who employed a continuous-flow system to generate (*E*/*Z*)-tamoxifen, an antibreast cancer drug, in 100% conversion and 84% yield (Fig. 7.12) [71]. The telescoped synthesis combined four chemically distinct transformations into one stream, using five pumps pumping at three different flow rates, five reactor coils, and four different temperature zones, to give sufficient material in in 80 min. This is a real demonstration in a research environment of the power of the continuous-flow approach for the synthesis of important functionalized molecules.

7.3.4 Library synthesis

In the drug discovery field, the need to find novel drug-like molecules faster than competitors has fostered the exploration of innovative enabling technologies for molecule production. In this respect, flow chemistry can be applied to reach chemical space that is new or is inaccessible by traditional batch approaches. In addition, common synthetic bench operations, such as reagent loading, mixing, workup, purification, analysis, and so on, can be automated and library building can be accelerated and platforms for lead discovery can be set up.

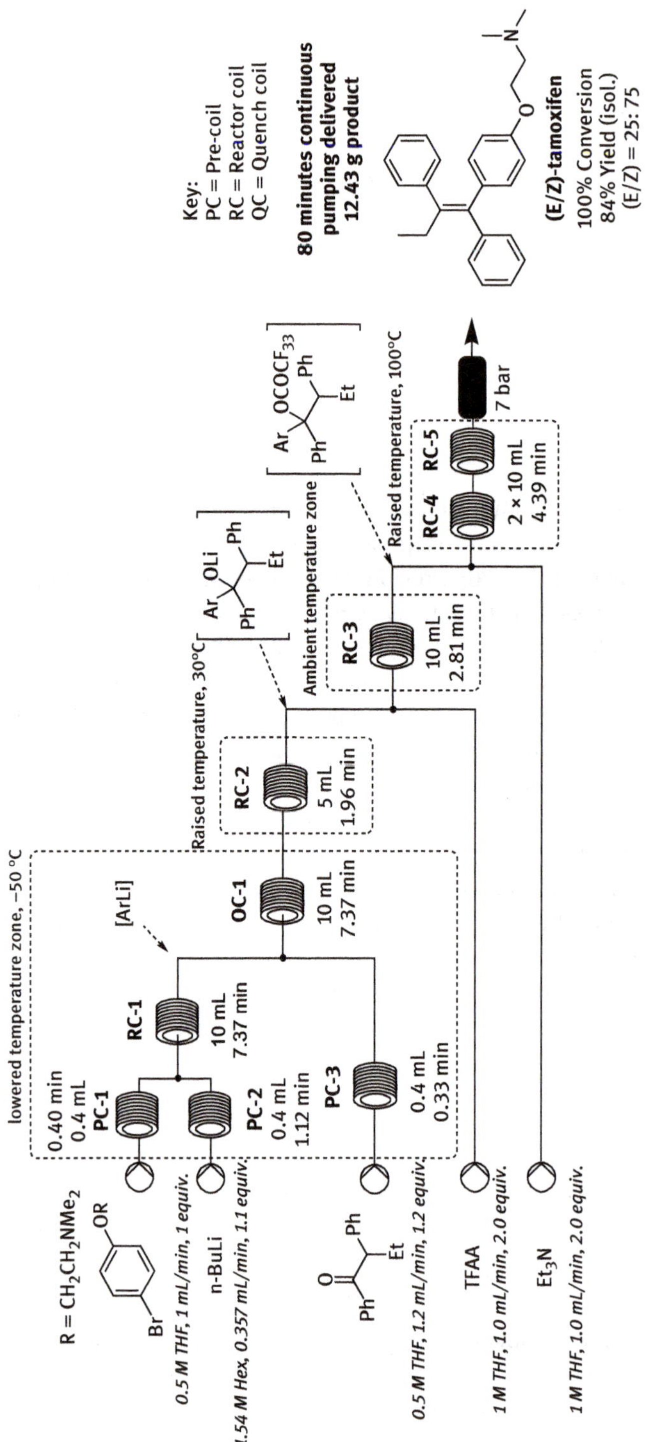

Fig. 7.12: Continuous-flow telescoped synthesis of (*E*/*Z*)-tamoxifen.

A wide range of examples of syntheses of compound libraries have been published in the literature over the years [72–77]. Recently, Ley and coworkers demonstrated a multistep continuous synthesis of a library of substituted pyrazoles from a series of aromatic diazonium compounds synthesized in situ from anilines and tBuONO or $NaNO_2$ [78]. This example demonstrates how unstable diazo compounds can be easily managed safely in a laboratory setting using flow. Unreacted materials are quenched and there is never an appreciable buildup of hazardous material. The reaction sequence included the corresponding reduction with ascorbic acid and hydrolysis and the authors obtained a library of pyrazoles generated by subsequent condensation with a series of 1,3-dicarbonyl compounds (Fig. 7.13).

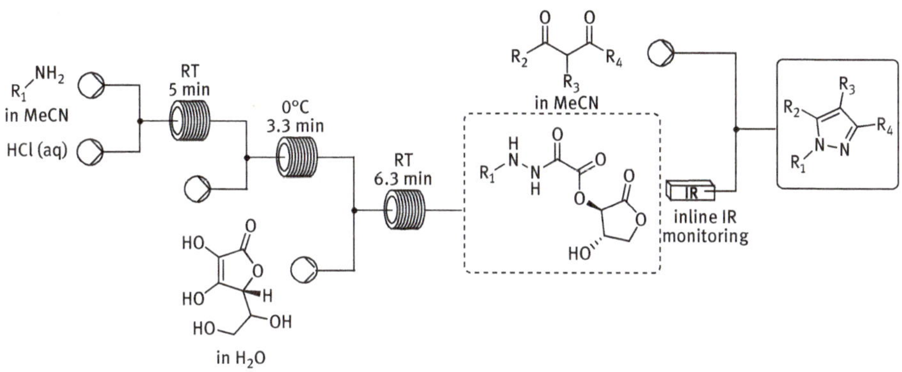

Fig. 7.13: Ley Group library synthesis using diazonium compounds.

Koolman and coworkers further expanded the scope of this tube-in-tube reactor by developing an automated diazomethane platform (DRIFT, diazomethane reactor in flow technology) that can be used for library generation or reaction scale-up [79]. As opposed to reactions that are run in the outer tube of the tube-in-tube reactor, DRIFT prepares an anhydrous solution of diazomethane in THF, which is mixed with a reagent stream to afford the desired products. It was noted that when reagent streams pass over the tube-in-tube reactor, palladium nanoparticle buildup over time was observed on the AF-2400 tubing and this led to inconsistent results. With DRIFT running in library production mode, a series of 15 arylcyclopropyl boronates was synthesized in modest yield. The automated scale-up mode was used to obtain up to 190 mg h^{-1} of the substrate (Fig. 7.14). At AbbVie, DRIFT has been applied to drug discovery in the context of building block and intermediate synthesis.

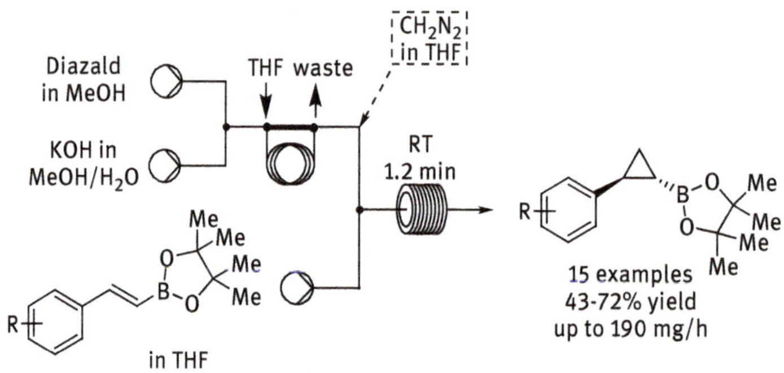

Fig. 7.14: Arylcyclopropyl boronate synthesis using tube-in-tube diazomethane reactor.

7.3.5 Other technologies applicable to drug discovery

Empowering technological chemistry advances employed for drug discovery can provide a substantial advantage. The objective is not only to reduce the time or cost of a process but also to increase the probability of success. The development of more secure, effective and green processes has been carried out by combining flow chemistry with other technologies to facilitate chemical transformations that are problematic or hazardous in batch procedures. In a typical continuous chemistry process, the narrow diameters and increased surface areas in a flow reactor, along with the possibility of immobilized beds, can be advantageous for photochemical, electrochemical and biocatalyzed reactions [80, 81].

7.3.5.1 Photochemistry

Photochemical changes are of great interest in drug discovery since the reactivity of singlet and triplet states regularly enhances a complex blend of compound structures and stereochemistry. Photochemical responses have been studied by applying innovations in microfluidic technology, for example, chip-based microreactors or clear tubing, to offer advantages over conventional batch forms. One advantage is the homogeneity of the illumination of the total response volume with the greatest penetration of light. The use of microfluidic systems also allows better temperature control, particularly in relation to heat dispersal and the minimization of hot spots. Finally, photochemistry in flow offers the extra advantage that its flow rate and reactor volume can be used to control UV reaction time, thus avoiding the under- and over-exposure to light found in batch reactors [82]. Moreover, visible-light photocatalysis has attracted significant attention recently. Well-known but costly ruthenium or iridium metal complexes have been widely used [83–85]. The use of metal-free sensitizers such as biologically derived catalysts of Rose Bengal, porphyrin, or riboflavin subsidiaries broadens the range of applications of flow photocatalysis, thus making the method much greener [86]. An excellent review on heterogeneous photocatalysis in flow chemical reactors and its application in environmental remediation has recently been published [87].

There are numerous examples in the literature of photocatalyzed reactions to obtain interesting products in drug discovery [88–91]. Most of these examples concerned to widely used ordinary pharmaceutical drugs, such as naproxen, ibuprofen, flurbiprofen, tolmetin, and fexofenadine, and these compounds share a common strategy for their formation by Negishi cross-coupling to achieve α-arylation of carbonyl compounds. However, this approach often fails when electron-rich heterocycles and chloro-derivatives are required. With the aim of overcoming this drawback and obtaining compounds of interest in drug discovery, Alcázar and coworkers carried out a visible-light-induced Negishi cross-coupling involving a Pd(0)-zinc complex, which enabled the reaction of zinc enolates with deactivated aryl halides (Fig. 7.15) [92].

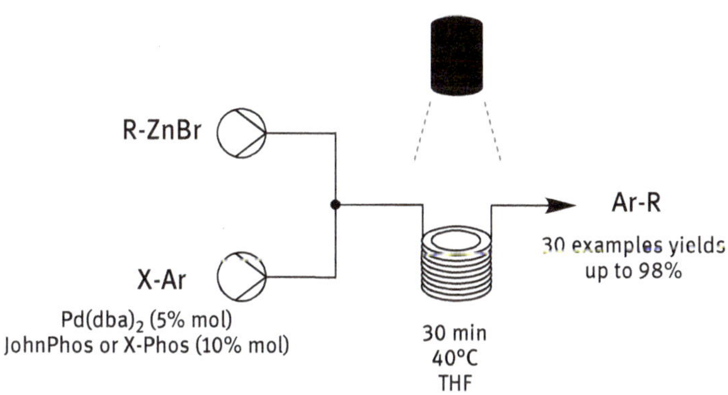

Fig. 7.15: Light-induced palladium-catalyzed Negishi reactions in flow conditions.

7.3.5.2 Electrochemistry

Electrochemical processes are inalienably green due to the replacement of hazardous chemical oxidants and the number of operators by electrical power, which can be created from renewable sources. However, the need for reproducibility, standardized equipment, and conventions has for many years hampered the application of this technique in standard chemical research [93]. Scaling electrochemistry in batch reactors is problematical due to the need for cathodes with large surface to volume ratios to create suitable electron flow, the presence of inhomogeneous electric areas, and energy loss owing to Joule heating [94].

These issues can be overcome by utilizing electrochemical flow cells. Microreactors provide high surface/volume ratios and allow appropriate control over temperature, reaction time, and flow rate. Furthermore, under flow conditions the possibility of overoxidation is diminished as the reaction mixture flows continuously outside the reactor volume, in contrast to batch electrolysis processes. In an electrochemical flow cell the reduced distance between the two electrodes decreases the ohmic resistance and, as a result, electrolysis can be carried out in the absence, or a low concentration, of supporting electrolyte. The combination of electroorganic reactivity and flow chemistry therefore provides a safer, more selective, controllable, economical, and eco-friendly chemical methodology [95].

A prime example of the use of electrochemistry in flow conditions is the synthesis of finerenone (BAY-94-8862). This compound is a nonsteroidal mineralocorticoid receptor antagonist developed by Bayer to slow down kidney disease derived from type 2 diabetes mellitus and diabetic kidney disease [96]. The product is prepared as a racemic mixture that is subsequently separated by chiral column chromatography to give the active *S* enantiomer. Bayer recently filed a patent to refeed the undesired *R* enantiomer through an electrochemical oxidation/reduction process to obtain a racemic mixture (Fig. 7.16) [97].

7.3.5.3 Biocatalysis

Biocatalysis concerns the use of biological frameworks (for the most part proteins) as catalysts and the scope and application of this approach has expanded significantly due to the advances in protein and metabolic design along with biocatalyst immobilization. In whole-cell catalysis the complete living species is employed, for example, *Escherichia coli*, whereas purified protein biocatalysis involves the use of a protein extracted from the cell [98]. However, biocatalysis does suffer from some synthetic and operational issues. The integration of biocatalyzed reactions with flow reactor innovation has led to sustainable and highly beneficial continuous processes. The foremost advantages of flow-based biocatalysis are restriction of substrate/product inhibition effects, in-line filtration with simple recovery of the product, and no mechanical blending. Flow systems accelerate biotransformations due to improved mass exchange, thus making large-scale generation more financially feasible with smaller equipment with a considerable reduction in the reaction time from hours to minutes. Automation for in-line product recovery is presently accessible at moderately low cost and this makes flow-based biocatalysis a viable innovation [99].

One clear example of biocatalysis is the three-step flow chemical synthesis of 2-aminophenoxazin-3-one. The first step is the zinc-catalyzed reduction of nitrobenzene to phenylhydroxylamine. The second reaction in the flow stream occurs with a hydroxylamine benzene mutase (HAB mutase) from *Pseudomonas pseudoalcaligenes* JS45 immobilized on silica to convert the hydroxylamine derivative to 2-aminophenol. In the final step, which also involves an enzyme immobilized on

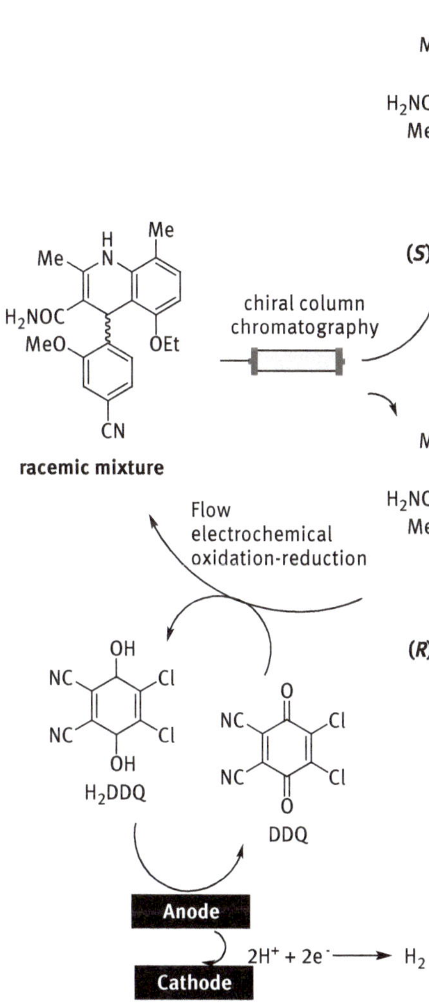

Fig. 7.16: Flow electrochemical oxidation/reduction of finerenone.

silica, soybean peroxidase (SBP) catalyzes the oxidation to render the dimerization product (Fig. 7.17). However, the low substrate concentration (1 mM), the overall yield (19%), and the productivity require further improvement.

7.3.5.4 Microwaves

Microwave (MW) radiation is an energy source that has markedly changed how chemical synthesis is performed. MW heating usually leads to shorter reaction times, simplified isolation procedures, increased product purities, higher yields,

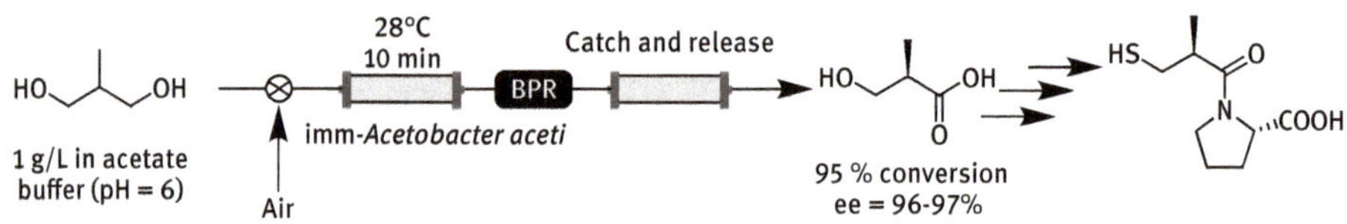

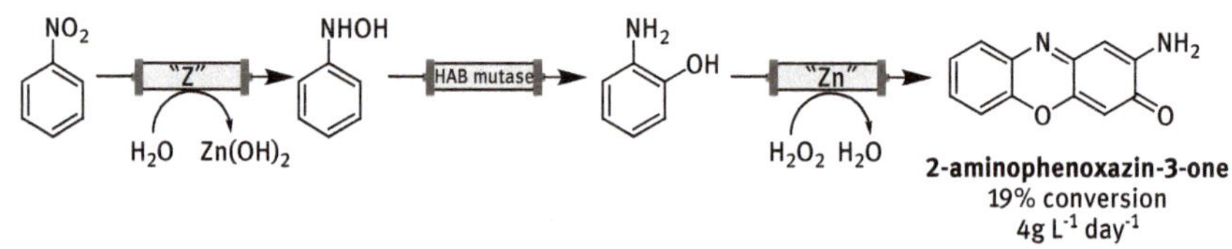

Fig. 7.17: Continuous synthesis of 2-aminophenoxazin-3-one.

and, in some cases, modifications in the selectivity. Furthermore, some reactions that do not proceed under conventional heating can be performed under MW irradiation [100]. MWs and flow chemistry have several factors in common. The combination of MW irradiation with flow conditions has been depicted as a better fit than for conventional heating conditions since heat exchange is exceptionally fast and overheating is possible when a process is performed under pressure [101].

MW irradiation and flow processes can be effortlessly performed at high temperatures and pressures and these techniques are complementary when scale-up is considered. One of the clearest advantages of flow responses is the ease with which processes can be scaled without the need to reoptimize the response conditions, even when working with small scale and mesoreactors. The small distance across of the flow reactors is perfect for the application of MW. This advantage overcomes the problem of the low penetration depth of MW when scaling MW-assisted processes [102]. As a result of the current synthetic utility of MW-assisted continuous-flow processes, nearly all MW manufacturers currently have systems that are suitable for flow conditions.

Some of the best examples of the versatility of MW reactions under flow conditions are found in the work of Ley and coworkers. One of the most recent examples concerns a single-mode bench-top resonator for the MW-assisted flow generation of primary ketenes by thermal decomposition of α-diazoketones at high temperature. Under these conditions the decomposition products in an in situ reaction with imines through a [2 + 2] Staudinger cycloaddition afforded the corresponding *trans*-configured β-lactams in good yields (Fig. 7.18) [103]. The authors also provided some insights into the mechanism of the reaction at high temperature.

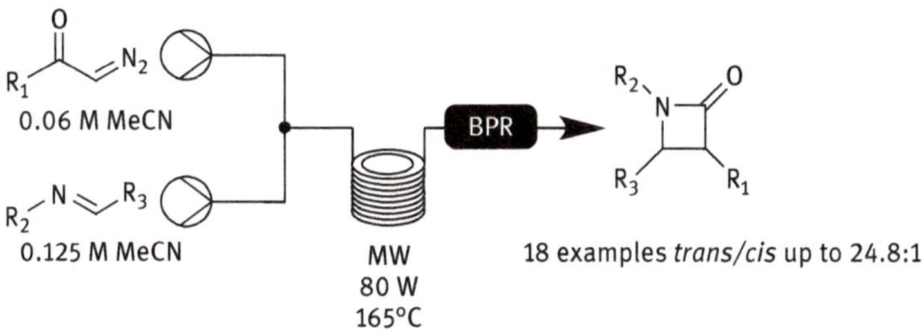

Fig. 7.18: Microwave-assisted Wolff–Staudinger flow strategy with the formation in situ of β-lactams.

7.4 Flow chemistry in fragrance and agrochemical production

In previous sections of this book it has been shown how pharmaceutical and medicinal chemists are leaning toward modular manufacturing and process intensification design concepts and they are not alone on this approach. Chemists and chemical engineers in the energy industry have performed chemical reactions in continuous flow for decades because of the economy of scale, chemical process safety, and the push for efficiency. Nowadays, fine chemicals manufacturing (such as agrochemicals) and fine commodities (such as the synthesis of fragrances or aromas), which have long been carried out using batch-wise operations, are now entering this area.

7.4.1 Fragrance production

Until 100 years ago, perfumes were made entirely from natural materials, whereas the modern perfumery industry makes extensive the use of synthetic chemicals, with the tonnage consumed running into six Figures. Today, it is estimated that fewer than 5% of the 3,000 fragrance ingredients available to the perfumers come directly from natural sources. In this field the exact molecules and ratios in flavor recipes are usually trade secrets [104]. Moreover, flavors are essential in driving consumption [105]. Odor characteristics play a crucial role in consumer preference and acceptance, thus making flavors essential components in the food industry [106]. In general, flavors are mixtures of aroma constituents blended with synthetic and natural raw materials [107].

One recent example of flow chemistry highlights the potential value of this technique when applied to target synthesis in the flavor and fragrance industries. This example is focused on the *gem*-dimethylcyclohexene unit that is particularly prevalent in carotenoids (retinal, vitamin A, β-carotene) and many benchmark fragrance classes such as the ionones and damascones [108]. The multistep sequence was robust and quickly scaled to allow the kg production of the target material and its intermediates, with more than one kilogram of the target compound eventually generated. Several options to improve the synthesis were identified during the course of the study and these were based on both reactor design and chemical synthesis (Fig. 7.19).

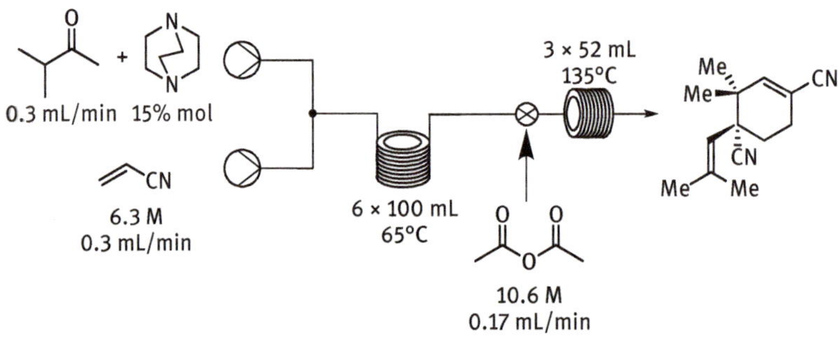

Fig. 7.19: Reactor setup for the dual-stage formation of *gem*-dimethylcyclohexene derivatives in fragrance synthesis.

Another recent strategy has seen the development of specific fragrance delivery systems, which are usually based on microencapsulation [109, 110] through either chemical (coacervation, cocrystallization, or molecular inclusion) or mechanical (spray-drying or chill drying) approaches [111]. Another way of tackling this problem is through the use of less volatile and odorless precursor molecules (profragrances) that release the fragrance under defined chemical conditions, which may include variations of temperature, enzymatic or pH-dependent hydrolysis, oxidation, or (ultraviolet) light [112, 113].

In addition, it can be pointed out that it is possible to employ flow continuous biocatalysis because of it allows to perform multiphase processes, like liquid–liquid reactions. This fact makes possible an efficient mass transfer, better compartmentalization and high local concentration of the catalyst. On the other hand, it is also possible to incorporate enzymatic immobilization in continuous reactors. Bearing in mind these evidences, both technologies can be combined. Contente and coworkers [114] developed an automated continuous biocatalytic process for the preparation of a wide range of flavor esters, employing only natural substrates with great yields in only 5 min. For this process, a transferase from *Mycobacterium smegmatis* was immobilized on beads of agarose.

Another successful strategy is the use of ionic liquids (ILs, defined as salts with melting points below 100 °C) [115], which thanks to their tunable properties have attracted a great deal of interest as alternative solvents and for material applications, among other uses [116, 117]. ILs have also been extensively studied in multiple fields of chemistry applications [118, 119] and scale-up synthesis [120].

Only a few examples can be found in the literature where ILs have been used either to slow down or accelerate the rate of evaporation of a perfume (volatile) component when used in formulations as fragrance dissolution media [121, 122]. The selection of the appropriate IL was, however, crucial to control the release of volatile components, especially considering that – depending on composition – the primary components of the ILs could be detected in the headspace of IL/fragrance mixtures [123, 124].

7.4.2 Agrochemical production

Agrochemicals are an essential part of current agriculture production systems around the world. At present, the use of agrochemicals in the form of fertilizers, fungicides, and pesticides is a common practice in many nations around the world [125]. Most pesticides used worldwide are focused on crop protection, and the agricultural market is estimated to have grown to some five billion worldwide by 2020. For this reason, numerous companies have used flow technology as an enabler for the preparation of intermediates of interest on kilogram scale in order to cover the expected market volume [126].

Some of the most interesting examples of agrochemical products in the literature are organophosphorus derivatives. These compounds are widely used in a number of commonly employed transformations such as the Arbuzov reaction, the Wittig and Horner–Wadsworth–Emmons olefinations, the Mitsunobu reaction, or the Staudinger reaction. Phosphorus compounds are also used as ligands in a large variety of organometallic complexes. The majority of the agrochemical procedures under flow conditions are under patent protection, but one clear example that is particularly useful in various processes for the synthesis of carotenoids and retinoids with applications rooted in agrochemical and pharmaceutical industry, is the isomerization reaction under continuous-flow conditions of the phosphonium salt patented by BASF [127]. In the continuous process the starting material (*E*,*E* enantiomer) is first treated with a 30 wt.% solution of sodium methoxide in methanol and then filtered. The filtrate is next injected into a continuous-flow reactor at 145 °C for a residence time of 25 min (Fig. 7.20). The filtrate is then concentrated and recrystallized from 2-propanol with seeding crystallization by the residual starting material in heptane. The final product (*Z*,*E* enantiomer) was obtained in 20% isolated yield after filtration.

Other examples of the application of flow chemistry in agrochemical compounds are the synthesis of two important fungicides: hymexazol and boscalid.

Hymexazol is a highly efficient broad-spectrum and low-toxicity fungicide which is used for soil sterilization and seed disinfection in agrochemistry. Furthermore,

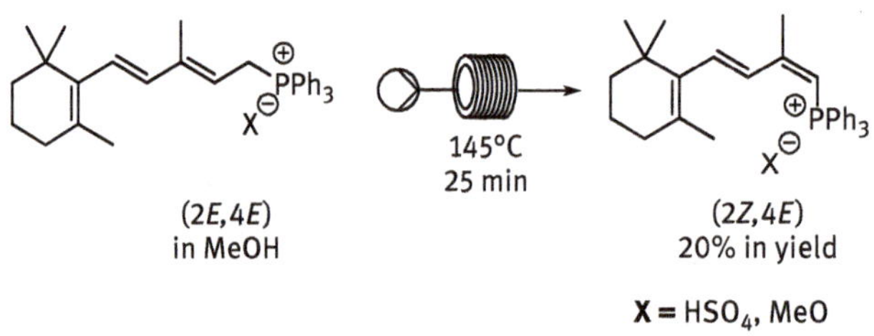

Fig. 7.20: Continuous-flow isomerization of a phosphonium salt patented by BASF.

it possesses important regulatory effects on plant diseases caused by different fungi such as *Rhizoctonia solani*, *Fusarium*, and *Mortierella*.

There are different conventional routes to the synthesis of hymexazol, but in 2019, flow chemistry was used for this goal reducing the reaction time and inhibiting the side reactions [128]. The first step for this process is the combination of ethyl acetoacetate and hydroxylamine hydrochloride to get a hydroxamic acid intermediate. Then, the reaction solution is quenched with concentrated hydrochloric acid to achieve the final product, hymexazol (Fig. 7.21). The process was successfully optimized, and in those conditions, it was possible to achieve a total yield of the target product around 86% being also able to scale the process to a kilogram scale.

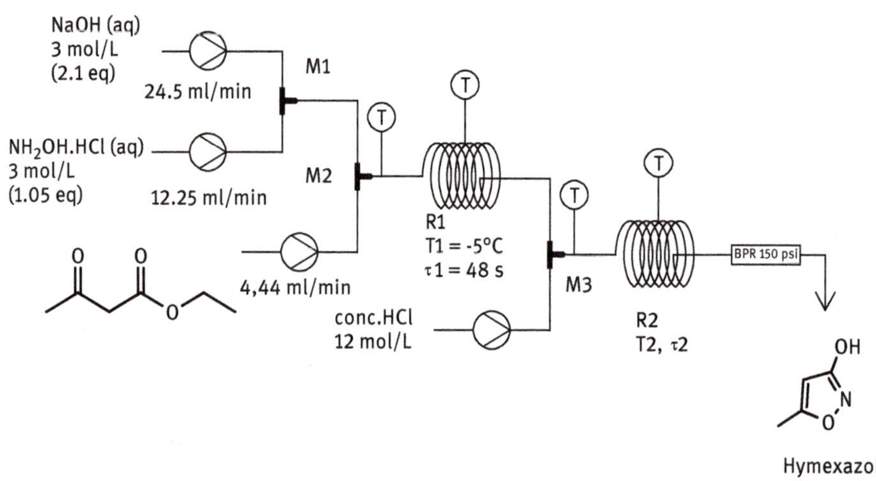

Fig. 7.21: Synthesis of hymexazol by flow chemistry.

Boscalid is another fungicide which enables the efficient control of *ascomycetes* on various fruits and vegetables.

Kappe and coworkers developed a novel two-step continuous-flow procedure for the synthesis of 2-amino-4′-chlorobiphenyl, an immediate precursor of Boscalid

(2-chloro-*N*-(4′-chlorobiphenyl-2-yl)nicotinamide) (Fig. 7.22) [129]. The first step is a Suzuki-Miyaura cross-coupling reaction between 1-chloro-2-nitrobenzene and 4-chlorophyenylboronic acid, performed in a microtubular flow reactor at 160 °C employing potassium *tert*-butoxide and the mixture *tert*-butanol/water. The mixture achieved the desired compound 4′-choro-2-nitrobiphenyl in high yield. Then, the second step is the hydrogenation of this compound to get the desired compound 2-amino-4′-chlorobiphenyl in great yield and high selectivity. This last reaction was *in line* scavenging of palladium metal with a thiourea-based resin.

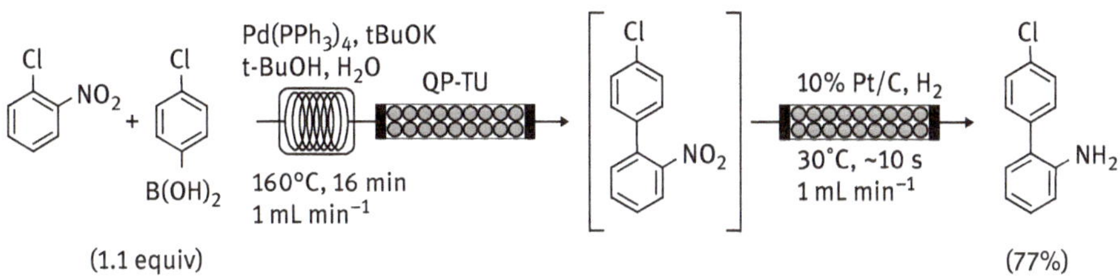

Fig. 7.22: Synthesis of 2-amino-4′-chlorobiphenyl, precursor of boscalid by a continuous-flow procedure.

(For related topics please see Volume 1, Chapter 2, Title: Principles of controlling reactions in flow chemistry, *Volume 2, Chapter 1,* Photochemical transformations in continuous-flow reactors, *and Chapter 2,* Electrochemical processes in flow*)*

7.5 Conclusions and outlook

The growth in the use of flow chemistry in the last few decades has been astounding, as evidenced by the articles and patents published in this time. This technique has had an enormous impact in synthesis for the design of drugs, fragrances, and agrochemical products, among other products.

The main goal of flow chemistry is to provide superior solutions for the chemical industry, while employing the best available technology to meet demanding processing needs today, tomorrow, and undoubtedly in the future. The development of new technologies that enable the competitive and environmentally friendly chemical manufacture of intermediates, building blocks, drug substances, and active ingredients by continuous flow is imperative for the continued growth of manufacturing industries. For industry, an initial focus on expediting development is the target of chemistries that are not usually accessible or scalable in batch conditions. There is a platform called Almac's Flow Assisted Synthesis Technology, whose goal is to facilitate the manufacture of new and highly valuable functional group interchanges using high energy, high pressure, oxidation, and photochemical transformations in a safe and scalable manner.

The enormous number of chemical reactions carried out with flow chemistry demonstrates the robustness and maturity of this exciting enabling technology. The limits and the potential of flow chemistry are still unknown, but this technique will surely become one of the fundamental components of chemistry and industry in the near future.

Study questions

7.1 Why is telescoped synthesis favored with flow methodologies?

7.2 How is it possible to minimize hazards exposure while in flow process?

7.3 How can clogging be minimized in flow reactions?

7.4 Why is flow chemistry an enabling technology for the pharmaceutical industry in research, development, and production?

7.5 Argue why the introduction of flow chemistry improves the continuous manufacturing of fragrances.

7.6 Why is the production of agrochemicals important in flow?

References

[1] Anastas PT, Warner JC, Green Chemistry, Frontiers, 1998, 640.

[2] Ley SV, On being green: can flow chemistry help?, Chem Rec, 2012, 12(4), 378–390. Doi: 10.1002/tcr.201100041.

[3] Bana P, Orkenyi R, Lovei K, Lako A, Turos GI, Eles J, Faigl F, Greiner I, The route from problem to solution in multistep continuous flow synthesis of pharmaceutical compounds, Bioorg Med Chem, 2017, 25(23), 6180–6189. Doi: 10.1016/j.bmc.2016.12.046.

[4] Baumann M, Integrating continuous flow synthesis with in-line analysis and data generation, Org Biomol Chem, 2018, 16(33), 5946–5954. Doi: 10.1039/c8ob01437j.

[5] Fleming GS, Beeler AB, Integrated drug discovery in continuous flow, J Flow Chem, 2017, 7(3–4), 124–128. Doi: 10.1556/1846.2017.00027.

[6] Gérardy R, Emmanuel N, Toupy T, Kassin V-E, Tshibalonza NN, Schmitz M, Monbaliu J-CM, Continuous flow organic chemistry: successes and pitfalls at the interface with current societal challenges, European J Org Chem, 2018(20-21), 2301–2351. Doi: 10.1002/ejoc.201800149.

[7] Dallinger D, Kappe CO, Why flow means green – Evaluating the merits of continuous processing in the context of sustainability, Curr Opin Green Sustain Chem, 2017, 7, 6–12. Doi: 10.1016/j.cogsc.2017.06.003.

[8] Souza JM, Galaverna R, Souza AAN, Brocksom TJ, Pastre JC, Souza R, Oliveira KT, Impact of continuous flow chemistry in the synthesis of natural products and active pharmaceutical ingredients, An Acad Bras Cienc, 2018, 90(1 Suppl 2), 1131–1174. Doi: 10.1590/0001-3765201820170778.

[9] Cole KP, Johnson MD, Continuous flow technology vs. the batch-by-batch approach to produce pharmaceutical compounds, Expert Rev Clin Pharmacol, 2018, 11(1), 5–13. Doi: 10.1080/17512433.2018.1413936.

[10] Hartman RL, McMullen JP, Jensen KF, Deciding whether to go with the flow: evaluating the merits of flow reactors for synthesis, Angew Chem Int Ed Engl, 2011, 50(33), 7502–7519. Doi: 10.1002/anie.201004637.

[11] Illg T, Lob P, Hessel V, Flow chemistry using milli- and microstructured reactors-from conventional to novel process windows, Bioorg Med Chem, 2010, 18(11), 3707–3719. Doi: 10.1016/j.bmc.2010.03.073.

[12] Vaccaro L, Lanari D, Marrocchi A, Strappaveccia G, Flow approaches towards sustainability, Green Chem, 2014, 16(8), 3680–3704. Doi: 10.1039/c4gc00410h.

[13] Nicolaou KC, Bulger PG, Sarlah D, Palladium-catalyzed cross-coupling reactions in total synthesis, Angew Chem Int Ed Engl, 2005, 44(29), 4442–4489. Doi: 10.1002/anie.200500368.

[14] Liu S, Fukuyama T, Sato M, Ryu I, Continuous microflow synthesis of butyl cinnamate by a mizoroki–heck reaction using a low-viscosity ionic liquid as the recycling reaction medium, Org Process Res Dev, 2004, 8(3), 477–481. Doi: 10.1021/op034200h.

[15] Ryu I, Fukuyama T, Rahman M, Sumino Y, 100 gram scale synthesis of a key intermediate of matrix metalloproteinase inhibitor in a continuous-flow system based on a copper-free sonogashira reaction using an ionic liquid as a catalyst support, Synlett, 2012, 23(15), 2279–2283. Doi: 10.1055/s-0031-1290456.

[16] Rahman MT, Fukuyama T, Kamata N, Sato M, Ryu I, Low pressure Pd-catalyzed carbonylation in an ionic liquid using a multiphase microflow system, Chem Commun, 2006, (21), 2236–2238. Doi: 10.1039/b600970k.

[17] Comprehensive Organic Transformations, Vol. 6, VCH Publishers, 1989, 989–990.

[18] Pattabiraman VR, Bode JW, Rethinking amide bond synthesis, Nature, 2011, 480(7378), 471–479. Doi: 10.1038/nature10702.

[19] Constable DJC, Dunn PJ, Hayler JD, Humphrey GR, Leazer JJL, Linderman RJ, Lorenz K, Manley J, Pearlman BA, Wells A, Zaks A, Zhang TY, Key green chemistry research areas—a perspective from pharmaceutical manufacturers, Green Chem, 2007, 9(5), 411–420. Doi: 10.1039/b703488c.

[20] Muñoz JDM, Alcázar J, De La Hoz A, Díaz-Ortiz Á, Alonso De Diego S-A, Preparation of amides mediated by isopropylmagnesium chloride under continuous flow conditions, Green Chem, 2012, 14(5). Doi: 10.1039/c2gc35037h.

[21] Yoshida J, Nagaki A, Yamada T, Flash chemistry: fast chemical synthesis by using microreactors, Chem Eur J, 2008, 14(25), 7450–7459. Doi: 10.1002/chem.200800582.

[22] Deadman BJ, Collins SG, Maguire AR, Taming hazardous chemistry in flow: the continuous processing of diazo and diazonium compounds, Chem Eur J, 2015, 21(6), 2298–2308. Doi: 10.1002/chem.201404348.

[23] Mastronardi F, Gutmann B, Kappe CO, Continuous flow generation and reactions of anhydrous diazomethane using a Teflon AF-2400 tube-in-tube reactor, Org Lett, 2013, 15(21), 5590–5593. Doi: 10.1021/ol4027914.

[24] Pinho VD, Gutmann B, Miranda LS, De Souza RO, Kappe CO, Continuous flow synthesis of alpha-halo ketones: essential building blocks of antiretroviral agents, J Org Chem, 2014, 79(4), 1555–1562. Doi: 10.1021/jo402849z.

[25] Znidar D, O'kearney-mcmullan A, Munday R, Wiles C, Poechlauer P, Schmoelzer C, Dallinger D, Kappe CO, Scalable Wolff–Kishner reductions in extreme process windows using a silicon carbide flow reactor, Org Process Res Dev, 2019, 23(11), 2445–2455. Doi: 10.1021/acs.oprd.9b00336.

[26] Anastas PT, Zimmerman JB, Design through the 12 principles of green engineering, Environ Sci Technol, 2003, 37(5), 94A–101A. Doi: 10.1021/es032373g.

[27] Anastas P, Eghbali N, Green chemistry: principles and practice, Chem Soc Rev, 2010, 39(1), 301–312. Doi: 10.1039/b918763b.

[28] Fish RH, Jaouen G, Bioorganometallic chemistry: structural diversity of organometallic complexes with bioligands and molecular recognition studies of several supramolecular hosts with biomolecules, alkali-metal ions, and organometallic pharmaceuticals, Organometallics, 2003, 22(11), 2166–2177. Doi: 10.1021/om0300777.

[29] Allardyce CS, Dorcier A, Scolaro C, Dyson PJ, Development of organometallic (organo-transition metal) pharmaceuticals, Appl Organomet Chem, 2005, 19(1), 1–10. Doi: 10.1002/aoc.725.

[30] Hessel V, Cortese B, De Croon MHJM, Novel process windows – Concept, proposition and evaluation methodology, and intensified superheated processing, Chem Eng Sci, 2011, 66(7), 1426–1448. Doi: 10.1016/j.ces.2010.08.018.

[31] Ehrfeld W, Hessel V, Löwe H (2000) Microreactors. doi:10.1002/3527601953

[32] Alcázar J, Sustainable Flow Chemistry in Drug Discovery, In: Sustainable Flow Chemistry, 2017, 135–164. Doi: 10.1002/9783527689118.ch6.

[33] Battilocchio C, Hawkins JM, Ley SV, A mild and efficient flow procedure for the transfer hydrogenation of ketones and aldehydes using hydrous zirconia, Org Lett, 2013, 15(9), 2278–2281. Doi: 10.1021/ol400856g.

[34] Moghaddam MM, Pieber B, Glasnov T, Kappe CO, Immobilized iron oxide nanoparticles as stable and reusable catalysts for hydrazine-mediated nitro reductions in continuous flow, ChemSusChem, 2014, 7(11), 3122–3131. Doi: 10.1002/cssc.201402455.

[35] Schröder F, Erdmann N, Noël T, Luque R, Van der eycken EV, Leaching-free supported gold nanoparticles catalyzing cycloisomerizations under microflow conditions, Adv Synth Catal, 2015, 357(14-15), 3141–3147. Doi: 10.1002/adsc.201500628.

[36] Tan LM, Sem ZY, Chong WY, Liu X, Hendra, Kwan WL, Lee CL, Continuous flow Sonogashira C-C coupling using a heterogeneous palladium-copper dual reactor, Org Lett, 2013, 15(1), 65–67. Doi: 10.1021/ol303046e.

[37] Martínez A, Krinsky JL, Peñafiel I, Castillón S, Loponov K, Lapkin A, Godard C, Claver C, Heterogenization of Pd–NHC complexes onto a silica support and their application in Suzuki–Miyaura coupling under batch and continuous flow conditions, Catal Sci & Technol, 2015, 5(1), 310–319. Doi: 10.1039/c4cy00829d.

[38] Egle B, Muñoz J, Alonso N, De Borggraeve W, De La Hoz A, Díaz-Ortiz A, Alcázar J, First example of alkyl–aryl negishi cross-coupling in flow: mild, efficient and clean introduction of functionalized alkyl groups, J Flow Chem, 2015, 4(1), 22–25. Doi: 10.1556/jfc-d-13-00009.

[39] Ricciardi R, Huskens J, Holtkamp M, Karst U, Verboom W, Dendrimer-encapsulated palladium nanoparticles for continuous-flow suzuki-miyaura cross-coupling reactions, ChemCatChem, 2015, 7(6), 936–942. Doi: 10.1002/cctc.201500017.

[40] Veerakumar P, Thanasekaran P, Lu K-L, Liu S-B, Rajagopal S, Functionalized silica matrices and palladium: a versatile heterogeneous catalyst for suzuki, heck, and sonogashira reactions, ACS Sustain Chem Eng, 2017, 5(8), 6357–6376. Doi: 10.1021/acssuschemeng.7b00921.

[41] Greco R, Goessler W, Cantillo D, Kappe CO, Benchmarking immobilized Di- and triarylphosphine palladium catalysts for continuous-flow cross-coupling reactions: efficiency, durability, and metal leaching studies, ACS Catal, 2015, 5(2), 1303–1312. Doi: 10.1021/cs5020089.

[42] Pavia C, Ballerini E, Bivona LA, Giacalone F, Aprile C, Vaccaro L, Gruttadauria M, Palladium supported on cross-linked imidazolium network on silica as highly sustainable catalysts for

the suzuki reaction under flow conditions, Adv Synth Catal, 2013, 355(10), 2007–2018. Doi: 10.1002/adsc.201300215.
[43] De M. Muñoz J, Alcázar J, De La Hoz A, Díaz-Ortiz A, Cross-coupling in flow using supported catalysts: mild, clean, efficient and sustainable suzuki-miyaura coupling in a single pass, Adv Synth Catal, 2012, 354(18), 3456–3460. Doi: 10.1002/adsc.201200678.
[44] Price GA, Hassan A, Chandrasoma N, Bogdan AR, Djuric SW, Organ MG, Pd-PEPPSI-IPent-SiO2: a supported catalyst for challenging negishi coupling reactions in flow, Angew Chem Int Ed Engl, 2017, 56(43), 13347–13350. Doi: 10.1002/anie.201708598.
[45] Briggs NEB, Schacht U, Raval V, McGlone T, Sefcik J, Florence AJ, Seeded crystallization of β-l-glutamic acid in a continuous oscillatory baffled crystallizer, Org Process Res Dev, 2015, 19(12), 1903–1911. Doi: 10.1021/acs.oprd.5b00206.
[46] Mallia CJ, Baxendale IR, The use of gases in flow synthesis, Org Process Res Dev, 2015, 20(2), 327–360. Doi: 10.1021/acs.oprd.5b00222.
[47] Yu T, Jiao J, Song P, Nie W, Yi C, Zhang Q, Li P, Cover feature: recent progress in continuous-flow hydrogenation (ChemSusChem 11/2020), ChemSusChem, 2020, 13(11), 2803–2803. Doi: 10.1002/cssc.202001206.
[48] Gardiner J, Nguyen X, Genet C, Horne MD, Hornung CH, Tsanaktsidis J, Catalytic static mixers for the continuous flow hydrogenation of a key intermediate of linezolid (Zyvox), Org Process Res Dev, 2018, 22(10), 1448–1452. Doi: 10.1021/acs.oprd.8b00153.
[49] Rahman MDT, Wharry S, Smyth M, Manyar H, Moody TS, FAST hydrogenations as a continuous platform for green aromatic nitroreductions, Synlett, 2020, 31(06), 581–586. Doi: 10.1055/s-0037-1610751.
[50] Thathagar MB, Ten Elshof JE, Rothenberg G, Pd nanoclusters in C-C coupling reactions: proof of leaching, Angew Chem Int Ed Engl, 2006, 45(18), 2886–2890. Doi: 10.1002/anie.200504321.
[51] Cantillo D, Kappe CO, Immobilized transition metals as catalysts for cross-couplings in continuous flow-a critical assessment of the reaction mechanism and metal leaching, ChemCatChem, 2014, 6(12), 3286–3305. Doi: 10.1002/cctc.201402483.
[52] Noel T, Buchwald SL, Cross-coupling in flow, Chem Soc Rev, 2011, 40(10), 5010–5029. Doi: 10.1039/c1cs15075h.
[53] Movsisyan M, Delbeke EI, Berton JK, Battilocchio C, Ley SV, Stevens CV, Taming hazardous chemistry by continuous flow technology, Chem Soc Rev, 2016, 45(18), 4892–4928. Doi: 10.1039/c5cs00902b.
[54] Yang H, Martin B, Schenkel B, On-demand generation and consumption of diazomethane in multistep continuous flow systems, Org Process Res Dev, 2018, 22(4), 446–456. Doi: 10.1021/acs.oprd.7b00302.
[55] Cossar PJ, Hizartzidis L, Simone MI, McCluskey A, Gordon CP, The expanding utility of continuous flow hydrogenation, Org Biomol Chem, 2015, 13(26), 7119–7130. Doi: 10.1039/c5ob01067e.
[56] Richardson P, Fluorination methods for drug discovery and development, Expert Opin Drug Discov, 2016, 11(10), 983–999. Doi: 10.1080/17460441.2016.1223037.
[57] Cantillo D, Kappe CO, Halogenation of organic compounds using continuous flow and microreactor technology, React Chem & Eng, 2017, 2(1), 7–19. Doi: 10.1039/c6re00186f.
[58] Colella M, Nagaki A, Luisi R, Flow technology for the genesis and use of (highly) reactive organometallic reagents, Chem Eur J, 2020, 26(1), 19–32. Doi: 10.1002/chem.201903353.
[59] Yoshida J, Takahashi Y, Nagaki A, Flash chemistry: flow chemistry that cannot be done in batch, Chem Commun, 2013, 49(85), 9896–9904. Doi: 10.1039/c3cc44709j.
[60] Wakami H, Yoshida J-I, Grignard exchange reaction using a microflow system: from bench to pilot plant, Org Process Res Dev, 2005, 9(6), 787–791. Doi: 10.1021/op0501500.

[61] Petersen TP, Becker MR, Knochel P, Continuous flow magnesiation of functionalized heterocycles and acrylates with TMPMgClLiCl, Angew Chem Int Ed Engl, 2014, 53(30), 7933–7937. Doi: 10.1002/anie.201404221.
[62] Kondo Y, Shilai M, Uchiyama M, Sakamoto T, TMP–zincate as highly chemoselective base for directed ortho metalation, J Am Chem Soc, 1999, 121(14), 3539–3540. Doi: 10.1021/ja984263t.
[63] Wunderlich SH, Knochel P, (tmp)(2)Zn x 2 MgCl(2) x 2 LiCl: a chemoselective base for the directed zincation of sensitive arenes and heteroarenes, Angew Chem Int Ed Engl, 2007, 46(40), 7685–7688. Doi: 10.1002/anie.200701984.
[64] Polyzos A, O'Brien M, Petersen TP, Baxendale IR, Ley SV, The continuous-flow synthesis of carboxylic acids using CO2 in a tube-in-tube gas permeable membrane reactor, Angew Chem Int Ed Engl, 2011, 50(5), 1190–1193. Doi: 10.1002/anie.201006618.
[65] Van Gool JJF, Van Den Broek SAMW, Ripken RM, Nieuwland PJ, Koch K, Rutjes FPJT, Highly controlled gas/liquid processes in a continuous lab-scale device, Chem Eng & Technol, 2013, 36(6), 1042–1046. Doi: 10.1002/ceat.201200553.
[66] Alonso N, Juan De MM, Egle B, Vrijdag JL, De Borggraeve WM, De La Hoz A, Díaz-Ortiz A, Alcázar J, First example of a continuous-flow carbonylation reaction using aryl formates as CO precursors, J Flow Chem, 2014, 4(3), 105–109. Doi: 10.1556/jfc-d-14-00005.
[67] Ueda T, Konishi H, Manabe K, Trichlorophenyl formate: highly reactive and easily accessible crystalline CO surrogate for palladium-catalyzed carbonylation of aryl/alkenyl halides and triflates, Org Lett, 2012, 14(20), 5370–5373. Doi: 10.1021/ol302593z.
[68] Britton J, Raston CL, Multi-step continuous-flow synthesis, Chem Soc Rev, 2017, 46(5), 1250–1271. Doi: 10.1039/c6cs00830e.
[69] Zhang P, Russell MG, Jamison TF, Continuous flow total synthesis of rufinamide, Org Process Res Dev, 2014, 18(11), 1567–1570. Doi: 10.1021/op500166n.
[70] Correia CA, Gilmore K, McQuade DT, Seeberger PH, A concise flow synthesis of efavirenz, Angew Chem Int Ed Engl, 2015, 54(16), 4945–4948. Doi: 10.1002/anie.201411728.
[71] Murray PRD, Browne DL, Pastre JC, Butters C, Guthrie D, Ley SV, Continuous flow-processing of organometallic reagents using an advanced peristaltic pumping system and the telescoped flow synthesis of (E/Z)-tamoxifen, Org Process Res Dev, 2013, 17(9), 1192–1208. Doi: 10.1021/op4001548.
[72] Baumann M, Baxendale IR, Kuratli C, Ley SV, Martin RE, Schneider J, Synthesis of a drug-like focused library of trisubstituted pyrrolidines using integrated flow chemistry and batch methods, ACS Comb Sci, 2011, 13(4), 405–413. Doi: 10.1021/co2000357.
[73] Lange PP, James K, Rapid access to compound libraries through flow technology: fully automated synthesis of a 3-aminoindolizine library via orthogonal diversification, ACS Comb Sci, 2012, 14(10), 570–578. Doi: 10.1021/co300094n.
[74] Bryan MC, Hein CD, Gao H, Xia X, Eastwood H, Bruenner BA, Louie SW, Doherty EM, Disubstituted 1-aryl-4-aminopiperidine library synthesis using computational drug design and high-throughput batch and flow technologies, ACS Comb Sci, 2013, 15(9), 503–511. Doi: 10.1021/co400078r.
[75] Gioiello A, Rosatelli E, Teofrasti M, Filipponi P, Pellicciari R, Building a sulfonamide library by eco-friendly flow synthesis, ACS Comb Sci, 2013, 15(5), 235–239. Doi: 10.1021/co400012m.
[76] Riesco-Domínguez A, Blanco-Ania D, Rutjes FPJT, Continuous flow synthesis of urea-containing compound libraries based on the piperidin-4-one scaffold, European J Org Chem, 2018, (11), 1312–1320. Doi: 10.1002/ejoc.201701539.
[77] Luque Navarro PM, Lanari D, Flow synthesis of biologically-relevant compound libraries, Molecules, 2020, 25, 4. Doi: 10.3390/molecules25040909.

[78] Poh JS, Browne DL, Ley SV, A multistep continuous flow synthesis machine for the preparation of pyrazoles via a metal-free amine-redox process, React Chem Eng, 2016, 1(1), 101–105. Doi: 10.1039/c5re00082c.

[79] Koolman HF, Kantor S, Bogdan AR, Wang Y, Pan JY, Djuric SW, Automated library synthesis of cyclopropyl boronic esters employing diazomethane in a tube-in-tube flow reactor, Org Biomol Chem, 2016, 14(27), 6591–6595. Doi: 10.1039/c6ob00715e.

[80] Bogdan AR, Dombrowski AW, Emerging trends in flow chemistry and applications to the pharmaceutical industry, J Med Chem, 2019, 62(14), 6422–6468. Doi: 10.1021/acs.jmedchem.8b01760.

[81] Gioiello A, Piccinno A, Lozza AM, Cerra B, The medicinal chemistry in the era of machines and automation: recent advances in continuous flow technology, J Med Chem, 2020, 63(13), 6624–6647. Doi: 10.1021/acs.jmedchem.9b01956.

[82] Vasudevan A, Bogdan AR, Koolman HF, Wang Y, Djuric SW, Enabling chemistry technologies and parallel synthesis-accelerators of drug discovery programmes, Prog Med Chem, 2017, 56, 1–35. Doi: 10.1016/bs.pmch.2016.11.001.

[83]. Prier CK, Rankic DA, MacMillan DW, Visible light photoredox catalysis with transition metal complexes: applications in organic synthesis, Chem Rev, 2013, 113(7), 5322–5363. Doi: 10.1021/cr300503r.

[84] Schultz DM, Yoon TP, Solar synthesis: prospects in visible light photocatalysis, Science, 2014, 343(6174), 1239176. Doi: 10.1126/science.1239176.

[85] Zeitler K, Photoredox catalysis with visible light, Angew Chem Int Ed Engl, 2009, 48(52), 9785–9789. Doi: 10.1002/anie.200904056.

[86] Rehm T, Flow photochemistry as a tool in organic synthesis, Chem Eur J, 2020. Doi: 10.1002/chem.202000381.

[87] Thomson CG, Lee AL, Vilela F, Heterogeneous photocatalysis in flow chemical reactors, Beilstein J Org Chem, 2020, 16, 1495–1549. Doi: 10.3762/bjoc.16.125.

[88] Levesque F, Seeberger PH, Continuous-flow synthesis of the anti-malaria drug artemisinin, Angew Chem Int Ed Engl, 2012, 51(7), 1706–1709. Doi: 10.1002/anie.201107446.

[89] Oelgemoller M, Solar photochemical synthesis: from the beginnings of organic photochemistry to the solar manufacturing of commodity chemicals, Chem Rev, 2016, 116(17), 9664–9682. Doi: 10.1021/acs.chemrev.5b00720.

[90] Abdiaj I, Alcazar J, Improving the throughput of batch photochemical reactions using flow: dual photoredox and nickel catalysis in flow for C(sp(2))C(sp(3)) cross-coupling, Bioorg Med Chem, 2017, 25(23), 6190–6196. Doi: 10.1016/j.bmc.2016.12.041.

[91] Hommelsheim R, Guo Y, Yang Z, Empel C, Koenigs RM, Blue-light-induced carbene-transfer reactions of diazoalkanes, Angew Chem Int Ed Engl, 2019, 58(4), 1203–1207. Doi: 10.1002/anie.201811991.

[92] Wei XJ, Abdiaj I, Sambiagio C, Li C, Zysman-Colman E, Alcazar J, Noel T, Visible-light-promoted iron-catalyzed C(sp(2))-C(sp(3)) Kumada cross-coupling in flow, Angew Chem Int Ed Engl, 2019, 58(37), 13030–13034. Doi: 10.1002/anie.201906462.

[93] Yan M, Kawamata Y, Baran PS, Synthetic organic electrochemistry: calling all engineers, Angew Chem Int Ed Engl, 2018, 57(16), 4149–4155. Doi: 10.1002/anie.201707584.

[94] Pletcher D, Green RA, Brown RCD, Flow electrolysis cells for the synthetic organic chemistry laboratory, Chem Rev, 2018, 118(9), 4573–4591. Doi: 10.1021/acs.chemrev.7b00360.

[95] Elsherbini M, Wirth T, Electroorganic synthesis under flow conditions, Acc Chem Res, 2019, 52(12), 3287–3296. Doi: 10.1021/acs.accounts.9b00497.

[96] Barfacker L, Kuhl A, Hillisch A, Grosser R, Figueroa-Perez S, Heckroth H, Nitsche A, Erguden JK, Gielen-Haertwig H, Schlemmer KH, Mittendorf J, Paulsen H, Platzek J, Kolkhof P, Discovery of BAY 94-8862: a nonsteroidal antagonist of the mineralocorticoid receptor for the

treatment of cardiorenal diseases, ChemMedChem, 2012, 7(8), 1385–1403. Doi: 10.1002/cmdc.201200081.
[97] Platzek J, Gottfried K, Assmann J, Lolli G (2017) Method For The Preparation Of (4s)-4-(4-Cyano-2-Methoxyphenyl)-5-Ethoxy-2,8-Dimethyl-1,4-Dihydro-1-6-Naphthyridine-3-Carboxamide And Recovery Of (4s)-4-(4-Cyano-2-Methoxyphenyl)-5-Ethoxy-2,8-Dimethyl-1,4-Dihydro-1-6-Naphthyridine-3-Carboxamide By Electrochemical Methods. Germany Patent WO2017032678,
[98] Britton J, Majumdar S, Weiss GA, Continuous flow biocatalysis, Chem Soc Rev, 2018, 47(15), 5891–5918. Doi: 10.1039/c7cs00906b.
[99] Tamborini L, Fernandes P, Paradisi F, Molinari F, Flow bioreactors as complementary tools for biocatalytic process intensification, Trends Biotechnol, 2018, 36(1), 73–88. Doi: 10.1016/j.tibtech.2017.09.005.
[100] De La Hoz A, Loupy A, Microwaves in Organic Synthesis, 2 Volume Set, Wiley, 2013.
[101] Glasnov TN, Kappe CO, The microwave-to-flow paradigm: translating high-temperature batch microwave chemistry to scalable continuous-flow processes, Chem Eur J, 2011, 17(43), 11956–11968. Doi: 10.1002/chem.201102065.
[102] De La Hoz A, Díaz-Ortiz A, Sustainable Flow Chemistry: Methods and Applications, Vaccaro L, ed, Wiley-VCH, Weinheim, 2017, 219–248.
[103] Ley SV, Musio B, Mariani F, Śliwiński E, Kabeshov M, Odajima H, Combination of enabling technologies to improve and describe the stereoselectivity of Wolff–Staudinger cascade reaction, Synthesis, 2016, 48(20), 3515–3526. Doi: 10.1055/s-0035-1562579.
[104] Epstein JL, Castaldi M, Patel G, Telidecki P, Karakkatt K, Using flavor chemistry to design and synthesize artificial scents and flavors, J Chem Educ, 2014, 92(5), 954–957. Doi: 10.1021/ed500615a.
[105] Murphy MM, Douglass JS, Johnson RK, Spence LA, Drinking flavored or plain milk is positively associated with nutrient intake and is not associated with adverse effects on weight status in US children and adolescents, J Am Diet Assoc, 2008, 108(4), 631–639. Doi: 10.1016/j.jada.2008.01.004.
[106] Zhu G, Xiao Z, Zhou R, Lei D, Preparation and simulation of a taro flavor, Chin J Chem Eng, 2015, 23(10), 1733–1735. Doi: 10.1016/j.cjche.2015.07.026.
[107] Zhu G, Xiao Z, Creation and imitation of a milk flavour, Food Funct, 2017, 8(3), 1080–1084. Doi: 10.1039/c7fo00034k.
[108] Surburg H, Panten J, Common Fragrance and Flavor Materials: Preparation, Properties and Uses, Wiley, 2006.
[109] Rogers K, Controlled release technology and delivery systems, Cosmet toilet, 1999, 114(5), 53–60.
[110] Gautschi M, Bajgrowicz JA, Kraft P, Fragrance chemistry — milestones and perspectives, CHIM Int J Chem, 2001, 55(5), 379–387.
[111] Madene A, Jacquot M, Scher J, Desobry S, Flavour encapsulation and controlled release - a review, Int J Food Sci Technol, 2006, 41(1), 1–21. Doi: 10.1111/j.1365-2621.2005.00980.x.
[112] Herrmann A, Controlled release of volatiles under mild reaction conditions: from nature to everyday products, Angew Chem Int Ed Engl, 2007, 46(31), 5836–5863. Doi: 10.1002/anie.200700264.
[113] Kuhnt T, Herrmann A, Benczédi D, Weder C, Foster EJ, Controlled fragrance release from galactose-based pro-fragrances, RSC Adv, 2014, 4(92), 50882–50890. Doi: 10.1039/c4ra07728h.
[114] Contente ML, Tamborini L, Molinari F, Paradisi F, Aromas flow: eco-friendly, continuous, and scalable preparation of flavour esters, J Flow Chem, 2020, 10(1), 235–240. Doi: 10.1007/s41981-019-00063-8.
[115] Wasserscheid P, Welton T, Ionic Liquids in Synthesis, John Wiley & Sons, 2008.

[116] Smiglak M, Metlen A, Rogers RD, The second evolution of ionic liquids: from solvents and separations to advanced materials–energetic examples from the ionic liquid cookbook, Acc Chem Res, 2007, 40(11), 1182–1192. Doi: 10.1021/ar7001304.

[117] Petrat F-M, Schmidt F, Stutzel B, Kohler G (2006) Fragrance composition comprising at least one ionic liquid, method for production and use thereof. Google Patents, WO/2004/035018.

[118] Watanabe M, Thomas ML, Zhang S, Ueno K, Yasuda T, Dokko K, Application of ionic liquids to energy storage and conversion materials and devices, Chem Rev, 2017, 117(10), 7190–7239. Doi: 10.1021/acs.chemrev.6b00504.

[119] Lei Z, Chen B, Koo YM, MacFarlane DR, Introduction: ionic liquids, Chem Rev, 2017, 117(10), 6633–6635. Doi: 10.1021/acs.chemrev.7b00246.

[120] García-Verdugo E, Altava B, Burguete MI, Lozano P, Luis SV, Ionic liquids and continuous flow processes: a good marriage to design sustainable processes, Green Chem, 2015, 17(5), 2693–2713. Doi: 10.1039/c4gc02388a.

[121] Berton P, Bica K, Rogers RD, Ionic liquids for consumer products: dissolution, characterization, and controlled release of fragrance compositions, Fluid Phase Equilib, 2017, 450, 51–56. Doi: 10.1016/j.fluid.2017.07.011.

[122] Holland L, Todini O, Eike D, Velazquez Mendoza J, Tozer S, Mcnamee P, Stonehouse J, Staite W, Fovargue H, Gregory J, Fragrance compositions comprising ionic liquids, PCT Int, Appl WO, 2017, A1.

[123] Ferrero Vallana FM, Holland LAM, Seddon KR, Todini O, Delayed release of a fragrance from novel ionic liquids, New J Chem, 2017, 41(3), 1037–1045. Doi: 10.1039/c6nj03200a.

[124] Cetti J, Eike D, Ferrero Vallana FM, Gunaratne HQN, Holland LAM, Puga AV, Seddon KR, Todini O, Modeling the vapor–liquid equilibria of ionic liquids containing perfume raw materials, J Chem Eng Data, 2017, 62(9), 2787–2798. Doi: 10.1021/acs.jced.7b00116.

[125] Carvalho FP, Agriculture, pesticides, food security and food safety, Environ Sci Policy, 2006, 9(7-8), 685–692. Doi: 10.1016/j.envsci.2006.08.002.

[126] Godineau E, Battilocchio C, Lal M, Building up a continuous flow platform as an enabler to the preparation of intermediates on kilogram scale, Chimia (Aarau), 2019, 73(10), 828–831. Doi: 10.2533/chimia.2019.828.

[127] Schaefer B, Siegel W (2014) Isomerization of olefinic compounds. US Patent, WO/2014/026896.

[128] Ma X-P, Chen J-S, Du X-H, A continuous flow process for the synthesis of hymexazol, Org Process Res Dev, 2019, 23(6), 1152–1158. Doi: 10.1021/acs.oprd.9b00047.

[129] Glasnov TN, Kappe CO, Toward a continuous-flow synthesis of boscalid®, Adv Synth Catal, 2010, 352(17), 3089–3097. Doi: 10.1002/adsc.201000646.

Kai Wang, Jian Deng, Chencan Du and Guangsheng Luo

8 Scale-up of flow chemistry system

8.1 Introduction of scale-up

8.1.1 Scale-up of chemical equipment

Today, there is a global market for chemical products consuming billions of tonnes of chemicals per year, such as refined oil, and many thousands of kilograms of chemicals through the generation of active pharmaceutical intermediates. Although the different chemical producers generate different products, the equipment used in their manufacture is similar. Large-scale chemical processes mainly use continuous production methods with high efficiencies, while fine chemical manufacturers often employ batch methods giving more flexibility owing to their greater flexibility. Irrespective of the production method adopted, the reactor, distillation device, extractor, crystallizer, and other core equipment of a production system must be scaled up from the laboratory scale to a plant scale. In addition to the larger equipment, adjustments of the corresponding processes and supporting systems must be carried out to adapt to the large-scale production. Therefore, the scale-up of a chemical process is done systematically, and is also one of the most difficult issues in chemical engineering technology.

For most chemical equipment used in the industry, the change of equipment geometry is preferred in the scaling-up process, as shown by the schematics of a tank reactor in Fig. 8.1. In this scale-up, the kinetics of the chemical process in the laboratory research are largely maintained. The most important feature is the mixing performance of the device, which is usually characterized by the substance mass transfer rate. Equations (8.1) and (8.2) are the fundamental convection diffusion equations for single-phase fluid, and mass transfer rate between phases, respectively:

$$\vec{N} = -D\nabla c + c\vec{u} \tag{8.1}$$

$$M = K_{\mathrm{m}} a(c_1 - c_2/m) \tag{8.2}$$

where N is the mass transfer flux, u is the velocity, M is the mass transfer rate between phases, D is the molecular diffusion coefficient, K_{m} is the mass transfer coefficient, a is the phase boundary area or interfacial area, c, c_1, and c_2 are concentrations, and m is the substance partition coefficient between phases. Substance concentration is usually fixed during the scale-up process; therefore, the flow condition should be carefully controlled, which determines the mass transfer area and coefficient. The fluid Reynolds number (Re) and the stirring speed of the tank reactor (for the mixing energy input), are mostly considered in the scale-up of mixing equipment [1]. Moreover, the mixing performance is also determined by the structure of the tank and stirrer.

https://doi.org/10.1515/9783110693690-008

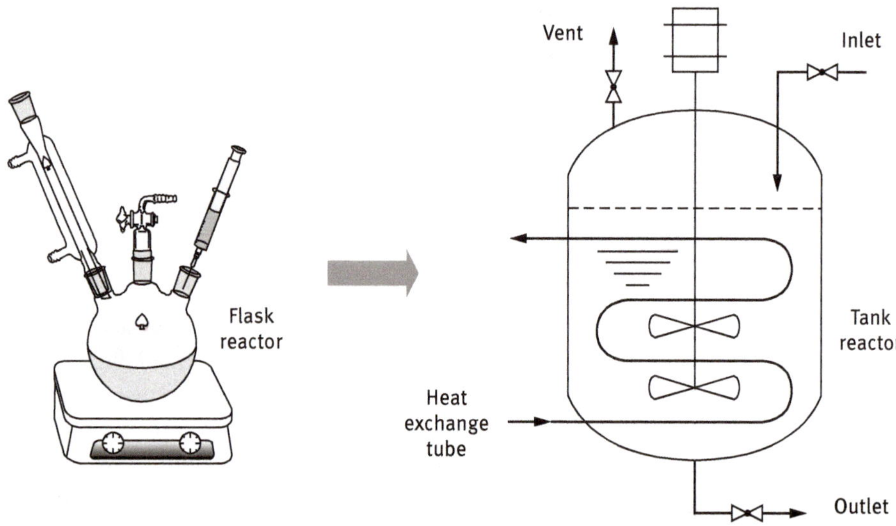

Fig. 8.1: Schematic of geometrical scale-up of a flask reactor to a tank reactor.

The heat transfer performance is another important feature that should be carefully designed for the enlarged chemical equipment, especially for the reactor, to contain strong exothermic reactions. For liquid reactions, heat transfer resistance often exists between the chemicals and the reactor cooling component. Flask reactors in laboratory research use glass walls to conduct heat, but the reactor specific surface area decreases greatly with enlargement of the geometry. Therefore, coiled heat exchange tubes are commonly employed in industrial vessel reactors, as shown in Fig. 8.1. Equation (8.3) shows the basic heat transport equation between the chemicals and the heat exchange medium (HEM) around the reactor:

$$Q_{\mathrm{h}} = K_{\mathrm{h}} A (T_{\mathrm{c}} - T_{\mathrm{HEM}}) \tag{8.3}$$

where Q_{h} is the heat transfer rate, A is the heat transfer area, K_{h} is the total heat transfer coefficient, T_{c} is the temperature of the chemical, and T_{HEM} is the temperature of the HEM. Adding more heat transfer surfaces in the reactor to increase A in equation (8.3), employing a higher temperature gradient between chemicals and HEM, would also enhance the heat transfer in the industrial reactors. However, locally extreme temperature gradient in the boundary layer would also influence the chemical process unexpectedly. The heat transfer coefficient is mainly affected by the flow conditions in chemical engineering devices, and can be represented by the Nusselt number (Nu) [2]. Usually, a high Reynolds number and Prandtl number (Pr) will lead to a higher Nusselt number.

In addition to the mixing and heat exchange performances, reaction kinetics are the third important feature in scaling-up of reactors. The intrinsic reaction kinetics shown in equation (8.4) do not vary with equipment, but the reaction rates in practice are usually determined by the apparent rates of reaction, which are affected by the mass or heat transfer rates:

$$-\mathrm{d}c_A/\mathrm{d}t = k_r c_A{}^{\alpha} c_B{}^{\beta} \cdots \tag{8.4}$$

where c_A and c_B are the reactant concentrations, k_r is the reaction rate factor, and α and β are the orders of the reaction. Understanding the reaction kinetics is very important for determining the residence time of a substance (reaction time) in an industrial reactor. Sometimes, the reaction time is manually slow down by gradually feeding the high-activity reactant or catalyst into the reactor, which are commonly used in the laboratory syntheses research. However, most industrialized reaction technologies still apply this reaction technology, which are time consuming and expensive for production.

In addition to these three important features, other performance parameters, such as the reaction time distribution, process safety, and automation, should also be carefully considered in the scale-up. Although the requirements for chemical equipment scale-up are well understood, it is difficult to satisfy all of them; this means that the demands of mixing, heat exchange, reaction time, and others cannot all be satisfied together. The decline in the dynamic performance of the equipment is commonly called to scaling-up effect. Therefore, the scale-up of traditional chemical equipment should be done step by step, constantly revising the equipment structure and the corresponding process. In addition to the scaling-up effect, another issue that poses difficulty in the scale-up of chemical processes is the variation of the production method. Most laboratory studies use batch production methods, but they would be turned to continuous methods in the industry if the chemicals being manufactured have to be made for a large market. Many unexpected problems, such as wide residence time distribution, device blocking, and device corrosion, combine the transformation of production methods. Therefore, the scale-up of chemical equipment still depends on the modeling of chemical equipment and the experience of engineers. Since this chapter does not aim to show all the details of equipment scale-up, we will not discuss the conventional chemical equipment scale-up here further.

(For further details on related issues please see Volume 1, Chapter 1, Title: *Fundamentals of Flow Chemistry)*

8.1.2 Principle of scale-up of flow chemistry system

For the scale-up of flow chemistry systems, the demands for maintaining reactant mixing, heat exchange, reaction time, and other essential parameters that we introduced above are basically the same. The difference between a flow chemistry system, and the traditional chemical system, is mainly in the form of the core equipment. In laboratory research, flow chemistry studies usually employ microstructured devices (simply called microdevices in this chapter), such as microchannels [3] and microtubes [4], to carry out reactions and separations, where the scale of mixing and mass transfer ranges from tens of micrometers to sub-millimeters. Passive mixing method without stirring is commonly employed, except for those with large amounts of crystallization or easily depositing solids [5]. When introducing the fundamentals of flow chemistry in the first volume of this book, we have seen that the microstructured equipment provides a short mass transfer distance, high specific surface area, and regular flow pattern, to enhance the mixing, heat exchange, and reaction time control of the reaction system. Therefore, their performance is geometrically dependent.

To break the hold of geometry on the production capacity, increasing the number of flow reactors had been promoted by academics initially. Parallel reaction channels or tubes are easy to fabricate and assemble with the aid of modern mechanical technology. However, simply a parallel scale-up of laboratory equipment will result in unacceptable costs for industrial applications. Thus, the scale-up of flow chemistry systems should be a hybrid method of numbering-up and traditional scale-up in equipment size. Figure 8.2 shows an example of a flow reaction system to prepare a crystal product. The reaction occurs between reactants A and B dissolved in a solvent:

$$A + B \rightarrow P \xrightarrow{\text{cooling}} P_{\text{crystal}} \tag{8.5}$$

In a laboratory, we can conclude that reactant mixing should be rapid, and the reaction heat should be quickly removed from the system. Therefore, scaling-up of the mixer and reaction tube with the heat transfer medium, are essential. To maintain the mixing and heat transfer performances, the numbering-up of the mixing channels in the mixer and the zigzag-shaped tube could be implemented, as shown in Fig. 8.2. However, The other components of the production system, including reactant B dissolution, product solution collection, crystallization, and filtration, can be entirely selected from mature chemical equipment, which does not affect the product properties, unless there is a special demand for the size of the reaction system, such as the vehicle-mounted mobile chemical system [6]. In the following sections, we will show that with pure numbering-up of microchannel, it is still difficult to meet the requirements of industrial production of general fine chemicals; therefore, similarity-up technology such as the membrane dispersion unit, is employed to assist in achieving large amounts of mixing for the gas–liquid and liquid–liquid systems [7, 8]. Nevertheless, the hybrid scaling-up method is an extension of the classical chemical process scale-up, which makes the flow chemistry technologies more feasible for industrialization. More details on the scale-up of reactant mixers and tubular reactors will be introduced in the following sections.

In addition to the example in Fig. 8.2, several other hybrid approaches for scaling-up flow chemistry systems will be discussed in the last section of this chapter. At the end of this section, we want to show that the scale-up of flow chemistry systems

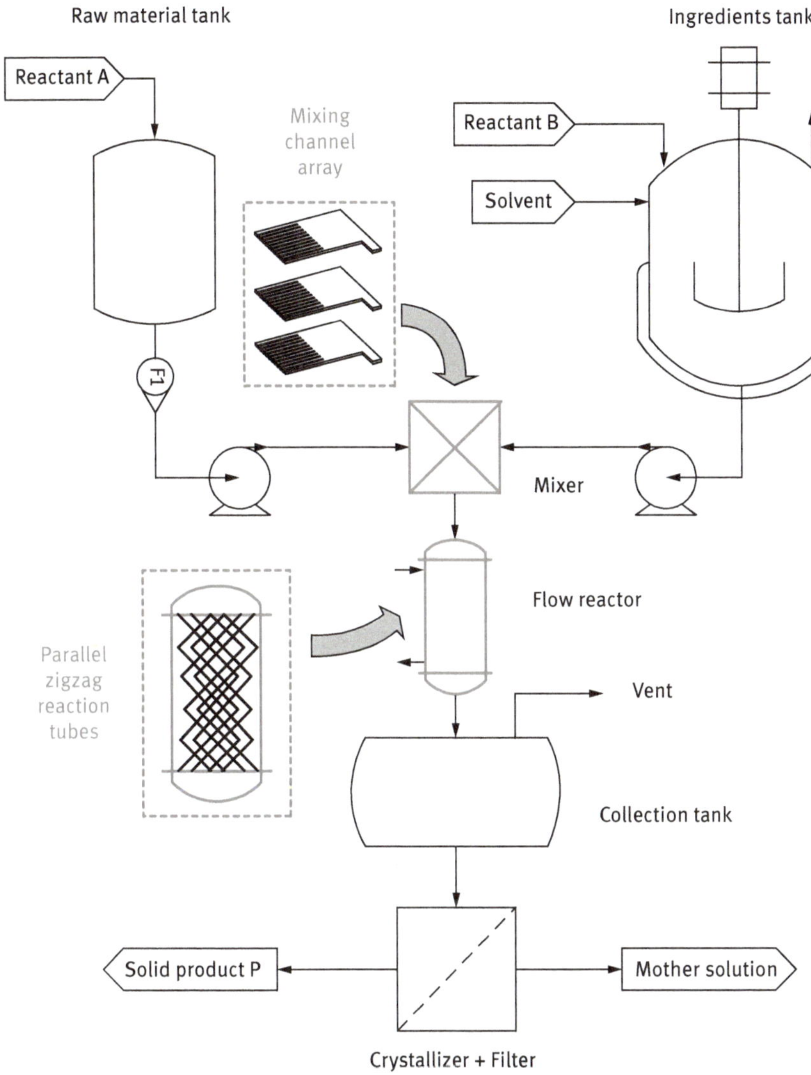

Fig. 8.2: Schematic of a pilot-plant flow reaction system with numbering-up core components.

is not as difficult as in traditional chemical engineering scale-up processes, especially for producing millions of tons of chemicals per year. We can also gain from the experience of petroleum processing, which have achieved "flow-production" for a long time now. Another advantage of flow chemistry is that the laboratory technical prototypes have achieved continuous operation; therefore, there is no need to worry about technology adjustment during scale-up. The debugging processes in laboratory research are also touchstones of process craft, helping to avoid technical lacunae in industrial processes.

8.2 Scale-up of mixing equipment

8.2.1 Numbering-up of mixing units

Typical mixing structures for flow chemistry systems are cross microchannels, such as T-junction, Y-junction, and cross-junction microchannels, which are curved in microchips, or commercial tube fittings such as T-junction, Y-junction, or cross-junction tube connectors. These structures seem to act like one mixing unit in the scaled-up mixing equipment. T-junction mixing devices are more widely used in laboratories because of their simple structure and ease of fabrication, making them more robust in industrial applications. In the T-junction mixing unit, one fluid flows in the horizontal direction and the other flows through the vertical channel. These fluid contact processes at the T-junction are called "droplet" or "bubble dispersions," for heterogeneous systems, and "cross-flow mixing" for homogeneous systems, both of which lead to micrometer scaled fluid contact or mass transfer of substance. To maintain the mixing performance of the T-junction mixing in the scale-up, the simplest method is to use multiple mixing units, which means that many identical mixing units are integrated in one device, similar to the method for manufacturing electronic chips. Figure 8.3 shows a typical multiple-units approach: parallel arrays, cycling arrays, and array stacks [9]. Generally, the mixer has two inlets and one or more outlets, and the inlet fluids are divided into many child fluids by the distributor, and then mixed at the junctions. Numbering-up is an efficient scaling-up strategy, especially for microfluidic applications, or pharmaceutical production with low throughput. However, with only numbering-up of mixing units it is difficult to meet the requirement for general fine chemicals. For example, for a chemical product with a throughput of 1 kt a^{-1} (working for 7,200 h a^{-1}), if the mass fraction of the product in the solvent is 10% and the solution density is 1,000 kg m^{-3}, a scale-up of a 1 mm^2 cross-section mixing unit with an inner flow rate of 0.1 m s^{-1} requires 3,858 mixing units, which are difficult to assemble. For a chemical that have a greater than 10 kt a^{-1} requirement, the numbering-up of mixing units is unacceptable. Therefore, other scaling-up methods, should be implemented to assist the numbering-up of the mixing units, necessary to meet the high requirements of the industry.

8.2.2 Similarity-up of T-junction mixing unit

Similarity-up of a mixing unit means the enlargement of the mixing unit geometry without affecting the mixing or mass transfer characteristic dimension of the mixing process. This method has been widely used in the scale-up of conventional chemical equipment, such as reaction tanks or columns. However, general methods cornering to maintain the energy input per unit equipment volume as in the laboratory experiment, or maintain the local Re number in the scaled-up device, cannot be directly

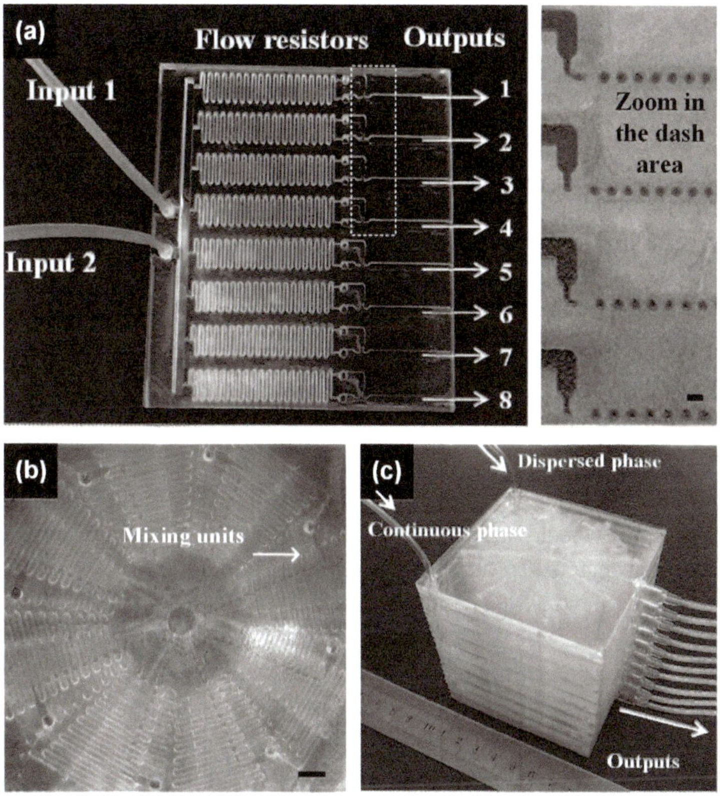

Fig. 8.3: Photos of numbering-up of T-junction mixing unit, which is adapted with permission [9]. Copyright 2017, Elsevier. (a) The numbering-up of T-junction microchannels with a parallel array and (b) a cycling array. (c) Stacking of cycling microchannel junction arrays.

used in flow chemistry processes because the droplet/bubble dispersion or substance mixing in the small space have their unique rules. Taking the mixing of heterogeneous fluids in the T-junction microchannel as an example, since the mixing performance is significantly related to the bubble or droplet size in the mixer, our aim is to maintain the bubble or droplet dispersion sizes in scaling-up the mixer. In our previous study [10], the average bubble or droplet diameters in similar T-junction mixing units with square cross-section straight mixing channels, and round cross-section branch channels, can be normalized and correlated to the following semiempirical equations, where the capillary number representing the competition between viscous force and interfacial force was employed:

$$\frac{d_{av}}{d_e}=0.33\left(\frac{\mu_D}{\mu_C}\right)^{0.1}\left(\frac{Q_D}{Q_C}\right)^{0.03}Ca_C^{-0.25}\ \text{at}\ 0.005<Ca_C<0.5, \mu_D/\mu_C<1 \quad \text{for liquid – liquid system} \tag{8.6}$$

$$\frac{d_{av}}{d_e}=0.26\left(\frac{\mu_D}{\mu_C}\right)^{-0.05}\left(\frac{Q_D}{Q_C}\right)^{0.03}Ca_C^{-0.25}\ \text{at}\ 0.008<Ca_C<0.2, \mu_D/\mu_C<200 \quad \text{for gas – liquid system} \tag{8.7}$$

$$Ca_C=\frac{\mu_C u_C}{\gamma} \tag{8.8}$$

where μ is the viscosity, u is the average velocity, γ is the interfacial tension, Q is the flow rate, and d_e is the diameter of the round cross-section branch channel. Subscripts C and D represent the continuous and dispersed phases, respectively. In the scale-up of the mixer, the fluid viscosity and phase ratio should be maintained; therefore, the fluid velocity and diameter of the round cross-section branch channel are the only parameters. To obtain the droplet or bubble diameters in the scaling-up process, we believe that the maintenance of Ca_C and d_e is the most important.

To break the geometry constraints of the mixing channel, we developed a microsieve dispersion unit, as shown in Fig. 8.4(a). In this mixing unit, multiple branch channels from the T-junction mixing unit are simplified by a microsieve pore array on the side of a millimeter-scale mixing channel [11]. Although there is approximately 15% error in determining the average sizes of bubbles and droplets in the increased mixing unit by using equations (8.6) and (8.7), we can still conclude that the similarity-up of the mixing unit is successful for the acceptable amplification effect. If people want to use a higher d_e to prevent channel blocking in industry, higher fluid velocity can be considered to reduce the bubble or droplet sizes. To further develop the microsieve dispersion mixing unit, the membrane dispersion unit, as shown in Fig. 8.4(b), was developed in Tsinghua, which has a higher treating capacity of the dispersed fluid for the high porosity of the membrane. The mixing or bubble/droplet dispersion rules in the membrane dispersion unit are similar to the microsieve dispersion unit, but the irregular pore shape will lead to a wider size distribution of bubbles and droplets. Therefore, it is usually used for a high flow rate mixing process that has a high mixing energy input, where the wide size distribution of the dispersed phase has a relatively weak effect on the total mass transfer rate in the working system. The pores in the membrane can range from 5 to 200 μm in size, which create tens to hundreds of micrometer bubbles and droplets in the mixing units [7, 8].

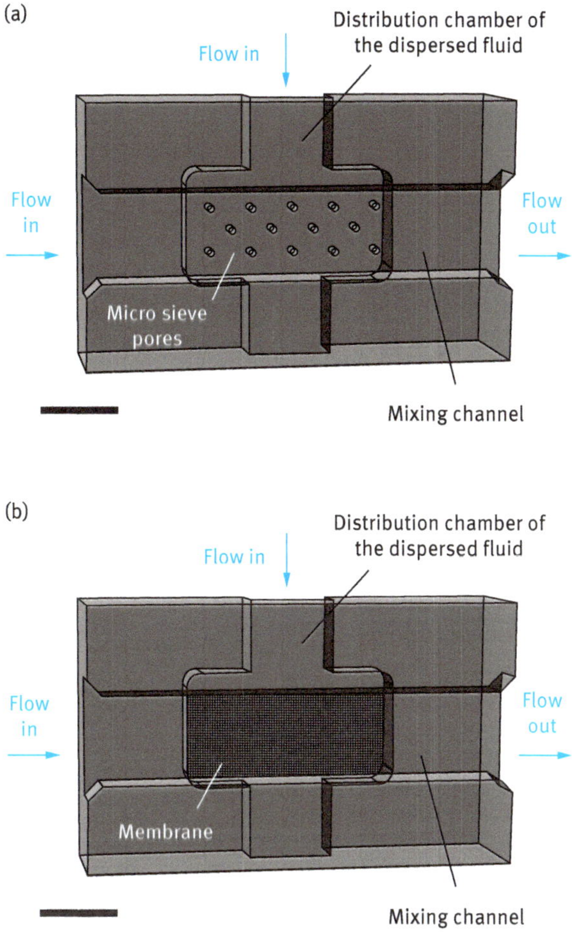

Fig. 8.4: Schematics of the mixing units with high working capacities. (a) Micro-sieve dispersion unit for fluid mixing in a channel. (b) Membrane dispersion unit for fluid mixing in a channel. Scale bars are 1 cm in length.

8.2.3 Fluid distributors in enlarged mixers

Fluid feed is crucial for reactant mixing in the flow chemistry process; however, it is uneconomical and impractical to replace many pumps in parallel for every individual mixing unit. Therefore, uniform fluid distribution is another core technology for the scale-up of flow chemistry equipment. The fluid distributors are typically designed to be of two types: wide distribution chambers, or branched feeding tubes, as shown in Fig. 8.5. The distribution chambers in Fig. 8.5(a) show a typical example of the flow fluid distribution between different T-junction mixing units. To obtain equal flow rates in all mixing units, the chamber volume and the resistances of the channels should be carefully considered. Figure 8.6(a) shows the resistance analysis for the mixing unit and the fluid distribution chamber. According to this analysis, we can find that the smaller the resistance in the chamber and the larger the flow resistance in the parallel mixing channels, the more uniform is the distribution of flow rates. An empirical equation that guides the design of the distributor is as follows [12]:

$$2n\frac{R_c}{R_u} < 0.01 \tag{8.9}$$

where R_c is the local resistance in the distribution chamber, R_u is the resistance through a mixing unit, and n is the number of mixing units. The risk of channel blockage can be avoided by using a wide distribution chamber; however, one fluid can flow into the chamber of the other fluid in an abnormal pressure situation, which can probably cause unpredictable reaction results.

The tree-like branched geometry in Fig. 8.5(b) is another commonly used distributor for fluids. To guarantee uniform fluid distribution and reduce the pressure drop as much as possible, Murray's law [13], which widely appears in the transport systems of plants and animals, can be used for the design of the tube radius. As shown in Fig. 8.6(b), for connecting a parent tube (radius r_0) to child tubes (radius r_i) without mass variation (equation (8.10)), a quantitative relationship between the radius of the parent tube and child pipes for laminar flow can be expressed by equation (8.11):

$$Q_0 = nQ_i \tag{8.10}$$

$$r_0^3 = nr_i^3 \tag{8.11}$$

where Q_0 is the total flow rate and Q_i is the flow rate in each mixing unit. This type of fluid distributor can evenly and symmetrically divide fluids into branched channels and does not need to consider the resistance of each mixing unit. However, significant space is required in the placement of the branching devices, and clogging of one channel in the branching structures will significantly affect the flow distribution due to the break in symmetry.

8.2.4 Package and connection of mixer

In addition to the parallel scale-up of core mixing units, packages and connections are necessary to apply the scaled-up micromixer in the industrial process. For hundreds of flow paths in a mixer, the mixing units must be stacked into arrays to make the device compact, and a shell has to be made to pack those arrays for sealing and withstanding the internal pressure of the fluid. This package process can be configured by borrowing the idea from commercial static mixers with packing and tubular shells. Figure 8.7

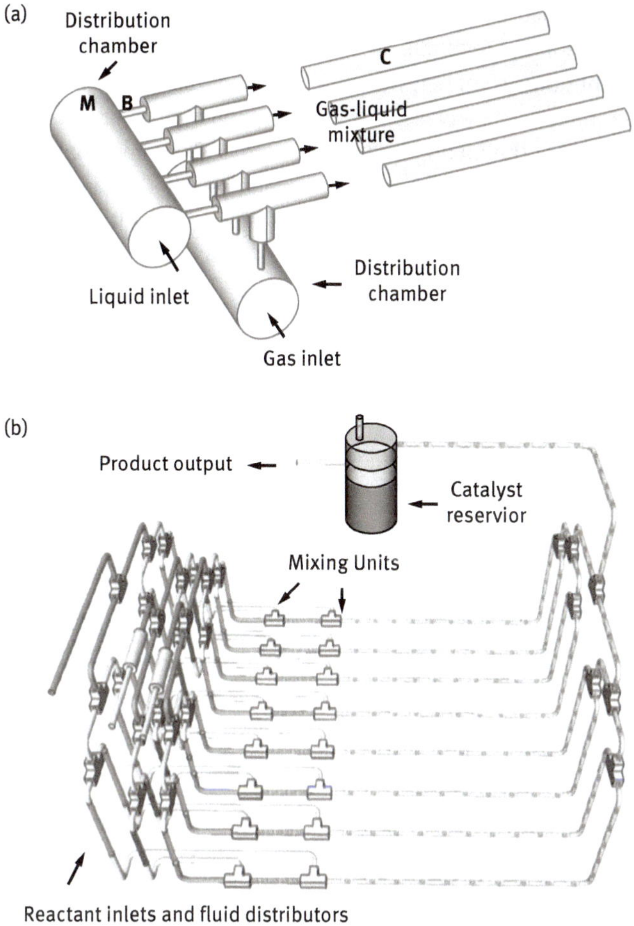

Fig. 8.5: Typical methods of fluid distribution between parallel mixing units in flow chemistry systems. (a) The method of using wide fluid distribution chambers for gas and liquid reactants to be mixed in four T-junction units. Figure is adapted with permission [11]. Copyright 2014, American Institute of Chemical Engineers. (b) The method of using branched tubes as the distributor of double T-junction mixing units, adapted with permission [12]. Copyright 2017, Elsevier.

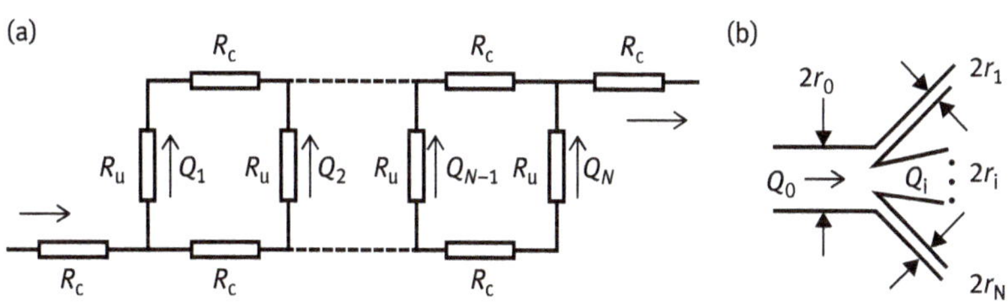

Fig. 8.6: Flow rate and resistance analysis for typical fluid distribution structures. (a) Resistance analysis of mixing units and the wide distribution chambers [12]. (b) Schematic of Murray's law [13].

shows a pilot-plant mixer developed by Tsinghua, which has a treatment capacity of 1 $m^3\,h^{-1}$. The microsieve dispersion mixing units are stacked in the stainless-steel shell, which integrates a wide distribution chamber. Standard DN 25 flanges are welded to the shell; therefore, the micromixer can be connected to other industrial chemical devices. Since there is a shell to protect the mixing units, both hard and soft materials can be chosen to make these mixing units, such as stainless steel, polyether-ether-ketone, polytetrafluoroethylene, and Teflon. The shell material is also multiselective. For example, steel-sprayed ethylene-tetrafluoroethylene is a good material for shell to resist corrosion.

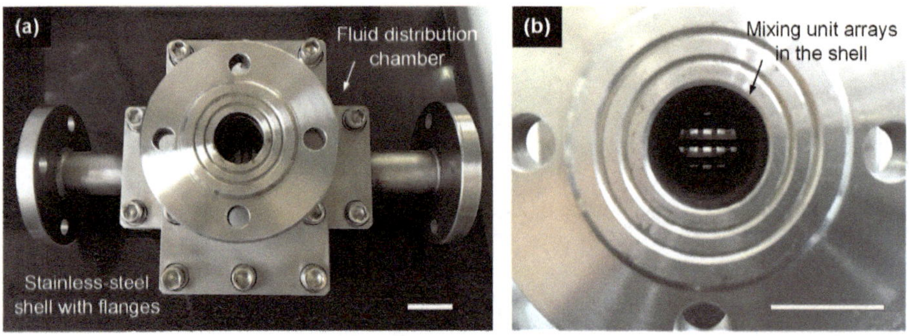

Fig. 8.7: Photos a micromixer from Tsinghua with a stainless-steel shell. (a) The outward of the mixer with DN 25 flanges. (b) Looking at the mixing unit arrays through a flange. Both scale bars are 25 mm in length.

8.3 Scale-up of reaction tubes and channels

In the above section, we find that the most important factors to maintain good mixing and mass transfer performances of flow chemistry equipment in their scaling-up process, is to maintain the dispersion scale of fluid elements, such as bubble or droplet average diameters, and similar flow conditions in differently sized devices. The principle for the reaction tube or channel scale-up follows the same method. In the scale-up of reaction tube, researchers focus more on the heat exchange performance between the reactor and the environment. Therefore, maintaining the characteristic scale, which determines the heat transfer area per unit volume of the reactor, is straightforward for the diameter of reaction tubes or channels. In addition to the heat transfer performance, the propagation distances of light in the photochemical reactors, and the resistance in electrochemical reactors, are also limiting factors for the increase in the reaction tube diameter. Thus, it is sometimes difficult to scale up reaction tube geometries. In this situation, parallel repeated reaction tubes or plates containing channels have to be applied.

8.3.1 Numbering-up of reaction tubes for exothermic reactions

The excellent heat transport ability is one of the biggest advantages of flow chemistry equipment in laboratory research. Many exothermic reactions that are usually carried out by batch reactors with slowly fed reactants are therefore continuously implemented by the flow chemistry method with high yield, and safety. With the enhancement of reactant mixing, the heat exchange demand for a flow reaction tube or capillary also becomes higher for the increased release of reaction heat by accelerating the apparent reaction rate:

$$\frac{dQ_h}{dL} = -\Delta H_r \cdot Q \cdot \frac{dc_A}{dL} \tag{8.12}$$

In equation (8.12), L is the reaction tube length, Q is the volume flow rate of the reacting solution, and ΔH_r is the reaction enthalpy. According to the convective heat transfer theory shown by equation (8.3) in the above section, the heat transfer rate Q_h is dominated by the heat transfer coefficient K_h and heat transfer area A at a constant temperature difference between the chemicals and HEM. Both parameters are determined by the tube or channel diameter, which is a characteristic of the scale in the investigations. Therefore, the reaction tube or channel diameter cannot be increased infinitely. Parallel scale-up is the best choice, but extremely small reaction tubes, such as capillaries, are difficult to mass-assemble for a large reaction capacity. Thus, heat transfer enhancement methods can be employed to enhance the heat exchange between parallel reaction tubes and HEM.

Using a coiled reaction tube is an effective method to enhance substance mixing, and it is also effective to enhance the heat transfer performance by employing Dean Flow. The heat transfer coefficient of the coiled reaction tube can be evaluated by the semiempirical models related to the Nusselt number. Equations (8.13) and (8.14) show models of the Nusselt numbers inside and outside of the coiled tubes provided by Nigam et al. [14]:

$$\mathrm{Nu_{in}} = \begin{cases} 0.08825\mathrm{Re}^{0.7}\ \mathrm{Pr}^{0.4}(D/d)^{-0.1} & \text{at Re} < 10{,}000 \\ 0.0271\mathrm{Re}^{0.85}\ \mathrm{Pr}^{0.4}(D/d)^{-0.1} & \text{at Re} > 10{,}000 \end{cases} \tag{8.13}$$

$$\mathrm{Nu_{out}} = 15.44\ \mathrm{Re}^{0.345}\ \mathrm{Pr}^{0.33} \quad \text{at } 1{,}000 < \mathrm{Re} < 3{,}000 \tag{8.14}$$

where Nu is the Nusselt number, Nu = hd/k, h is the heat transfer film coefficient, d is the tube diameter, D is the curvature diameter of the coiled tube, and k is the thermal conductivity. Figure 8.8(a) shows a schematic of the parallel scale-up of coiled reaction tubes assembled in a vessel containing HEM. The magnified tubular reactor can be operated like a heat exchanger with co-current or counter-current heat transfer process. In 2006, Yoshida's group reported a prototype of parallel scale-up of reaction tubes for methyl methacrylate polymerization, as shown in Fig. 8.9(a), which had a capacity of 4.0 kg polymethyl methacrylate per week [15]. In the parallel scale-up of the reaction tube, the heat transfer rate should be fit for the reaction heat release rate, which is determined by the reaction kinetics. To carry out fast exothermal reactions, the tube should be thin enough to prevent the high hot-point temperature along the tube. To carry out relatively slow reactions, the tube diameter is allowed to be increased to a few millimeters. In this situation, some commercialized heat exchangers with millimeter-sized zigzag-shaped tubes, as shown in Fig. 8.8(b), can be used instead of the coiled tubes. Photos of these heat exchange tubes can be found on the reference website [16].

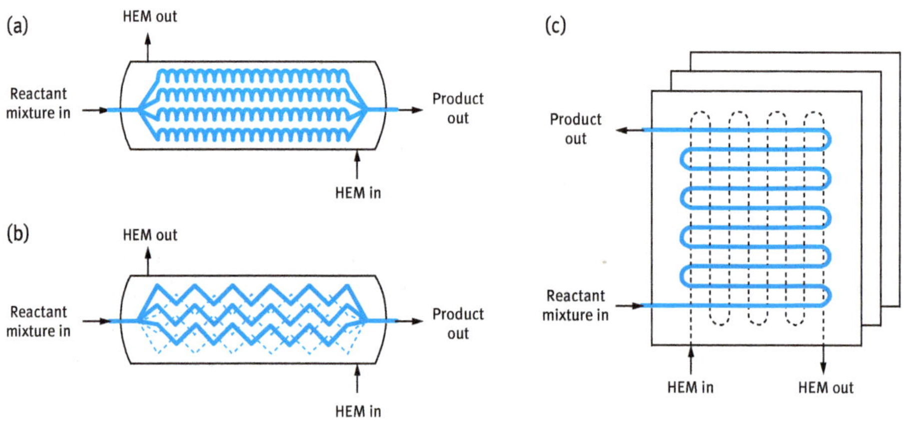

Fig. 8.8: Schematic of different parallel scale-up methods for reaction tubes and channels. (a) Parallel coiled tubes in a heat exchanger. (b) Parallel zigzag tubes in a heat exchanger. >(c) Numbering-up of reaction channel plates and heat exchange channel plates with sandwich structures.

In addition to tubular reactor structures, parallel plates with reaction channels and heat exchange channels have also been considered in scale-up research. In contrast to the reaction tubes with round cross sections, the cross sections of reaction and heat exchange channels as well as their patterns can be customized with different aspect ratios and structures to fit the demand of reaction kinetics and the heat transfer rate. Another advantage of the sandwich structure of the parallel plates, as shown in Fig. 8.8(c), is that they are easy to repair and clean with removable parts. Figure 8.9(b) shows a pilot-plant reactor from Kockmann's group [17], to understand the mixing and heat transfer characteristics of stacked reactor plates. A modified Gnielinski correlation was then proposed by them [18], as follows:

$$\mathrm{Nu_{Gu}} = \begin{cases} \dfrac{(f/8)(\mathrm{Re}-1{,}000)\,\mathrm{Pr}}{1+12.7(f/8)^{1/2}\left(\mathrm{Pr}^{2/3}-1\right)} & \text{at } 3{,}000 \le \mathrm{Re} \le 5\times10^{6}, 0.5 \le \mathrm{Pr} \le 2{,}000 \\[2ex] \dfrac{(f/8)(\mathrm{Re}-141)\,\mathrm{Pr}}{1+18.5(f/8)^{1/2}\left(\mathrm{Pr}^{2/3}-1\right)} & \text{at } 440 \le \mathrm{Re} \le 800, \text{for } 5-90\ ^\circ\mathrm{C water} \end{cases} \tag{8.15}$$

where f is the friction factor. Currently, parallel scale-up reaction channels have been commercialized. Corning Inc. reported their G5 Advanced-Flow™ Reactor in 2020, which has an annual treating capacity of 10,000 tons reacting solution for diazotization reaction [19].

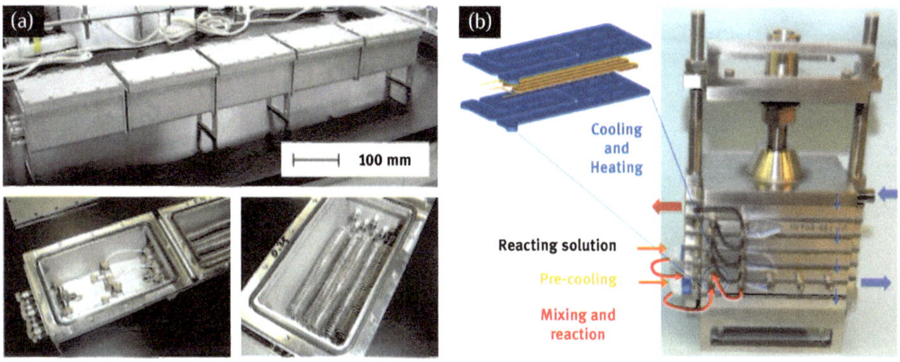

Fig. 8.9: Photos of parallel scale-up for coiled tube reactor and plate structures reactor. (a) Methyl methacrylate polymerization reactor from Yoshida's group, adapted with permission [15]. Copyright 2006, American Chemical Society. (b) Pilot-plant investigation reactor from Kockmann's group, adapted with permission [17]. Copyright 2011, Elsevier.

8.3.2 Numbering up of photochemical and electrochemical flow reactors

Similar to the above tubular flow reactors for exothermal reactions, photochemical and electrochemical flow reactors, which are the new technologies for sustainability and environmental friendliness of chemical processes, are also channel size controlled equipment. One of the most important reasons for applying the flow chemistry method in photochemical reactions is the Lambert–Beer law, which exhibits the variation of light absorbance with the light incidence depth, as follows:

$$A = \log\left(\frac{I_0}{I_t}\right) = K \cdot c \cdot l \tag{8.16}$$

where A is the light absorbance, I_0 is the incident light intensity, I_t is the transmitted light intensity, K is the absorption coefficient, c is the substance concentration, and l is the light path length. According to this law, the diameter of the reaction tube or the depth of the reaction channel should be confined by l to prevent low light intensity in the deep solution. Parallel transparent reactor plates made from glass or quartz have been used for the commercial scale-up of photochemical reactors. In 2017, Su and Noël reported a novel scale-up method for photochemical reactors using coiled reaction tubes, as shown in Fig. 8.10 [20]. This method is easy to run with commercial LED light sources and has the advantage of lower production cost of the scaled-up reactor. A similar scale-up method has also been used for the parallel increase of other chemical reactions [21].

Fig. 8.10: Photos of parallel scale-up for a photochemical reactor from Noël's group, adapted with permission [20]. Copyright 2017, the Royal Society of Chemistry.

For the scale-up of electrochemistry reaction equipment, ensuring that the distance between the electrodes is at a suitable value, is also important. Most electrosynthesis reactions are implemented in weakly polar solutions, with poor ionic transport ability. The low electrical conductivity causes high resistance in the electrical loop, which increases the energy cost and heat loss. According to the theory of ionic transport in solution, the electrical potential required to overcome the cell resistance is expressed as follows:

$$U = \frac{I \cdot d_e}{A_e \cdot \kappa} \tag{8.17}$$

where I is the electrical current, d_e is the distance between electrodes, A_e is the surface area of the electrode, and κ is the solution conductivity [22]. In addition, the distance between electrodes also dominates the specific surface areas of the electrodes in the electrochemical reactor. In many electrochemical processes, supporting electrolytes, such as Bu_4NBF_4 and Bu_4NPF_6, are used to increase the conductivity, which, in turn, increase the difficulty of product purification after reaction. However, the electrical resistance is automatically reduced by decreasing the electrode distance. To maintain the electrically conductive state of the reactor, parallel scale-up is the best choice. As we write this chapter, there is still inadequate data on the industrial applications of electrosynthesis flow devices. However, a similar scale-up method in the flow battery has been applied in the industrial process. Figure 8.11 shows photos of 100 and 240 kW for all vanadium flow battery energy storage systems developed by Baoguo Wang from Tsinghua University. This type of flow battery is a good template for scale-up of flow electrosynthesis devices.

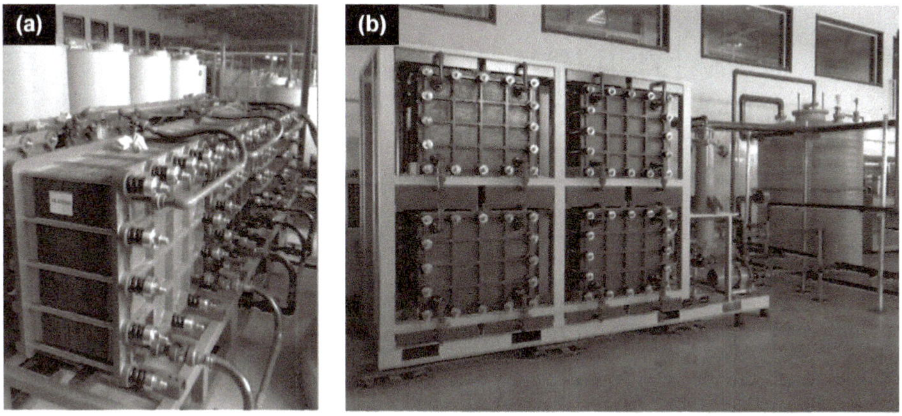

Fig. 8.11: All vanadium redox flow battery energy storage systems: (a) 100 kW and (b) 240 kW. Pictures are authorized by Baoguo Wang from Tsinghua University.

8.3.3 Fluid distributors of reaction tubes and channels

The fluid distributors for parallel increased reaction tubes and channels are basically the same as we have discussed in the second section for the numbering-up of mixing units. Both the methods of employing fluid distribution chambers and branched fractal pipelines, as shown in Fig. 8.12, can be used for diversion and convergence of fluids. In contrast to the mixing units in the micromixers, the reaction tubes or channels can be as long as a few meters; therefore, the flow resistance reduces on tubes or channels, which is beneficial for the uniform distribution of reactants. For the parallel scale-up of reaction units, monitoring the working status of each reaction tube and channel is still challenging. Tonomura at Kyoto University has done a lot of work for the blocking diagnostic analysis of reaction channels based on system engineering models and CFD simulations [23].

8.4 Coupling of microequipment and conventional equipment

In the above sections, the parallel scale-up of the core components in flow chemistry system was introduced. More methods can be found in the literature [26, 27], or created, based on researchers' innovation. In addition to the core equipment to enhance the reaction or separation processes, the combination of microequipment and conventional equipment is also important for the scale-up of flow chemistry processes. As mentioned in the first section, hybrid scale-up is more realistic for achieving continuous chemical

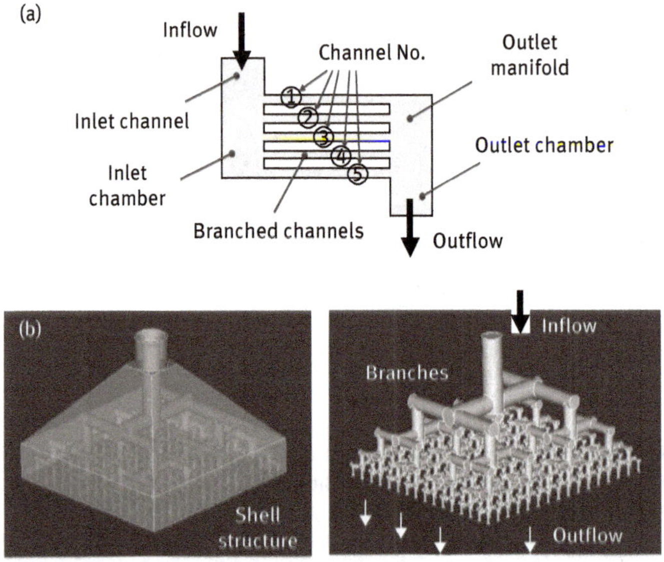

Fig. 8.12: Schematics of the distribution structures of reaction tubes and channels. (a) Fluid distributor by manifolds, adapted with permission [24]. Copyright 2004, Elsevier. (b) Branched fluid distributor, adapted with permission [25]. Copyright 2004, Elsevier.

syntheses in industrial processes with hundreds of tons of products per year. Conventional chemical equipment is a good assistant for microequipment. It can help the microdevices in extending residence time, providing continuous mixing, storing solid components, or realizing phase separation. From a process point of view, conventional equipment is also crucial for substance dissolution, feeding, distillation, filtration, and centrifugation, which have been widely used in traditional unit operations. Since the application of these chemical equipment is mainly based on manufacturing technology selection, and they have been commercialized for a long time, we will not provide their introduction in this chapter. In the following section, we will focus on the coupling of microdevices and conventional equipment or components, which helps to expand the application of flow chemistry technology in the industrial process.

8.4.1 Integration of micromixer with tubular reactor

In laboratory research, small tubes or capillaries are commonly used to provide sufficient reaction time after mixing the reactants. In industrial applications, tubular reactors work in the same way. Only ultra-fast reactions can be finished in micromixers, such as those in flash chemistry processes. Therefore, the combination of a mixer and tubular reactor can also be used in industrial processes. If the reaction is strongly exothermic, parallel reaction tubes have to be employed, as mentioned in the above sections. However, if the reaction is not exothermic, such as the halogenation reaction of butyl rubber [28] that we will introduce in the next section, or can be implemented as a self-heated process, such as the synthesis of the *trans*-2-hexenal intermediate that we have mentioned [29], one thick reaction tube can be applied downstream of the micromixer instead of numbering-up the reaction tube. Figure 8.13 shows a schematic of this combination. This type of flow reaction system is more useful for homogeneous reaction systems that require minute reaction times. If the dispersed state of the multiphasic reaction system is stable at this time, such as emulsified oil–water mixtures or nanoparticle suspensions, the integration of the micromixer and tubular reactor is also the most convenient method for scale-up. In this kind of integrated reaction system, the substance residence time can be strictly controlled by the tube volume and flow rate of the reactant, as shown by equation (8.18), with narrow distributions for the negligible volume of the micromixer:

$$t_r = \frac{\pi D^2 L}{4\beta \sum Q} \tag{8.18}$$

where t_r is the average residence time, D is the tubular reactor diameter, and β is the volume expansion coefficient of the reaction system. In many cases, the combination of micromixer and tubular reactor can be repeated in order to carry out multiple feeds of reactants, or quench reactions.

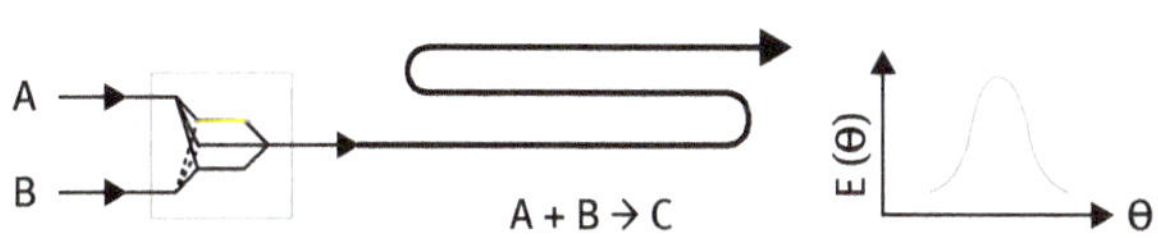

Fig. 8.13: Schematic of the combination of parallel scaled reactant mixer and a reaction tube for the reaction to generate substance C. A schematic diagram for the residence time distribution is shown on the right side with normalized residence time $\theta = t/t_r$.

8.4.2 Integration of micromixer with packed bed reactor

Similarly, the combination of a micromixer and packed bed reactor is also based on the same principle of the integration of micromixers and tubular reactors, but it is more useful for heterogeneous systems, such as the liquid–liquid Beckmann rearrangement reaction of cyclohexanone oxime [30], which can easily appear as phase separation in the tubular reactor, or if the reaction process requires a packed catalyst, such as the hydrogenation reaction in pharmaceutical applications [31]. Bead materials are more widely used in lab-scaled packed bed reactors, but they can be turned into other shapes during the scaling-up. The packing materials are mainly used to maintain the mixing or dispersion states in the reaction system, if they do not work as catalysts. Therefore, the random, or structured packages, in traditional extraction or absorption columns, can be used and improved, based on the concepts of bubble or droplet break-up in confined spaces, as shown in Fig. 8.14. In addition, to narrow the residence time distributions, the packed bed reactor is usually vertically hold, which is different from the straight or coiled reaction tubes used for homogeneous reactions.

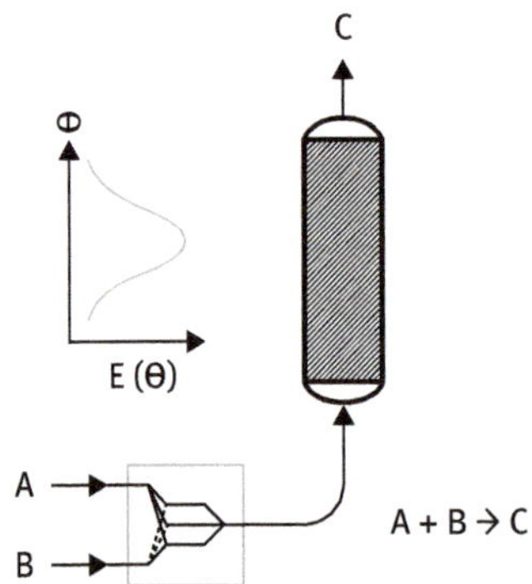

Fig. 8.14: Schematic of the combination of a parallel scaled reactant mixer and a packed bed reactor for the reaction of A + B → C.

8.4.3 Integration of micromixer with stirred tank reactor

In contrast to the combination of micromixer and tubular reactor, or the combination of micromixer and packed bed reactor, the coupling of a micromixer and a stirred tank reactor brings a wider residence time distribution, as shown in Fig. 8.15. The strong back mixing in the tank makes the unturned reactant and reaction product fully blended, which is sometimes harmful to the yield or selectivity of the reaction system containing competing side reactions. However, the stirred tank reactor can provide a large volume for the reaction system and continuous energy input of mixing, which is beneficial for the precipitation reactions or aging process for some organic reactions that require a long time. In the next section, we present an example of the preparation of $CaCO_3$ nano particles in a flow reaction system with a micromixer and a stirred tank reactor. The micromixer is used to mix $Ca(OH)_2$ solution and CO_2 quickly to realize explosive nucleation, and the following tank is used to complete the crystal growth and $Ca(OH)_2$ dissolution. In addition, the combination equipment of the micromixer and stirred tank reactor can be operated in a semicontinuous manner. In this process, the reactant can be fed into the micromixer quickly, and the reactant mixture then finishes the reaction slowly with a long, aging reaction, similar to the conventional batch reactor. The application of such a reaction system is relatively special. We developed a reaction system several years ago in the synthesis of polyvinyl butyral [32]. As a fast precipitation reaction between polyvinyl alcohol and *n*-butyraldehyde, a micromixer was used to enhance the contact of reactants. After the precipitation of polyvinyl butyral with a low conversion of hydroxyls, 4 h aging reaction was implemented in the stirred tank reactor for the reaction between polyvinyl butyral particles and *n*-butyraldehyde. The combination of the micromixer and stirred tank adapted well with the kinetic characteristics of the polyvinyl butyral synthesis reaction, and it is easy to implement a scale-up.

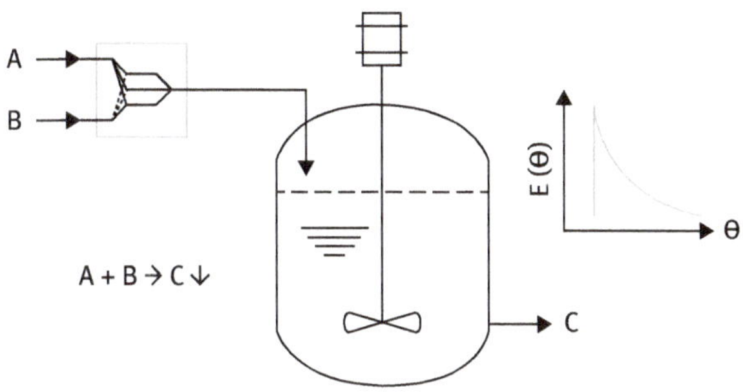

Fig. 8.15: Schematic of a parallel scaled reactant mixer and a continuous stirred tank reactor combination for the reaction of A + B → C with C as a solid.

8.5 Examples of flow chemistry systems in industry or pilot plant

At the end of this chapter, let us see some examples of flow chemistry systems that have been used in the industry, or as pilot-plant reaction systems. These examples basically cover the concepts we have discussed above, but they are still being improved and finalized.

8.5.1 Butyl rubber bromination microreaction system

Bromobutyl rubber is an airtight material for rubber products such as radial tires and bottle stoppers. They are prepared from the reaction between butyl rubber and bromine in *n*-alkane solutions [28], as shown in Fig. 8.16. In the preparation of bromobutyl rubber, a large stirring tank containing bromine and alkane solvent is the source of risk to the plant; therefore, it is best to continuously carry out the butyl rubber bromination reaction in a compact reaction system with a small reactor volume. In 2019, the research group form Tsinghua University developed a continuous flow reaction platform in their collaborating enterprise, which had a capacity of 4 tons bromobutyl rubber per hour. The reaction system was started from a micromixer containing 288 mixing units, which achieved a sandwich-type multilayer flow for mixing the high-viscosity butyl rubber solution and bromine solution. A photograph of the industrial micromixer is shown in Fig. 8.17(a). Since the homogeneous bromination reaction does not have an obvious exothermic effect, a tubular reactor was employed downstream of the micromixer to complete the reaction. After the bromination reaction, the generated HBr in the reaction solution was further neutralized by NaOH aqueous solution, which was achieved by the second micromixer connected to the tubular reactor, as shown in Fig. 8.17(b). The micromixer had 320 mixing units with microsieve pore dispersion structures, which generated well-dispersed NaOH droplets in the bromobutyl rubber solution. The mixing solution from the second micromixer was finally collected in a stirring vessel, and further treated for waste and solvent removal.

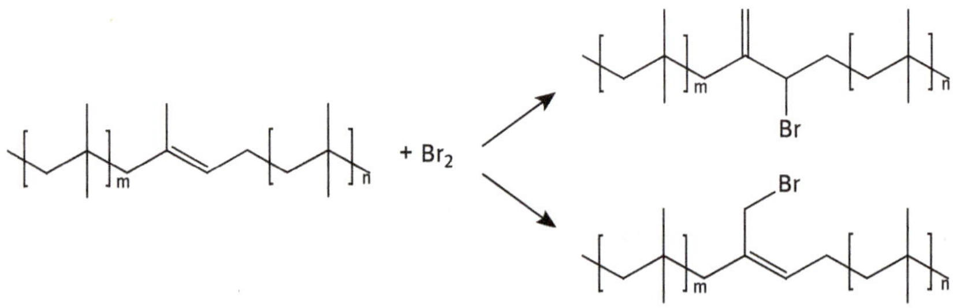

Fig. 8.16: Bromination reaction of butyl rubber.

Fig. 8.17: Photos of the micromixers for (a) butyl rubber and bromine solutions; and (b) bromobutyl rubber and base solutions (photo is from the research group of Tsinghua University in 2014).

8.5.2 Nano-calcium carbonate powder preparation reactor

Nano-calcium carbonate is a powdery material for paper whitening. It is prepared from a simple reaction between carbon dioxide and a calcium hydroxide suspension. However, mixing carbon dioxide with the calcium hydroxide suspension should be fast enough to enhance the nucleation of calcium carbonate, to create 40–60 nm average diameter particles. In 2007, we developed a flow reaction system with a membrane dispersion mixer to achieve nano-calcium carbonate preparation in a reaction system with submillimeter bubbles dispersed in calcium hydroxide suspensions [7]. Since the reaction between carbon dioxide and calcium hydroxide is ultrafast, the reaction is almost complete after bubble generation. Therefore, the membrane dispersion mixer is the core equipment of the reaction platform. Figure 8.18(a) shows the industrial mixers with more than 200 membrane dispersion mixing units in a one-layer mixer. To achieve a capacity of 10 tons $CaCO_3$ per day, six layers of mixer with independent flow rate and CO_2 pressure control systems were combined as a mixer group. Figure 8.18(b) shows a TEM image of the prepared nano-calcium carbonates. Although the parallel mixers are large enough to finish the calcium carbonate preparation reaction, they had difficulty in fully converting all the calcium hydroxide suspensions in one pass because of the low solubility of calcium hydroxide in water. Therefore, we designed the entire reaction into a semicontinuous process with recycled slurries, as shown in Fig. 8.18(c). The stirring tank was used to dissolve calcium hydroxide into water and collect the product of the reaction. After the pH in the stirred tank drops to less than 8, the reaction is believed to be completed and the nano-calcium carbonate suspension is exhausted to downstream treatment. It should be noted that the solid calcium carbonate can slowly deposit in the mixers, connection tubes, and the tank. Therefore, the entire system is cleaned with dilute acid regularly.

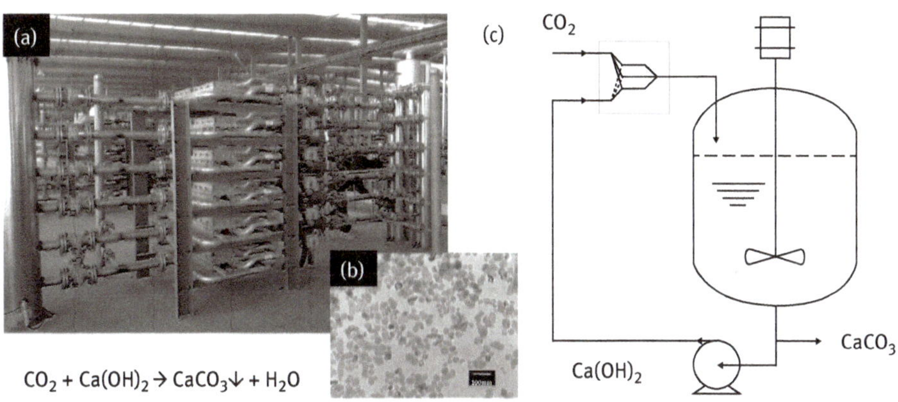

Fig. 8.18: Photos and schematic of the semicontinuous reaction system for nano-calcium carbonates. (a) The parallel increased micromixer for carbon dioxide and calcium hydroxide suspension. Copyright 2020, Wiley. (b) TEM of prepared nano-calcium carbonates. (c) Schematic to show the cycling reaction craft (photos are from the research group of Tsinghua University in 2009).

8.5.3 Cyclohexanone-oxime Beckman rearrangement reactor

ε-Caprolactam (CPL), the monomer of nylon-6, is widely used in the manufacture of fibers and resins. At present, CPL is mainly produced by the Beckmann rearrangement of cyclohexanone oxime (CHO) catalyzed by concentrated sulfuric acid or oleum. Since the reaction is rapid and highly exothermic, a large amount of product circulation is needed to control the temperature. To overcome the local over-temperature and for faster removal of the reaction heat, a heterogeneous process involving an inert solvent was developed. In this process, a solvent containing CHO is employed as the continuous phase and oleum is used as the dispersion phase. Meanwhile, owing to the high differences in density and viscosity of two-phase fluids and large interfacial tension, the dispersion phase tends to coalesce along the reaction tube. In this context, a micromixer combining a packed bed reactor was proposed for this reaction. The micromixer is used for the mixing of two phases by generating tiny oleum droplets in an organic solvent, and the packed bed is used to break up the coalesced oleum droplets to enhance the mass transfer [30]. Figure 8.19 shows a photo of the pilot-plant reaction system designed by Tsinghua, which has a capacity of 100 tons ε-caprolactam per year. The micromixer has 81 mixing units with microsieve pores, and 3 mm stainless steel beads are used to feed the packed bed reactor. Experimental results show that the micromixer and the packed bed reactor completed the reaction in less than 40 s residence time with more than 99% selectivity.

Fig. 8.19: Photo of the micromixer and packed bed reactor for the cyclohexanone-oxime Beckmann rearrangement reaction (photo is from the research group of Tsinghua University in 2019).

8.5.4 Bromo-3-methylanisole synthesis reaction system

The above examples are all chemical processes with single step reactions. Nowadays more and more flow chemistry methods have been used to carry out multistep reactions [33]. Here, we introduced a pilot-plant research of the continuous synthesis of bromo-3-methylanisole, which is an intermediate in the synthesis of heat- and pressure-sensitive dye 2-anilino-3-methyl-6-dibutylaminofluorane [34]. The reaction shown in Fig. 8.20 includes a methylation reaction and a bromination reaction from 3-methylphenol. Both reactions are fast reactions; therefore, the combination of micromixer and reaction tube is a good choice to carry out the reaction.

Figure 8.21 shows the pilot-plant flow reaction system developed by the research group from Tsinghua University and collaborators. The system has a production capacity about 100 kg bromo-3-methylanisole per hour. For the methylation reaction, the reaction between 3-methylphenol sodium solution and dimethyl sulfate was carried out via microsieve dispersion mixer, which had 12 mixing units and a coiled reaction tube. The reaction was controlled at free temperature rising condition for its relatively low reaction enthalpy and the yield of 3-methyl anisole was higher than 94%. Almost pure 3-methyl anisole was obtained from the flow reaction after alkaline washing of the organic phase product, and then the 3-methyl anisole was reacted with Br_2-HBr aqueous solution in the second microsieve dispersion mixer with almost the same structure as the previous one. After the micromixer a following packed bed reactor was employed to finish the reaction at low temperature in a few minutes. The yield of bromo-3-methylanisole (including about 95% 4-bromo-3-methylanisole and 5% 6-bromo-3-methylanisole) was higher than 98%. Since the both the methylation reaction and the bromination reaction, as well as the alkaline washing reaction were fast processes, we continuous operated those reactions in the lab research as shown in Fig. 8.21; however, they were independently operated in the pilot-plant process for the robustness of whole system. Intermediate storage tanks were employed between each step for storing the reactant solutions in temporary, which was important to balance the flow rate fluctuation in the industrial process.

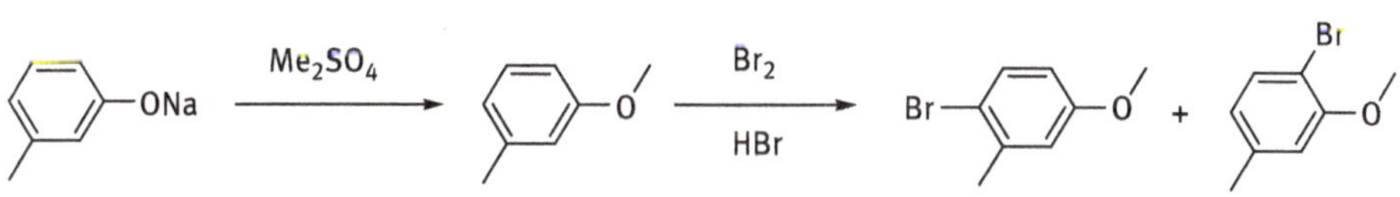

Fig. 8.20: Two-step reaction for bromo-3-methylanisole synthesis.

8.5.5 Food-grade phosphoric acid purification equipment

As the last example in this chapter, an industrial system for substance purification is shown here. This application is for the purification of phosphoric acid from the raw acid material, which is obtained from the sulfuric acid extract of phosphate rock. The aim of this purification is to prepare food-grade phosphoric acid, which strictly limits the contents of calcium, magnesium, silicon, iron, and other impurities at very low levels. Therefore, liquid–liquid phosphoric acid extraction and back-extraction are important unit operations of the total purification process. Fortunately, the extraction and back-extraction of phosphoric acid do not require many stages, but have to be enhanced in mass transfer kinetics for reducing the amount of solvent in extraction devices. Thus, the research group from Tsinghua University and collaborator designed a

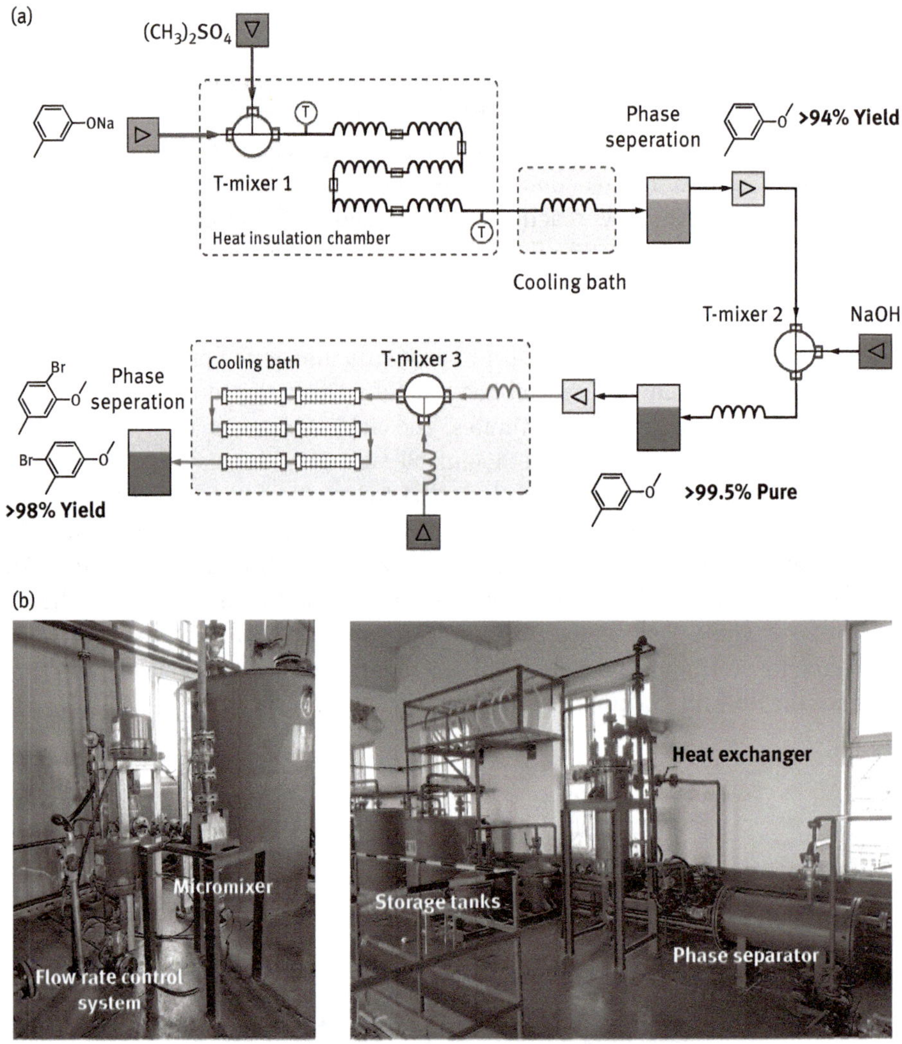

Fig. 8.21: The continuous synthesis of bromo-3-methylanisole with a flow reaction system. (a) Schematics of the continuous flow system in laboratory research. (b) Photos of part of the pilot-plant reaction system (photos are from the research group of Tsinghua University in 2020).

compacted extraction system based on a scaled-up micromixer and a traditional phase separation tank. The extractor and back-extractor have almost the same structure, and each of them employs 600 microsieve dispersion units for the dispersion of phosphoric acid aqueous solution as droplets into the solvent. Both the extractor and back extractor have the capacity to treat 50–70 $m^3\ h^{-1}$ oil–water mixtures, which meet the demand of 75 kt a^{-1} food-grade phosphoric acid. Photographs of the micromixer and back-extraction system are shown in Fig. 8.22.

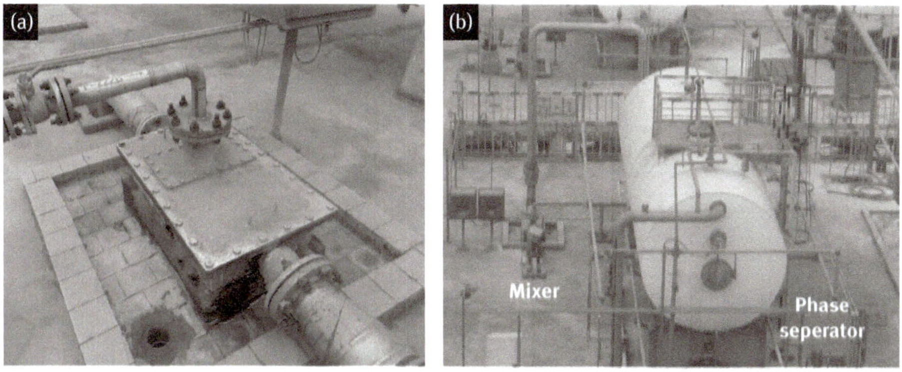

Fig. 8.22: Photos of the micromixer and the corresponding back-extraction device. (a) Micromixer containing 600 microsieve dispersion units. (b) Back-extraction system including the micromixer and phase separation tank (photos are from the research group of Tsinghua University in 2012).

8.6 Summary

This chapter shows the scale-up of flow chemistry equipment and systems, aiming for the industrial application of flow chemistry technology. As the core components of flow chemistry systems, micromixers, and reaction tubes are suggested to be numbered up using the process intensification method, the hybrid method with the combination of microdevices and traditional auxiliary equipment is highlighted as a feasible and reliable scale-up method from today's engineering point of view. Typical examples that have been studied by the authors are introduced in this chapter. There are also other types of flow chemical equipment that others have developed, such as micropacked bed reactors and flow photochemical reactors. These methods can be found in the literature for the reported flow chemical methods being increasingly tested in pilot plant or industrial stages. Therefore, we strongly believe that the industrial scaled flow chemistry system will be more popular, and finally change the R&D and production models of chemical production.

Further readings

- Annual Review of Chemical and Biomolecular Engineering: Zhang, J.S., Wang, K., Teixeira, A.R., Jensen, K.F., Luo, G.S., Design and Scaling Up of Microchemical Systems: A Review. Vol. 8, Annual Reviews. Palo Alto, 2017.
- Microreactors: New Technology for Modern Chemistry: Ehrfeld, W., Hessel, V., Löwe, H. State of the Art of Microreaction Technology. Weinheim, WILEY-VCH, 2000.
- Microreactor Technology and Process Intensification: Wang, Y., Holladay, J. D. Washington, DC, American Chemical Society, 2005.
- Basics of Flow Microreactor Synthesis: Yoshida, J. Kyoto, Springer, 2015.

Study questions

8.1 What is the challenge of employing microstructure devices in chemical engineering separation processes?

8.2 Is a flow chemistry system workable for preparing different chemical products?

8.3 How long does a flow chemistry system run continuously?

References

[1] Paul EL, Atiemo-Obeng VA, Kresta, S. M., Handbook of Industrial Mixing, Science and Practice, John Wiley & Sons, Hoboken, New Jersey, 2004.

[2] Incropera F, Lavine A, DeWitt D, Fundamentals of Heat and Mass Transfer, John Wiley & Sons, Hoboken, New Jersey, 2011.

[3] Marre S, Adamo A, Basak S, Aymonier C, Jensen KF, Design and packaging of microreactors for high pressure and high temperature applications, Ind Eng Chem Res, 2010, 49(22), 11310–11320.

[4] Shen Y, Weeranoppanant N, Xie L, Chen Y, Lusardi MR, Imbrogno J, Bawendi MG, Jensen KF, Multistage extraction platform for highly efficient and fully continuous purification of nanoparticles, Nanoscale, 2017, 9(23), 7703–7707.

[5] Okafor O, Weilhard A, Fernandes JA, Karjalainen E, Goodridge R, Sans V, Advanced reactor engineering with 3D printing for the continuous-flow synthesis of silver nanoparticles, React Chem & Eng, 2017, 2(2), 129–136.

[6] Kralisch D, Kockmann N, Noël T, Wang Q, Novel process windows for enabling, accelerating, and uplifting flow chemistry, ChemSusChem, 2013, 6(5), 746–789.

[7] Wang K, Wang YJ, Chen GG, Luo GS, Wang JD, Enhancement of mixing and mass transfer performance with a microstructure minireactor for controllable preparation of CaCO3 nanoparticles, Ind Eng Chem Res, 2007, 46(19), 6092–6098.

[8] Luo H, Chen GS, G. G., Tan B, Mass transfer performance and two-phase flow characteristic in membrane dispersion mini-extractor, J Memb Sci, 2005, 249(1–2), 75–81.

[9] Han T, Zhang L, Xu H, Xuan J, Factory-on-chip: modularized microfluidic reactors for continuous mass production of functional materials, Chem Eng J, 2017, 326, 765–773.

[10] Wang K (2010). Miniaturization of multiphase reaction processes with micro-structured chemical systems, Ph. D. thesis of Professor Kai Wang from Tsinghua University.

[11] Al-Rawashdeh M, Fluitsma LJM, Nijhuis TA, Rebrov EV, Hessel V, Schouten JC, Design Criteria for a Barrier-based Gas–liquid Flow Distributor for Parallel Microchannels, Chem Eng J, 2012, 181, 549–556.

[12] Yap SK, Wong WK, Ng NXY, Khan SA, Three-phase microfluidic reactor networks – design, modeling and application to scaled-out nanoparticle-catalyzed hydrogenations with online catalyst recovery and recycle, Chem Eng Sci, 2017, 169, 117–127.

[13] Murray CD (1926). The physiological principle of minimum work I: The vascular system and the cost of blood volume. Proceedings of the National Academy of Sciences of the United States of America, 12, 207–214.

[14] Mandal MM, Kumar V, Nigam KDP, Augmentation of heat transfer performance in coiled flow inverter vis-à-vis conventional heat exchanger, Chem Eng Sci, 2010, 65(2), 999–1007.

[15] Iwasaki T, Kawano N, Yoshida J, Radical polymerization using microflow system: numbering-up of microreactors and continuous operation, Org Process Res Dev, 2006, 10(6), 1126–1131.

[16] http://www.chemdodgen.com/page/list-33.html
[17] Kockmann N, Roberge DM, Scale-up concept for modular microstructured reactors based on mixing, heat transfer, and reactor safety, Chem Eng Process, 2011, 50(10SI), 1017–1026.
[18] Mielke E, Plouffe P, Koushik N, Eyholzer M, Gottsponer M, Kockmann N, Macchi A, Roberge DM, Local and overall heat transfer of exothermic reactions in microreactor systems, React Chem & Eng, 2017, 2(5), 763–775.
[19] https://www.corning.com/worldwide/en/about-us/news-events/corning- advances-flow-reactor-technology-for-industrial-chemical-production.html
[20] Kuijpers KPL, Van Dijk MAH, Rumeur QG, Hessel V, Su Y, Noël T, A sensitivity analysis of a numbered-up photomicroreactor system, React Chem & Eng, 2017, 2(2), 109–115.
[21] Qiu M, Zha L, Song Y, Xiang L, Su Y, Numbering-up of capillary microreactors for homogeneous processes and its application in free radical polymerization, React Chem & Eng, 2019, 4(2), 351–361.
[22] Noël T, Cao Y, Laudadio G, The fundamentals behind the use of flow reactors in electrochemistry, Acc Chem Res, 2019, 52(10), 2858–2869.
[23] Tanaka Y, Tonomura O, Isozaki K, Hasebe S, Detection and diagnosis of blockage in parallelized microreactors, Chem Eng J, 2011, 167(2–3), 483–489.
[24] Tonomura O, Tanaka S, Noda M, Kano M, Hasebe S, Hashimoto L, CFD-based optimal design of manifold in plate-fin microdevices, Chem Eng J, 2004, 101(1–3), 397–402.
[25] Tondeur D, Luo L, Design and scaling laws of ramified fluid distributors by the constructal approach, Chem Eng Sci, 2004, 59(8–9), 1799–1813.
[26] Shen Q, Zhang C, Tahir MF, Jiang S, Zhu C, Ma Y, Fu T, Numbering-up strategies of micro-chemical process: uniformity of distribution of multiphase flow in parallel microchannels, Chem Eng Process – Process Intensif, 2018, 132, 148–159.
[27] Rossetti I, Compagnoni M, Chemical reaction engineering, process design and scale-up issues at the frontier of synthesis: flow chemistry, Chem Eng J, 2016, 296, 56–70.
[28] Xie P, Wang K, Wang PJ, Xia Y, Luo GS, Synthesizing bromobutyl rubber by a microreactor system, AIChE J, 2017, 63(3), 1002–1009.
[29] Deng J, Zou PC, Wang K, Luo GS, Continuous-flow synthesis of (E)-2-Hexenal intermediates using a two-stage microreactor system, 2020, 10(4), 661–672.
[30] Du CC, Wang PC, Hu YP, Zhang JS, Luo GS, Liquid-liquid mass transfer enhancement in milliscale packed beds, Ind Eng Chem Res, 2020, 59(9), 4048–4057.
[31] Tu J, Sang L, Cheng H, Ai N, Zhang JS, Continuous hydrogenolysis ofN-diphenylmethyl groups in a micropacked-bed reactor, Org Process Res Dev, 2019, 24(1), 59–66.
[32] Lin XY, Wang K, Zhang JS, Luo GS, Process intensification of the synthesis of poly(vinyl butyral) using a microstructured chemical system, Ind Eng Chem Res, 2015, 54(14), 3582–3588.
[33] Noël T, Hessel V, Continuous-flow multistep synthesis of cinnarizine, cyclizine, and a buclizine derivative from bulk alcohols, ChemSusChem, 2016, 9(1), 67–74.
[34] Xie P, Wang K, Deng J, Luo GS, Continuous, homogeneous and rapid synthesis of 4-bromo-3-methylanisole in a modular microreaction system, Chin J Chem Eng, 2020, 28(8), 2092–2098.

Sara Miralles-Comins, Elena Alvarez, Pedro Lozano and
Victor Sans

9 Exothermic advanced manufacturing techniques in reactor engineering: 3D printing applications in flow chemistry

9.1 Introduction to 3D printing applied to flow chemistry

Additive manufacturing (AM), also known as 3D printing technologies (3DP), is a set of techniques that build 3D objects by successively adding materials typically in a layer-by-layer fashion [1]. The 3D models are generally designed with computer-aided design (CAD) software for the creation, modification, analysis, or optimization of a design [2].

The recent growth of AM techniques is attributed to several advantages including the rapid fabrication of complex geometries (i.e., internal channels or undercut features) with high precision and resolution, unprecedented flexibility in design, waste minimization and personalised customization reducing the time and cost of the process, as well as human interaction [3]. Whereas traditional fabrication techniques have limitations in controlling the geometry and architecture of macroscopic components without compromising the optimal performance, 3DP techniques are an efficient strategy to engineer well-controlled functional materials across the scales (from nano to macroscale) with an accurate control over geometry (dimension, porosity, and morphology) and structure. In addition, 3DP expands the boundaries of materials science and provides an exciting opportunity for interdisciplinary research as it allows the incorporation of multiple nanomaterials in the same printing process to obtain multifunctional devices. Therefore, AM technologies have rapidly developed providing new solutions in a broad range of industrial sectors [4].

AM technology can play a significant role in the development of sustainable flow processes. Minimizing or eliminating heat and mass transport limitations is a key challenge to optimize the efficiency of chemical transformations. An approach is to develop equipment and techniques that lead to compact, safe, energy-efficient, and environment-friendly processes. These developments are focused on the term of Process intensification (PI), which has been defined as "the development of innovative apparatuses and techniques that offer drastic improvements in chemical manufacturing and processing, substantially decreasing equipment volume, energy consumption, or waste formation and ultimately leading to cheaper, safer, sustainable technologies" [5–7]. PI implies the improvement in process efficiency by reducing the scale and by

https://doi.org/10.1515/9783110693690-009

integrating unit operations. Hence, it involves the development of smaller equipment that led to enhancements in the process efficiency, including improvements on the control of the reactor kinetics, providing higher selectivity through the reduction of by-products; higher energy efficiency on the process; reduction in the capital and raw material costs, and safer processes. In this sense, continuous-flow reactors (CFRs) represent a practical, safer, faster, and flexible alternative to traditional batch manufacturing where the mass and heat transfer phenomena play a key role in the performance of chemical reactions [8].

Among the different techniques used to intensify mass transfer (i.e., rotation, vibration, or mixing), the in-line mixer design has become very popular recently and is improving by leaps and bounds. In this context, the first requirement for a mixer is to reduce the length of the mixing path. Thus, a conventional reactors, where a large pot with a mixing path length equivalent to its radius is used.

Micro- and millimeter-scale reactors can be employed to improve the heat transfer phenomena, specially, for fast and highly exothermic reactions. Without an efficient heat transfer management, the heat generated can reduce the selectivity of the process due to side reactions or even present safety threats, such as runaway reactions. These problems are eliminated mainly due to the reduction in the hydraulic diameter, together with an increase in the heat transfer area [9]. The appropriate selection of the material should be determined by an equilibrium between: a) the availability, cost and ease of handling material and manufacturing process; b) the roughness of the material; c) the thermal conductivity; and d) the operational conditions [10].

It is important to note that due to the small characteristic length scales in compact reactors, which lead to a faster heat transfer, it is possible in many cases to carry out the reaction in the absence of any solvent, highlighting the green aspect of continuous-flow processing in microreactors [11, 12]. In this sense, the use of intensified technology for estimating reaction kinetics has been widely demonstrated, providing better accuracy than those obtained in conventional batch reactors [13–15]. Moreover, the small volumes of the reactors lead to a reduction in chemical consumption, making the use of microreactors more suitable with green techniques, as well as more cost-effective.

Emerging trends in this field include the development of more robust reactors, using polymers, metals, and ceramics. Recently, thermally and mechanically robust as well as chemically stable polymers have been printed to be used as highly resistant flow reactors [16]. Likewise, metal-printed reactors have been manufactured, enabling challenging reaction conditions in terms of temperature, pressure, and chemicals used [17, 18]. Ceramics are harder materials to manufacture with precision due to the required post-manufacturing thermal treatment to obtain the finished structures. However, it has been reported the possibility to print ceramic microreactors by a digital light processing (DLP) technique [19].

Additionally, 3DP flow reactors can be easily manufactured to add functionality to the reactors by integrating enabling technologies, such as catalysis, biocatalysis [20], microwave dielectric heating [21], electrochemistry, or photochemistry [22]. The post-modification of printed devices is an attractive approach to achieve this goal. For instance, the efficient immobilization of enzymes in a 3DP reactor has a big potential in green chemistry because it combines the efficiency of flow transformations with the selectivity and mild conditions achieved with biocatalysts [23, 24]. Nevertheless, not always a post-modification is necessary. A printing hydrogel containing thermostable enzymes has been used to form multienzyme flow reactors for cascade transformations [25–27]. Electrochemical flow reactors are also gaining interest for the development of sustainable chemical transformations. The employment of 3DP facilitates the manufacture of parts of electrochemical devices [28]. In addition, electroreactors between the electrodes and a high electrode surface [29]. Furthermore, the intensification of the electrochemical processes by improving mixing can have positive effects on activity and selectivity. In this regard, it has been demonstrated the improved mass transfer properties of 3DP electrochemical mixing electrodes [30].

The ability to integrate multiple enabling technologies, together with the freedom of design, which enables the operator to finely tune the reactor geometry are key advantages over the employment of simpler technologies, like tubular reactors or packed bed columns, which can work with standard fittings, T-mixers, and heating elements. In this regard, 3D printing should be considered as an alternative technology to the use of conventional reactors when the application sought will require enhanced features, which would be hard to integrate employing traditional manufacturing methods. Examples will be shown later in this chapter.

9.2 Classification of 3DP techniques

3DP techniques can be divided into seven areas, according to the way the materials are deposited, the form of the starting material or the method of assembling in the additive fashion, opening new pathways for creating devices with outstanding performance at fine resolutions [31]. These techniques cover a broad range of scales from micro- to macroscale devices and materials available. The first rapid prototyping system "Stereolithography Apparatus" (SLA) was patented by Charles Hull in 1986 [32]. Since then, many different types of rapid prototyping processes have been developed and commercialized (Tab. 9.1). Depending on the technique employed, the parts may be post-processed to improve the strength or surface quality. There are many factors which have a direct influence while manufacturing flow reactors that must be considered when using the different AM categories, hence the method of fabrication plays a major role. This section provides a brief overview about the state of the art of AM technology [33].

Tab. 9.1: AM classification: vat photopolymerization, powder bed fusion, direct energy deposition, material extrusion, material jetting, binder jetting, and sheet lamination [34].

Process principle	AM technology	Materials
Vat photopolymerization	Stereolithography Dynamic light processing Continuous liquid interphase production	Photopolymer Ceramic
Powder bed fusion	Selective laser sintering Selective laser melting Electron beam melting	Metal Polymer Ceramic
Material extrusion	Fused filament fabrication Direct ink writing	Polymer Ceramic and Biomaterials
Material jetting	Ink-jetting Thermojet Polyjet	Photopolymer Wax
Binder jetting	Ink-jetting	Metal Polymer Ceramic
Direct energy deposition	DMD Laser deposition Laser consolidation EMD	Metal Powder Wire
Sheet lamination	Ultrasonic consolidation Laminated object	Hybrids Metallic Ceramic

Most commonly employed 3DP techniques for applications in flow chemistry are vat photopolymerization (VP), extrusion, laser sintering/melting, and jetting techniques [35], which are represented in Fig. 9.1 [1].

9.2.1 Vat photopolymerization

VP produces macroscopic solids from liquids by selective photopolymerization with UV light (or electron beam) inside a bath which contains UV-active monomers that instantly activate the polymerization. Each photopolymer layer is cured using a UV light source, producing fully cured models that can be handled and used immediately, without post-curing, although a post-process treatment like photo-curing or heating is usually employed for some printed parts in order to achieve the desired mechanical performance. Flash technology devices arise from the necessity to reduce the lead time

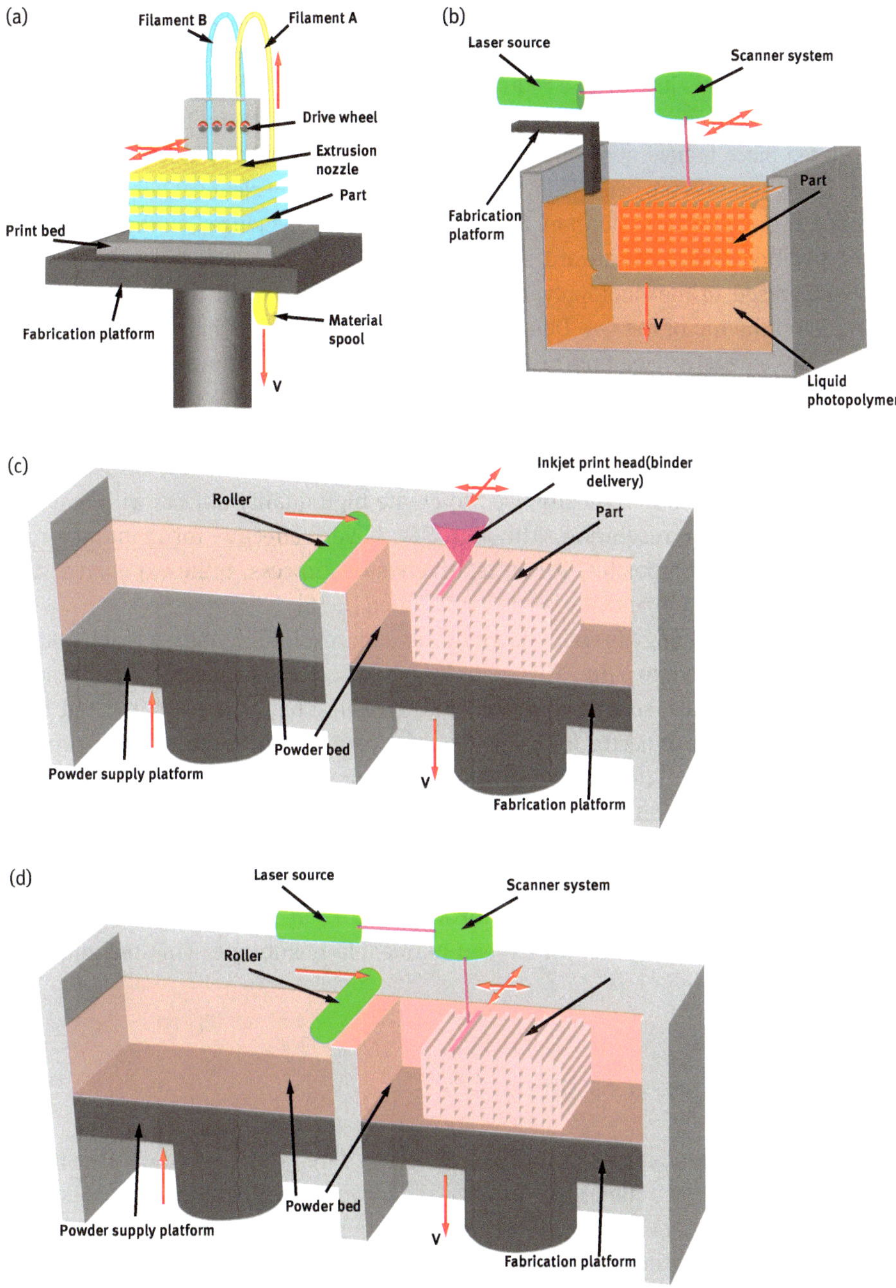

Fig. 9.1: Schematic representation of 3DP techniques. (A) Filament extrusion, commonly known as fused filament fabrication (FFF) or fused deposition modelling (FDM). (B) Vat photopolymerization. (C) Inkjet binder technique. (D) Selective laser sintering/melting technique. Adapted from reference [1] with permission.

and increase in build speed obtained with SLA, to produce 3D models. However, typically acrylate-based formulations result in materials that are soluble in some solvents. That is why its use is limited to a reduced range of solvents in flow reactors. There are two different processes of fabrication, by vat or bath configuration. In the former one, the printed piece is held above the vat of resin and, fixed in the platform, moves upward. The UV light is projected from underneath the optically clear vat window. On the contrary, in the bath configuration, the layers are photopolymerized on the vat surface as the platform moves downward toward the bottom of the platform so the piece is fully submersed in the photopolymer resin during the build. In this case, the light source is on the top of the vat. Due to several disadvantages of this methodology, including the extensive cleaning process normally required, vat systems are gaining more and more prominence. SLA, SLASLA-DLP, and continuous liquid interface production (CLIP) methods are common vat photopolymerization techniques.

a) **SLA:** SLA is a well-established AM technology as it was one of the earliest methods implemented. It is mostly used to create high-quality objects with features <10 μm at a fine resolution. Although SLA is very effective for manufacturing complex nanocomposites, it is a relatively slow process, quite expensive, and the range of printable materials is very limited [1, 36].
b) **DLP:** DLP technology is well known for its high resolution, able to reach a layer thickness of down to 5 μm. This technique is based on the use of UV photopolymerized materials, so as a film is coated in resin. Then, the resin is cured by a UV flash from a projector for each slice of product. Contrary to 3D laser printer, that project only of lines or points, DLP projector projects the entire layer. Therefore, the substitution of scanning time of a laser results in a methodology that allows building much quicker than other rapid prototype methods [4].
c) **CLIP:** CLIP is a new type of AM technology, based on photopolymerization principle. This technique uses a projector which works in continuous by controlling oxygen levels throughout an oxygen-permeable membrane. This technique is much faster than SLA [4].

9.2.2 Powder-based technologies

Powder bed fusion consists in thin layers of very fine powders, which are spread and closely packed on a platform, and then the powder in each layer is consolidated by heating with a laser or an electron beam over the photopolymer surface. PBF does not require support structures as the parts are supported by unfused powder in the build area and presents a fine resolution and high quality of printing, the process is slow with high cost [1, 36]. As a post-processing step, the material can be heated to improve physical and mechanical properties, although it can lead to shrinkage of parts. Moreover, cleaning the powder from flow reactor channels is a challenge, particularly in designs with narrow path and tight curves. This technique has been widely applied to directly make ceramics. For example, Griffith et al. developed ceramic powder (i.e., silica and alumina) and it was dispersed in a fluid UV monomer to prepare a ceramic-UV monomer suspension [37]. Shortly after this development, a widely laser-based AM technology was developed:

a) **Selective laser melting (SLM).** SLM is based on the application of a thin layer of the powder material, which will be beamed layer by layer. The laser will selectively melt a layer of powder with a controlled morphology, which is controlled by the slicer software. Then, the platform is lowered and another layer of powder is applied. Thereby, during the process, successive layers of metal powder are fully melted and consolidated on top of each other [4].
b) **Selective laser sintering (SLS).** SLS is a technology based on layer processes that use a high-power laser as energy source for sintering the powder into a single part in a layer. The process formation of the layer in this technique is similar to SLM. What make the difference is that in SLS is that the sintering processes do not fully melt the powder, but heat it to the point in which the powder can fuse together on a molecular level, reaching a great dimensional accuracy in the prototypes. Moreover, SLS can be used for a variety of polymers, metals, and alloy powders while SLM can only be used for certain metals [4].
c) **Electron beam melting (EBM).** EBM is another PBF method where materials are placed under vacuum and fused together by direct heating using a high-powder electron beam as source of energy reaching superior accuracy [38].

9.2.3 Extrusion technologies

Extrusion-based 3D-printing methods consist of the deposition of a model or support material directly from a nozzle head dispenser, after material pre-treatment (i.e., liquefaction). It is a low cost, high speed, and simple process technology [1, 36].

a) **Fused fabrication filament (FFF).** AM technique based on the extrusion of thermoplastic filaments. In this technique, the filament is extruded through a nozzle to print one cross section of an object. Then, the process is repeated to produce a new layer. Parts can be built with single or multiple materials with the option of a dual extruder nozzle on some platforms. In FFF technology, the knowledge of the material properties of the raw is crucial, as well as the effect that FFF builds parameters have on anisotropic material properties. In this sense, the most used materials are polycarbonates, polylactic acid, and acrylonitrile butadiene styrene. The problematics of FFF are the poor mechanical properties, the layer-by-layer appearance, low quality (resolution between 0.2 and 0.05 mm), and the limited number of thermoplastic materials, especially, the difficulty of creating a good seal between the layers to avoid leaks makes the development of flow reactors a challenge.

b) **Direct ink writing (DIW):** DIW extrudes a paste through a thin needle allowing the processing of relatively dense and viscous materials compared to other AM techniques. It can print the electrode with a high active material loading and with much less clogging risk [1, 36]. Creating leak-free reactors is dependent on multiple factors such as the selected material, fabrication conditions, reactor geometry, and layer height.

9.2.4 Jetting technologies

In **material jetting (MJ)**, a print head, similar to a standard inkjet printer head, is used to dispense materials in a continuous or drop on demand approach onto a platform. Droplets are released on demand due to the pressure change in the nozzle by thermal or piezoelectric actuators. The latter permits the use of a wider range of materials. Commercial MJ devices allow a layer resolution of micron size layer resolution [39], and multiple nozzles can be used in order to incorporate multiple materials in a single build. As a post processing step, the pieces can be cured under UV light to enhance their physical properties. However, it still faces challenges related to the limited solvent resistivity of the materials as it happens with VP methodology [1, 36].

Inkjet printing is one of the main methods for printing complex and advanced ceramics, consisting in pumping and depositing the stable ceramic suspensions in the form of droplets via the injection nozzle onto the substrate. This is a fast and efficient method able to print complex structures with high resolution and flexibility in design. While direct writing needs a high viscous ink, that is, non-Newtonian paste with shear-thinning behavior, inkjet printing requires low-viscosity Newtonian liquid-like ink. Another similar technology to inkjet printing is called contour crafting which is widely used for large building structures. This method is capable of extruding concrete paste or soil by using larger nozzles and high pressure [1, 36].

9.2.5 Other AM technologies

Binder jetting (BJ) is similar to powder bed fusion method since parts are built up from layers of powder but they are consolidated with binder agent jetted on demand form a nozzle head. BJ methods pose similar advantages in terms of resolution. Nevertheless, when the powder is fused with a binder, the porosity is generally higher compared to laser because it can print denser parts. Moreover, the binder-powder combination may not be suitable for a wide range of solvents.

On the other hand, **direct energy deposition (DED)** is frequently used for engineering high-performance super-alloys. DED also uses a source of energy (laser or electron beam) which melts a small region of the substrate and the feedstock material simultaneously. The melted material is then deposited and fused into the melted substrate and solidified after movement of the laser beam. The main difference between DED and SLM methods is that in DED no powder bed is used and the feedstock is melted before deposition in a layer-by-layer fashion similar to FFF but employing a higher amount of energy for melting the metals. It has low accuracy and quality just as FFF and can be used for printing large objects due to its high speed [1, 36].

The last method is **sheet lamination** where sheets of material are bonded together using an adhesive or welding and laser cut. A common methodology is laminated object manufacturing (LOM), a roll-to-roll-based process used in various industries such as paper manufacturing, foundry industries, electronics, and smart structures. LOM consists in layer-by-layer cutting and lamination of sheets or rolls of materials and results in a reduction of tooling cost and manufacturing time, becoming in one of the best methods for big structures. But it is not recommended for complex shapes due to its inferior surface quality and resolution [1, 36]. The resolution of DED and sheet lamination platforms are not suitable for flow reactors as they will require post machining to achieve the surface finish and feature sizes required.

9.3 Applications of 3D printing in flow chemistry

CFRs can be classified according its dimension into microscale reactors and milliscale reactors. Micro-scale tubular flow reactors have cross-sectional diameters below 1 mm while milliscale reactors have up to several millimeters of characteristic length [40]. Depending on to the scale of the reactor, multiple materials as well as different methods of manufacture can be employed to build them.

In recent years, researchers have been focused on advanced manufacturing technologies to construct more complex and customized CFRs. In this sense, it has been incorporated into these devices several structural elements (i.e., mixing structures, residence time channels, separation units, or interfaces).

In this context, the digitalization of the manufacturing procedures is becoming a very valuable tool to fabricate CFRs with innovative internal architectures as conventional techniques require high equipment and labor costs [41–43]. The use of 3DP to manufacture CFRs was first reported in 2012. The field has rapidly grown to develop a broad range of applications in the synthesis of chemicals, materials [44], and crystallization processes [45]. 3DP has a big potential for the development of sustainable flow processes, reducing heat and mass transfer limitations. Besides, this technology can integrate with multiple technologies, such as electrochemistry, biocatalysis, or chemocatalysis, to perform multiple transformations, improving reaction yields and minimizing the waste generation through efficient use of reactants [46].

Since Kitson et al. employed 3DP to manufacture continuous-flow reactors in 2012 [47], this technology has grown to develop a wide range of applications, including the synthesis of chemical and materials, as well as the fabrication of optimized reactor geometries. In the first work it was reported the versatility, feasibility, and configurability of this technology in the fabrication of micro- and millifluidic devices for organic, inorganic, and material synthesis. The simple and rapid design of a two-inlet device and a three-inlet reactor, composed by two initial inlets joined by a third inlet after the reaction, was demonstrated. A third part developed allowed the deposition of solid reagents within the fabricated "silos," during the manufacturing process. It is important to point out that the fabrication of these reactors is very difficult employing traditional methods. However, with this technology it was easier and cost-effective to alter the design of the devices in terms of geometry, inlets, outlets, and sizes of channels.

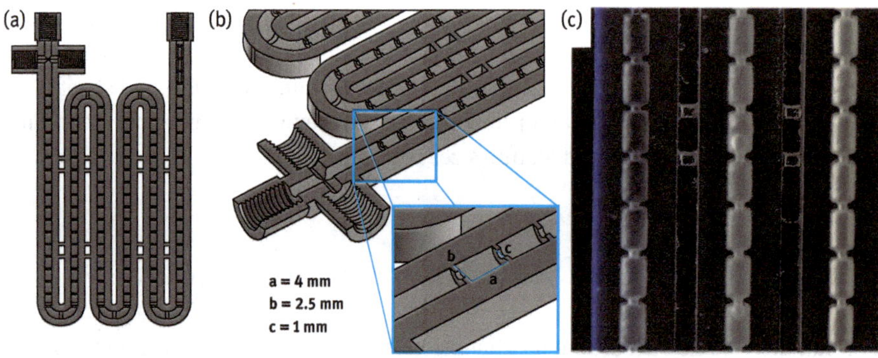

Fig. 9.2: Highly complex reactor architecture that can be easily fabricated with 3DP.
(A) Representative view of a compact scale baffled reactor with three inlets, periodic constrictions, and one outlet. (B) Detail of the dimensions of the baffles. (C) A detail of a 3D-printed version of the baffled reactor employing a Form2 printer from Form Labs with commercially available clear resin.

A key advantage of 3DP flow reactors is the possibility to integrate advanced functionalities capable to improve the performance of the reactors in terms of improved mixing, heat and mass transfer, catalyst contact with substrates, thus contributing to an effective PI. An example of the design of advanced reactor through 3DP to optimize mixing at the milli- and mesoscale is the continuous oscillatory baffled reactor (COBR) shown in Fig. 9.2 [48]. A COBR combines periodic constriction (baffles) along the reactor path with a complex flow behavior composed of a linear flow and a mechanically generated oscillatory flow [49]. This leads to an improved mixing at low Reynolds numbers (Re). This represents an advantage over traditional reactors, where at low Re laminar flow limits efficiency due to inefficient diffusion-based mixing. Okafor et al. reported a miniaturized COBR (mCOBR) based on 3DP technology for the continuous-flow synthesis of silver nanoparticles (Ag-NPs) [48]. Ag nanoparticles have a local plasmon of resonance (LSPR) that can be easily monitored by UV–vis spectroscopy. The LSPR is strongly dependent on size, shape, and environment of the NPs. They are challenging to synthesize under continuous flow due to its tendency to foul in the reactor walls [50], and their sensitivity to the reaction conditions, which means that an inefficient mixing profile within the reactor would lead to broad size distributions. The employment of an mCOBR is an ideal system to overcome these limitations. An mCOBR and a control tubular reactor of similar geometry were designed and manufactured using low cost SLA. The optimization of internal geometry of the mCOBRs (i.e., baffle open area, baffle spacing, or thickness) was easily achieved employing digital designs manufactured with 3DP and the performance was evaluated employing residence time distribution experiments (RTD). The mCOBR reactors were used for the continuous synthesis of silver nanoparticles (Ag-NPs) under continuous-flow. Figure 9.3A shows a scheme of the reaction platform employed. Two high pressure liquid chromatography (HPLC) pumps were employed to pump the reagents into the reactor. A programmable syringe pump filled with immiscible oil was employed to generate the controlled oscillations in the system. An in-line UV–vis flow cell was employed to monitor and characterize the Ag-NPs in real time. As it can be seen in Fig. 9.3B and C, a comparison between the extinction spectra as function of time was carried out in both, tubular and mCOBr, showing more stability over time of the NPs synthesized in the mCOBR reactor. Figure 9.3C shows a detail of characteristic parameters of the LSPRs of the Ag-NPs synthesized in both re-

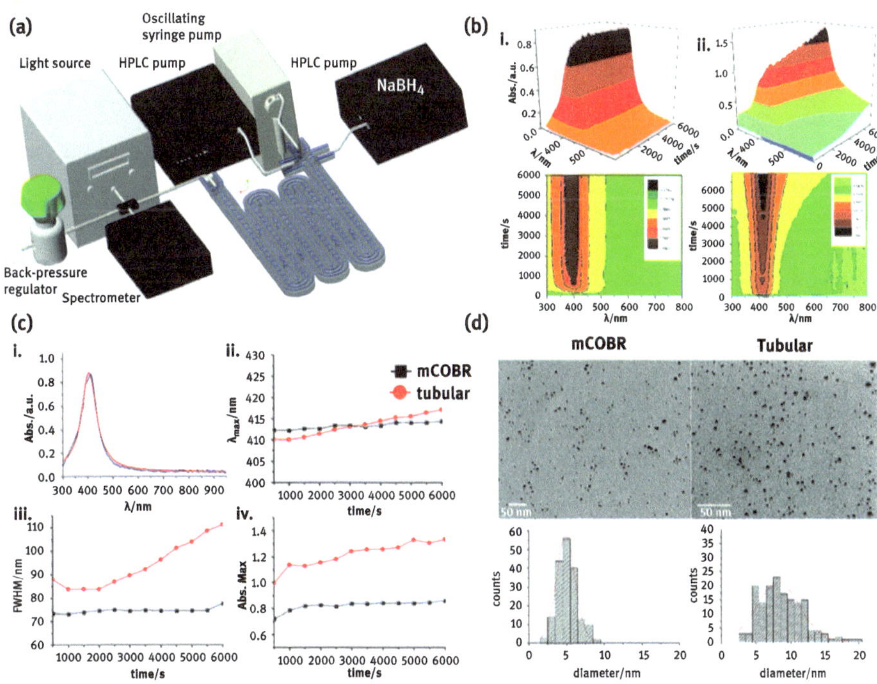

Fig. 9.3: (A) Example of the schematic representation of the platform employed to synthesize Ag-NPs under continuous-flow conditions. (B) Times series extinction spectra corresponding to the Ag-NPs synthesis from a side view and top view for an mCOBR (i) and tubular reactor (ii). Comparison of the LSPR profiles as a function of time in mCOBR and tubular reactor. Reproduced from reference 48 with permission from the Royal Society of Chemistry.

actors, including the maximum wavelength, the full width at half maximum, and the absorbance values. All these parameters are related to the particle size, the size distribution of the nanomaterials, and they are more constant in the case of an mCOBR. Transmission electron microscopy images confirmed that the Ag-NPs synthesized in the mCOBR were smaller and with a narrower particle size distribution (Fig. 9.3D).

In order to maximize the potential advantages of 3DP for the construction of CFRs, it is important to develop them in materials other than polymers, which exhibit low stability against a range of reagents and common organic solvents. Therefore, the use of other techniques, such as SLM, the printing of reactors from a variety of metal is possible, providing a large number of advantages for organic synthesis (i.e., thermal conductivity or mechanical and thermostability).

The design and manufacturing of a millifluidic stainless-steel reactor for continuous exothermic fluoromethylation reactions (Fig. 9.4) was recently reported [17]. The main advantage of stainless-steel reactor is that improve the chemical stability in organic reactions. It is also important to point out that stainless steel enables a high-pressure

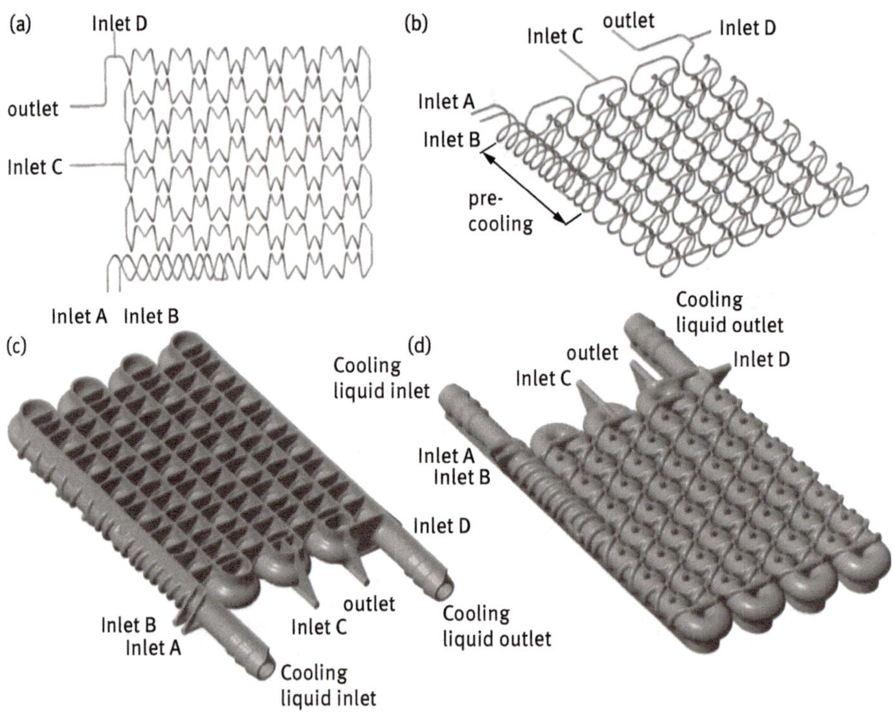

Fig. 9.4: Schematic example of the CAD drawings of the flow reactor designed by B. Gutman et al. for continuous fluoromethylation reactions. (A) Top view on the reaction channels. (B) Perspective view on the reaction channels. (C) View from below of the reaction channel attached onto the cooling outlet. (D) View from above of the reaction channel attached onto the cooling outlet. Reproduced from [17] from RSC under CC-BY 3.0.

resistance, as well as a fast heat transfer, allowing the performance of exothermic reactions under cryogenic temperatures and high pressures [17]. The reactor was designed with meandering cannels and their ability to provide adequate mixing was validated by computational fluid dynamics (CFD), providing a curvature-based geometry (see Fig. 9.4A). It has been demonstrated that the zig-zag nature of the channels enhances advective mixing through the stretching and folding of the flow stream. Besides, in Fig. 9.4B it can be seen the 3D channel geometry generated by the meandering channels, drawn onto the cooling element. The heat transport distances and cooling are easily achieved by the attachment of the reaction channel onto the cooling element. Figure 9.4C and D shows the simple connection between the end of the cooling tube with a circulation cryostat. The same group demonstrated the possibility to apply this methodology to generate optimized reactors for continuous-flow oxidations of Grignard reagents with in-line oxygen sensors to monitor the reaction progress [18].

9.4 Future directions: digitalization of reactor design and manufacturing

The possibility to generate complex geometries with high resolution offered by 3DP offers a new dimension to reactor engineering, where the designs can be based on finely and tailored calculations by means of CFD [41].

CFD methodologies employ finite element simulations to solve complex equations (e.g., Navier–Stokes equation) in geometries that have been modelled into discrete parameters, typically points or volumes defined with a mesh applied over the geometry of interest. Parra-Cabrera et al. published recently a review covering aspects of CFD applied to microreactors and its potential coupling to AM [41].
CFD allows the modeling of flow dynamics within the selected geometry, enabling a visualization of the flow within the reactor, and also to model parameters of interest, such as heat and mass transport, RTDs, and pressure drops. Figure 9.5 shows an example reported by Wu et al. where single and multiphase flows were simulated in a commercial AFR Corning reactor [51] and the models were validated with experimental measurements [52].

The advantages of combining CFD and CAD with 3DP to manufacture devices with advanced features are just starting to be realized [17, 18, 53]. CFD allows a fine understanding of the mixing ability within the reactor, hence enabling a fine tuning of the geometry to be manufactured. 3DP enables the realization of these complex geometries, which is the missing link to enable the next generation reactor technology. An example of the ability was very recently reported by Bettermann et al., where a 3DP reactor employed for emulsion polymerization [54] was analyzed by CFD calculations, to have an enhanced understanding of the flow patterns was recently demonstrated (Fig. 9.6) [53]. The CFD calculations of a flow channel indicated

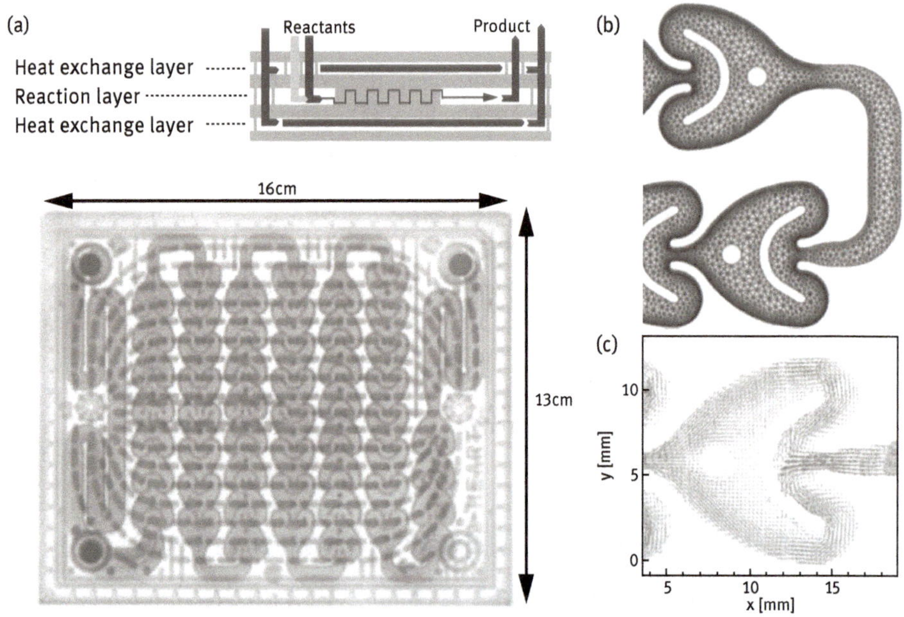

Fig. 9.5: Example of the use of CFD to model a commercial Corning continuous-flow chemical reactor with passive mixing features. (A) Overview of the reactor geometry including heat exchange channels. (B) Mesh of a selected section of the reactor geometry. (C) CFD modeled velocity profile within the reactor channel. Adapted from [52] from ACS under CC-BY 4.0.

that in this geometry, mixing occurred mainly in the curves, where Dean vortices were alterating the otherwise laminar flow patterns. The CFD analyses allowed the optimization of parameters including reactor wall thickness, radii of the bending areas to develop a lean manufacturing approach.

9.5 Conclusions

3DP is having a significant impact across a large number of manufacturing sectors due to the potential to realize complex geometries and to finely tune the design to the concrete application. In this way 3DP is expected to play a role in the development of advanced chemical reactors for flow chemistry applications. A number of representative examples have been presented in this chapter. The effective digitalization of the reactor design process, combining CAD, CFD, and 3DP opens the door to the development of advanced reactor geometries. This will play an increasing role in the field of flow chemistry, especially when robust reactors made in metal will become more widely employed.

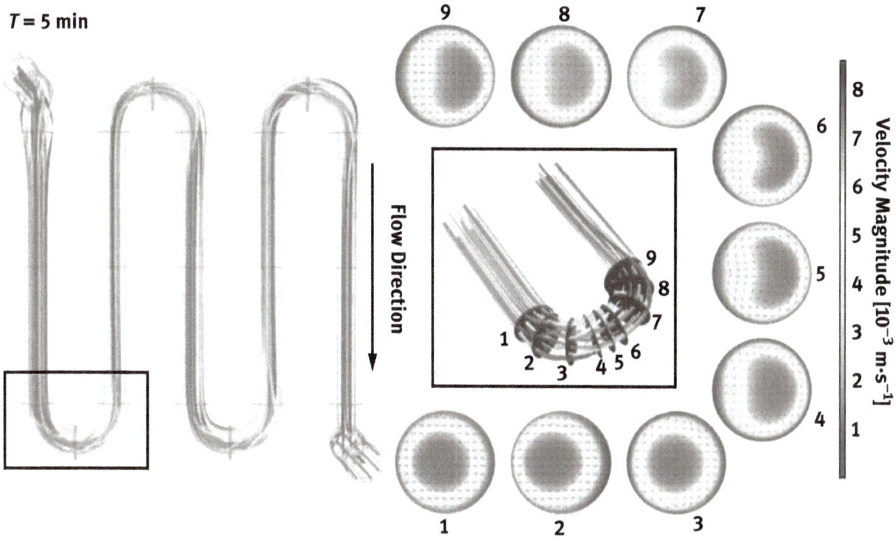

Fig. 9.6: CFD simulation of single-phase flow of a simple geometry, with a highlight of dean vortices in the curves. Reproduced from [53] with permission from Elsevier.

References

[1] Ngo TD, Kashani A, Imbalzano G, Nguyen KTQ, Hui D, Additive manufacturing (3D printing): a review of materials, methods, applications and challenges, Compos B Eng, 2018, 143, 172–196.

[2] Gao W, Zhanga Y, Ramanujan D, Ramania K, Chen Y, Williams CB, Wange CCL, Shina YC, Zhanga, S, Zavattieri PD The status, challenges, and future of additive manufacturing in engineering. Comput Aided Des, 2015, 69, 65–89.

[3] Ashley S, Rapid prototyping systems, Mech Eng, 1991, 113, 34.

[4] Gardan J, Additive manufacturing technologies: state of the art and trends, Int J Prod Res, 2015, 54.

[5] Hessel V Process intensification–engineering for efficiency, sustainability and flexibility. Green Process Synth, 2012, 1.

[6] Keil FJ, Process intensification, Rev Chem Eng, 2018, 34, 135–200.

[7] Re-Engineering the Chemical Processing Plant: Process Intensification, New York/USA, Marcel Dekker Inc., 2004.

[8] Noël T, Su Y, Hessel V, Beyond Organometallic Flow Chemistry: The Principles Behind the Use of Continuous-Flow Reactors for Synthesis, Springer International Publishing, 2015, 1–41.

[9] Gutmann B, Cantillo D, Kappe CO, Continuous-flow technology – a tool for the safe manufacturing of active pharmaceutical ingredients, Angew Chem Int Ed, 2015, 54, 6688–6728.

[10] Brandner JJ, Anurjew E, Bohn L et al., Concepts and realization of microstructure heat exchangers for enhanced heat transfer, Exp Therm Fluid Sci, 2006, 30, 801–809.

[11] Borukhova S, Noël T, Metten B, De vos E, Hessel V, Solvent- and catalyst-free huisgen cycloaddition to rufinamide in flow with a greener, less expensive dipolarophile, ChemSusChem, 2013, 6, 2220–2225.
[12] Snead DR, Jamison TF, A three-minute synthesis and purification of ibuprofen: pushing the limits of continuous-flow processing, Angew Chem Int Ed, 2015, 54, 983–987.
[13] Yue J, Schouten JC, Nijhuis TA, Integration of microreactors with spectroscopic detection for online reaction monitoring and catalyst characterization, Ind Eng Chem Res, 2012, 51, 14583–14609.
[14] Noël T, Hessel V, Membrane microreactors: gas–liquid reactions made easy, ChemSusChem, 2013, 6, 405–407.
[15] Zhou X, Medhekar R, Toney MD, A continuous-flow system for high-precision kinetics using small volumes, Anal Chem, 2003, 75, 3681–3687.
[16] Harding MJ, Brady S, O'Connor H et al., 3D printing of PEEK reactors for flow chemistry and continuous chemical processing, React Chem Eng, 2020, 5, 728–735.
[17] Gutmann B, Köckinger M, Glotz G et al., Design and 3D printing of a stainless steel reactor for continuous difluoromethylations using fluoroform, React Chem Eng, 2017, 2, 919–927.
[18] Maier MC, Lebl R, Sulzer P et al., Development of customized 3D printed stainless steel reactors with inline oxygen sensors for aerobic oxidation of Grignard reagents in continuous flow, React Chem Eng, 2019, 4, 393–401.
[19] Gyak KW, Vishwakarma NK, Hwang YH, Kim J, Yun HS, Kim DP, 3D-printed monolithic SiCN ceramic microreactors from a photocurable preceramic resin for the high temperature ammonia cracking process, React Chem Eng, 2019, 4, 1393–1399.
[20] Tamborini L, Fernandes P, Paradisi F, Molinari F, Flow bioreactors as complementary tools for biocatalytic process intensification, Trends Biotechnol, 2018, 36, 73–88.
[21] Glasnov TN, Kappe CO, Microwave-assisted synthesis under continuous-flow conditions, Macromol Rapid Commun, 2007, 28, 395–410.
[22] Su Y, Straathof NJW, Hessel V, Noël T, Photochemical transformations accelerated in continuous-flow reactors: basic concepts and applications, Chem Eur J, 2014, 20, 10562–10589.
[23] Peris E, Okafor O, Kulcinskaja E et al., Tuneable 3D printed bioreactors for transaminations under continuous-flow, Green Chem, 2017, 19, 5345–5349.
[24] Ye J, Chu T, Chu J, Gao B, He B, Versatile A, Approach for enzyme immobilization using chemically modified 3D-printed scaffolds, ACS Sust Chem Eng, 2019, 7, 18048–18054.
[25] Maier M, Radtke CP, Hubbuch J, Niemeyer CM, Rabe KS, On-demand production of flow-reactor cartridges by 3D printing of thermostable enzymes, Angew Chem Int Ed, 2018, 57, 5539–5543.
[26] Peng M, Mittmann E, Wenger L et al., 3D-printed phenacrylate decarboxylase flow reactors for the chemoenzymatic synthesis of 4-hydroxystilbene, Chem Eur J, 2019, 25, 15998–16001.
[27] Schmieg B, Schimek A, Franzreb M, Development and performance of a 3D-printable poly (ethylene glycol) diacrylate hydrogel suitable for enzyme entrapment and long-term biocatalytic applications, Eng Life Sci, 2018, 18, 659–667.
[28] Van Melis CGW, Penny MR, Garcia AD et al., Supporting-electrolyte-free electrochemical methoxymethylation of alcohols using a 3D-printed electrosynthesis continuous flow cell system, ChemElectroChem, 2019, 6, 4144–4148.
[29] Folgueiras-Amador AA, Wirth T, Perspectives in flow electrochemistry, J Flow Chem, 2017, 7, 94–95.
[30] Lölsberg J, Starck O, Stiefel S, Hereijgers J, Breugelmans T, Wessling M, 3D-printed electrodes with improved mass transport properties, ChemElectroChem, 2017, 4, 3309–3313.

[31] Zhu C, Liu T, Qian F et al., 3D printed functional nanomaterials for electrochemical energy storage, Nano Today, 2017, 15, 107–120.
[32] Hull CV Apparatus for production of three-dimensional objects by stereolithography. 1986, Patent version number: US4575330 A.
[33] Horn T, Harrysson O, Overview of current additive manufacturing technologies and selected applications, Sci Prog, 2012, 95, 255–282.
[34] Lee J-Y, An J, Chua CK, Fundamentals and applications of 3D printing for novel materials, Appl Mater Today, 2017, 7, 120–133.
[35] Capel AJ, Edmondson S, Christie SDR, Goodridge RD, Bibb RJ, Thurstans M, Design and additive manufacture for flow chemistry, Lab Chip, 2013, 13, 4583–4590.
[36] Zhang F, Wei M, Viswanathan VV et al., 3D printing technologies for electrochemical energy storage, Nano Energy, 2017, 40, 418–431.
[37] Griffith ML, Halloran JW, Freeform fabrication of ceramics via stereolithography, J Am Ceram Soc, 1996, 79, 2601–2608.
[38] Rajaguru K, Karthikeyan T, Vijayan V, Additive manufacturing – state of art, Mater Today Proc, 2020, 21, 628–633.
[39] Karjalainen E, Wales DJ, Gunasekera DHAT et al., Tunable ionic control of polymeric films for inkjet based 3D printing, ACS Sustain Chem Eng, 2018, 6, 3984–3991.
[40] Vikram A, Kumar V, Ramesh U et al., A millifluidic reactor system for multistep continuous synthesis of InP/ZnSeS nanoparticles, ChemNanoMat, 2018, 4, 943–953.
[41] Parra-Cabrera C, Achille C, Kuhn S, Ameloot R, 3D printing in chemical engineering and catalytic technology: structured catalysts, mixers and reactors, Chem Soc Rev, 2018, 47, 209–230.
[42] Rossi S, Puglisi A, Benaglia M, Additive manufacturing technologies: 3D printing in organic synthesis, ChemCatChem, 2018, 10, 1512–1525.
[43] Capel AJ, Rimington RP, Lewis MP, Christie SDR, 3D printing for chemical, pharmaceutical and biological applications, Nat Rev Chem, 2018, 2, 422–436.
[44] Okafor O, Weilhard A, Fernandes JA, Karjalainen E, Goodridge R, Sans V, Advanced reactor engineering with 3D printing for the continuous-flow synthesis of silver nanoparticles, React Chem Eng, 2017, 2, 129–136.
[45] Okafor O, Robertson K, Goodridge R, Sans V, Continuous-flow crystallisation in 3D-printed compact devices, React Chem Eng, 2019, 4, 1682–1688.
[46] Sans V, Emerging trends in flow chemistry enabled by 3D printing: robust reactors, biocatalysis and electrochemistry, Curr Opin Green Sustain Chem, 2020, 25, 100367.
[47] Kitson PJ, Rosnes MH, Sans V, Dragone V, Cronin L, Configurable 3D-Printed millifluidic and microfluidic 'lab on a chip' reactionware devices, Lab Chip, 2012, 12, 3267–3271.
[48] Okafor O, Weilhard A, Fernandes JA, Karjalainen E, Goodridge R, Sans V, Advanced reactor engineering with 3D printing for the continuous-flow synthesis of silver nanoparticles, React Chem & Eng, 2017, 2, 129–136.
[49] Harvey AP, Mackley MR, Stonestreet P, Operation and optimization of an oscillatory flow continuous reactor, Ind Eng Chem Res, 2001, 40, 5371–5377.
[50] Poe SL, Cummings MA, Haaf MP, McQuade DT, Solving the clogging problem: precipitate-forming reactions in flow, Angew Chem Int Ed, 2006, 45, 1544–1548.
[51] Advanced-Flow™ Reactors (AFR). Accessed 20-12-20. https://www.corning.com/cala/es/innovation/corning-emerging-innovations/advanced-flow-reactors.html.
[52] Wu K-J, Nappo V, Kuhn S, Hydrodynamic study of single- and two-phase flow in an advanced-flow reactor, Ind Eng Chem Res, 2015, 54, 7554–7564.

[53] Bettermann S, Kandelhard F, Moritz H-U, Pauer W, Digital and lean development method for 3D-printed reactors based on CAD modeling and CFD simulation, Chem Eng Res Des, 2019, 152, 71–84.
[54] Bettermann S, Schroeter B, Moritz H-U, Pauer W, Fassbender M, Luinstra GA, Continuous emulsion copolymerization processes at mild conditions in a 3D-printed tubular bended reactor, Chem Eng J, 2018, 338, 311–322.

Francesca Paradisi and László Poppe

10 Continuous-flow biocatalysis with enzymes and cells

The advantages of synthetic chemistry performed in continuous flow synergize with the benefits conferred by biocatalysis, including reactions with greener, milder, lower temperatures, and aqueous conditions. Furthermore, the fine control over reaction conditions in continuous flow can solve inherent challenges associated with catalysis by enzymes, such as substrate and product inhibition. Cells and enzymes also benefit from the improved mixing, mass transfer, thermal control, pressurized processing, decreased variation, automation, along with in-line product analysis and purification conferred through continuous flow. Thus, the combination of continuous flow and biocatalysis has emerged as a highly effective approach for creating diverse synthetic targets. Examples include immobilized enzymes and whole cells in continuous flow for the synthesis of pharmaceuticals, chemicals, and materials. Challenges to continuous-flow biocatalysis surveyed in this chapter include immobilization methods, the necessity to regenerate cofactors for cell-free biocatalysis, and optimization of biosynthetic steps with highly divergent conditions in flow.

10.1 Introduction

The use of enzymes in continuous flow, known as flow biocatalysis, is rapidly growing in popularity due to the many advantages, particularly in multistep reactions, with respect to batch conditions. Whether the enzyme(s) is immobilized or used in solution, the mass transfer is greater in flow with respect to batch, generally favoring the reaction rate. When the enzyme is immobilized and compartmentalized in a tube reactor, the catalyst loading accumulated in a small space (ratio of enzyme per substrate molecule at any given time) is exceptionally high, again dramatically increasing the conversion. Of course, the increase in reaction rate has no effect on the equilibrium of the reaction; therefore, if the catalyzed reaction has an unfavorable equilibrium due to a fast reverse reaction, in flow this will be simply reached sooner. In a flow setup, issues linked to product inhibition are also eliminated as the flow itself constantly removes the newly formed product. It is therefore easy to imagine how extremely beneficial flow biocatalysis can be when several enzymes are used in a cascade approach, where the product of the first reaction feeds directly into the second reactor containing the second enzyme, and so on (Fig. 10.1). No negative feedback would take place and, if the system is fully optimized, the conversion rate of each reactor can be synchronized to achieve maximum yields.

https://doi.org/10.1515/9783110693690-010

Interestingly, it is also often noted that the (immobilized) biocatalyst used continuously presents a longer working stability when compared to its shelf life under static conditions, with packed enzyme reactors reusable for several weeks.

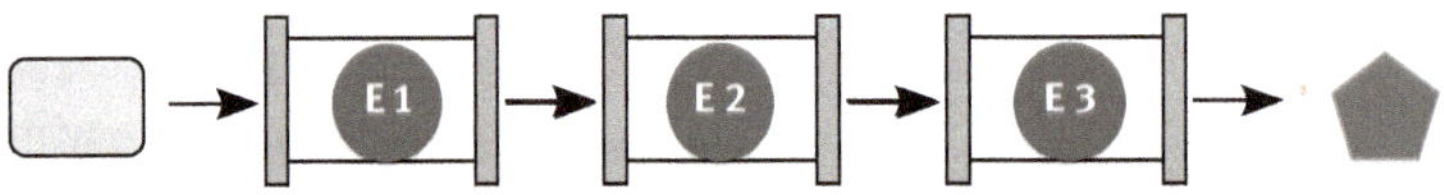

Fig. 10.1: General sketch of sequential reactors in a flow setup. The substrate is transformed into the product via and enzymatic cascade where the enzymes (E1, E2, etc.) are compartmentalized into separate reactors.

As for any technology, however, the application of enzymes in continuous flow presents some challenges and limitations [1–12]. While a single simple biotransformation (i.e., lipase-mediated ester hydrolysis) is relatively straightforward, as soon as the biotransformation requires, for example, a cofactor, therefore, a cofactor recycling strategy, or when multiple enzymes are assembled in line, the system will require extensive investigation of all variables and careful considerations of the benefits versus drawbacks of such a setup. A key issue is represented by the necessity of establishing a "compatibility window" of the enzymes involved in the cascade. Factors such as pH and buffer would ideally be kept constant throughout the cascade, though pH adjustments are possible with inlets upstream of each reactor. Therefore, it is possible that one or more enzyme could be utilized under suboptimal conditions in the cascade, limiting the overall efficiency of the system. Another potential problem, especially in the case of immobilized biocatalysts, is the solubility of the starting material. While slurries can be utilized (normally with flow reactors equipped with peristaltic pump systems), heterogeneous catalysts are inefficient and conversion yields tend to be poor, often worse than in batch.

The high degree of modularity of a flow apparatus opens up the possibility to utilize biocatalyst-filled reactors in new reaction pathways which are now almost exclusively dominated by standard chemical synthesis. It is important to mention that biocatalysis is new to the flow chemistry scene and has only relatively recently discovered the benefits offered by a continuous-flow setup. With respect to the abundance of systems, including industrial applications, reported for traditional chemical synthesis biocatalytic examples are still very limited. However, researchers have shown that flow biocatalysis enables multigram syntheses of high-value products with an exceptional ability to recover and reuse solvents and water wastes, reducing drastically the environmental impact.

10.2 Considerations for the design of CF biocatalysis procedures

10.2.1 Choice of biocatalysts

When performing a biocatalytic transformation, whether in batch or in flow, the choice of the catalyst is clearly important. Generally, key features such as the substrate scope, stability under operational conditions, and tolerance to suboptimal reaction environments (presence of solvents, shifts in pH, temperature, etc.) are good indicators of the robustness of the selected enzyme. In addition, the ease of expression yield and ease of purification, if needed, should be considered. Depending on the type of catalyst and its chemistry, considerations for the requirement of cofactors (and how to recycle), oxygen-rich or oxygen-free media, reaction equilibrium, and so on are also relevant to establish the preferred form of the enzyme (isolated or in the cell) and to evaluate the overall complexity of the reaction setup. On an analytical scale, for example to establish a proof of concept, the assembly of an elaborate system can be achieved more easily than on larger scale, and often this will impact the real implementation potential of a biocatalytic reaction.

10.2.1.1 Single-enzyme or multienzyme-based processes

When biocatalysis is carried out in flow, the first simple distinction is as to whether the reaction is a single step mediated by a single enzyme, or where a multienzyme system must be employed (Fig. 10.2).

Even a single step reaction may require a multienzyme system. This is the case for example when a cofactor is present and needs a coupled enzyme for its recycling, or if the equilibrium of the reaction is unfavorable and a coupled enzymatic reaction is used to shift it toward the formation of the product.

Clearly a single enzyme reaction, as mentioned before, will require only limited optimization, as the procedure will be simple. When more than one enzyme is needed however, the complexity of the system rapidly increases, with a decision making process that starts from the very basics, such as if the enzymes will be co-located in the same reactor (possibly co-immobilized on the same solid support), or if it is better for the overall process to keep them in different reactors, or whether the individual characteristics of one of the enzymes will yield to a hybrid system with some enzymes compartmentalized and others (generally only one) which may be added in soluble form with other reagents. Commonly, redox systems must be set up with an in situ recycling of the cofactor to minimize costs (this consideration applies to redox batch reactions as well), and while it is possible in some cases to exploit the main redox enzyme to recycle the cofactor as well as driving the intended reaction [13]. In other systems a second

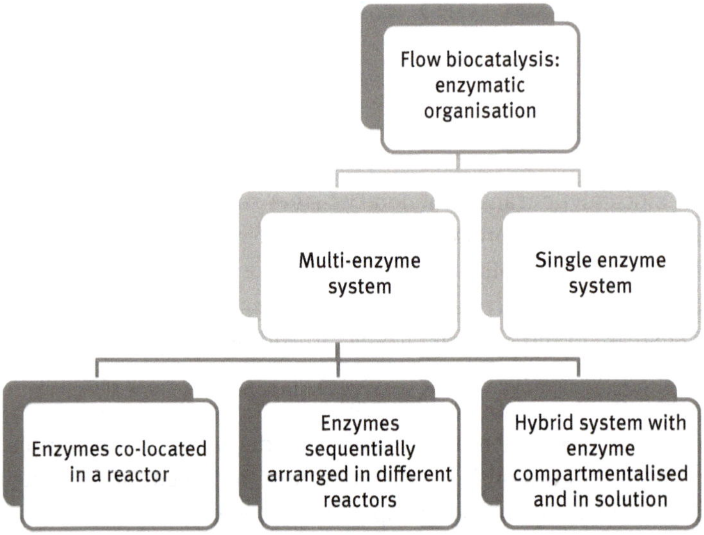

Fig. 10.2: Chart summarizing possible enzyme-based systems in a flow setup.

enzyme–substrate pair must be provided. Intuitively, the second enzyme in this case should be co-located in the same reactor to enable the use of catalytic amounts of cofactor. Co-location of catalysts can also be advantageous in the case of sequential reactions which may benefit from rapid transformation of an intermediate with poor stability, or in the case of two redox reactions which share the same cofactor [14].

10.2.1.2 Cell-free or cell-based: advantages and disadvantages

Both cell-free and cell-based flow systems have been successfully developed [9, 15], with advantages and disadvantages in both cases. These are summarized in Tab. 10.1.

Cells-based systems, normally bacterial or yeast cells, must be set up in such a way that the cells expressing the catalyst must be contained (either in membrane reactors, or immobilized) to enable recovery and simplify product purification. Cell-free systems, on the other hand, allow the catalyst to be added to the reaction bulk, either at the beginning or at some point in the cascade in the case of multiple reactors, but likewise they can be compartmentalized. A more substantial difference among the two systems is the access of the substrate to the biocatalyst. With a cell-free enzyme there is no physical barrier between the bulk of the reaction and the catalyst, in fact, the mass transfer is even facilitated under flow conditions with respect to batch. In whole-cell systems, the integrity of the cell membrane is essential for the retention of the biocatalyst(s) within the cellular space and diffusion across the membrane must be possible for both substrates and products. To enable such diffusion and then the enzymatic

Tab. 10.1: Advantages and disadvantages of cell-free and cell-based flow reactors[a].

Type	Advantages	Disadvantages
Cell free	– Cleaner system – Ready access to the catalyst – Faster flow rate – Higher substrate concentration	– Cofactors (if required) must be added in the media – Possible purification steps needed – Stability of the free form may be limited
Cell based	– Multiple enzymes can be expressed and contained in the same cell system – Cofactors are available within the cell system – Rapid preparation of the catalyst(s)	– Flow rate achievable is generally low – Permeability of the cell membrane/wall may be limited – Integrity of the cell structure and cell viability could be affected over time with the consequent loss of the catalyst(s) – Product recovery may be difficult – Side reactions due to cell metabolism

[a]Adapted from Pinto et al. [15].

reaction to take place, the flow rate that can be achieved is often very low. However, a clear advantage of whole cells is that they can harbor more than one recombinant enzyme, and once good expression levels are achieved, the cells can be simply spun down and utilized. If the reaction is carried out in buffer, this can also be supplemented with small amounts of nutrients to preserve cell viability. Enzymes within a cellular environment are also generally more stable, and this is particularly true for those enzymes that are membrane-bound; therefore, if the cell itself withstands the reaction media, reagents, and so on, it can be greatly advantageous not to extract the catalyst at all. In addition, whole cells are often preferred when the biocatalyst of interest require a cofactor, because these natural products are always endogenously available, and the addition of such expensive reagents can be avoided. In fact, whole cells spontaneously recycle cofactors which again streamline the setup.

In general, however, cell-free systems are more popular among researchers as they offer a higher degree of control with less variables. The quantification of the catalyst, for example, can be very precise, because this is treated like any other reagent, and therefore it is independent of the natural variability which can be observed in a living organism such as whole cells. Through immobilization (see later), the stability of the biocatalyst can be increased, and its retention within a reactor can be almost guaranteed.

10.2.1.3 Immobilized or not?

Immobilization strategies to anchor enzymes or whole cells expressing enzymes to a solid matrix are extensively used in flow biocatalysis because they provide the easiest

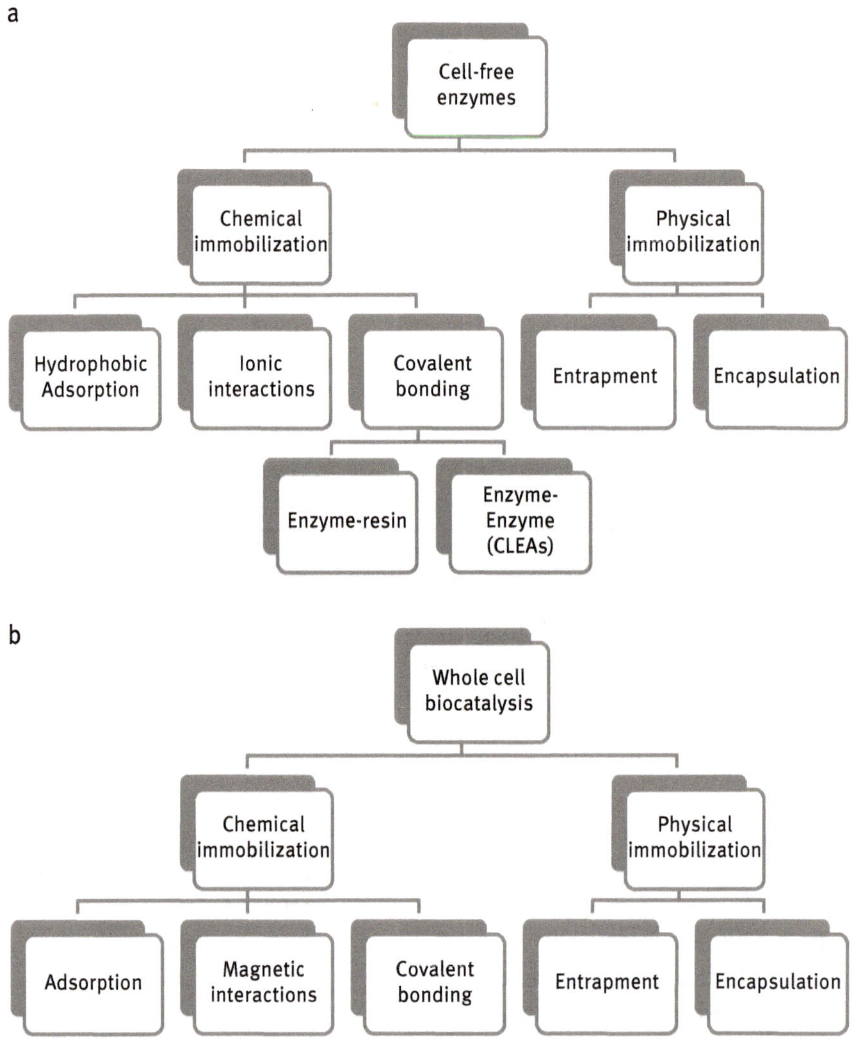

Fig. 10.3: Immobilization strategies for cell-free systems (a) and whole-cell biocatalysts (b).

way to contain a biocatalyst within the defined space of a reactor (Fig. 10.3, panels A and B) [16, 17].

A number of techniques have been developed over the last few years, in particular for cell-free enzyme systems. Enzymes can be covalently or noncovalently attached to a support which is then used to pack a reactor, or on the surface of the reactor itself (normally this approach is suitable for microfluidic systems where the narrow channels would not accommodate a solid resin), or on magnetic nano- and microparticles [16]. Whole cells can also be immobilized, the most common approach is to trap them in hydrogels which preserve their viability and, as it is done for immobilized enzymes, the matrix can be used in a packed-bed reactor. The ability of whole cells to form biofilms can also be exploited in microfluidic devices where the channel can be coated with a cell layer yielding a large active surface area which favor access to the biocatalyst.

Cell-free biocatalysts greatly benefit from immobilization because the close contact with a matrix (which can be carefully selected to offer a beneficial micro-environment), independently of the immobilization method, confers enhanced stability to the enzyme which can be used for longer period of times, often improving also thermal properties and solvent tolerance [18]. No optimal strategy for immobilization is known however, and each catalyst requires generally a series of trials before a suitable approach, better than alternative protocols, is selected, and this can be quite time consuming. Immobilization of cell-free enzymes has also drawbacks such as the possible obstruction of the active site if the anchoring is nondirectional, reduced activity of the catalyst if excessive rigidification is imposed (typical of covalent immobilization strategies), and possible interaction of the matrix itself with the reagents (often observed between hydrophobic resins and aromatic, hydrophobic reagents). Finally, if the immobilization is not covalent, the possibility of slow leaching of the catalyst, promoted overtime even by the sheer action of the flow itself, and consequent loss of activity in the reactor, must be considered. In whole-cell immobilization, these drawbacks are much less relevant as the catalyst is protected within the cell itself.

Enzyme loading: amount of biocatalyst added to the reactor (for immobilized biocatalysts is better to specify also the matrix loading: $mg_{biocatalyst}\ g_{matrix}^{-1}$).

Enzyme specific activity: U mg_{enzyme}^{-1}, where the enzyme is in solution, normally under its optimal conditions.

Activity of the immobilized enzyme: specific activity of the support following enzyme immobilization, expressed per gram of matrix (U g_{matrix}^{-1}).

Enzyme productivity: g of product synthesized per g of biocatalyst employed in the flow reaction.

Bioreactor stability: biocatalyst activity over time

Depending on the specific requirements of a reaction, it may in fact be preferable to keep the catalyst in solution and recover it (via liquid–liquid extraction or affinity column) downstream of the reactor [9]. Likewise, whole cells, provided they can be efficiently separated at the end of the process, can be used in "soluble" form because this increases the access to the cytoplasm and to the catalyst [19].

10.2.2 Key design criteria

Designing a synthetic process involving biocatalytic steps under continuous-flow conditions require multiple considerations [3, 4, 6, 9, 11, 12]. Typically, a synthetic process targeting a product (TM) involves several steps which might be performed with the aid of biocatalysts and by other chemical means (Fig. 10.4). Different parts of the whole process can be performed by biocatalysis and classical chemical methods and various (and not necessarily the same) segments of the full synthesis can be carried out under continuous-flow conditions. Consequently, designing and optimization of such a complex system is a demanding task. Even if considering only the biocatalytic parts, the increase in number of the enzymes involved in the process boosts the complexity of the system rapidly [20].

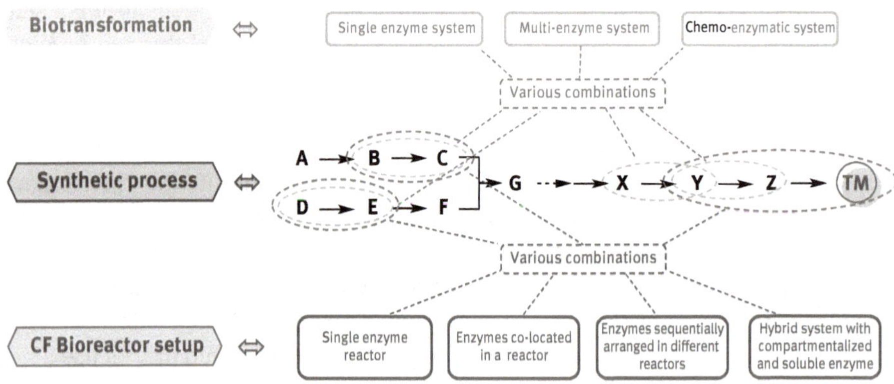

Fig. 10.4: Implementation of biocatalysis and continuous-flow bioreactor systems within a multistep synthetic process.

Even for optimization a single enzyme–based biotransformation, multiple issues should be considered. For example, remarkable differences exist between biotransformation with enzymes requiring no external cofactors (requiring no cofactor or acting with tightly bound and autocatalytically regenerating cofactor) and biotransformation with enzymes requiring external cofactors which should be retained and regenerated [21].

Before discussing the important optimization parameters of continuous-flow biocatalysis in detail, the differences between whole-cell and cell-free enzyme biocatalysis must be understood. Whole-cell biocatalysis uses the entire organism (such as *E. coli*) for biotransformation – even in continuous-flow mode [15] – whereas cell-free enzyme biocatalysis uses partially or fully purified enzyme recovered from the cell in soluble or in immobilized form. Understandably, influence of the operational parameters on the actual biotransformation depends not directly on the inherent properties of the single soluble enzyme but rather on the effective properties (kinetic parameters) of the enzyme in the actual form applied as a biocatalyst (Fig. 10.5) [22, 23].

Logically, optimization of a complex multienzyme system is challenging [24–28], because even in the case of a single-step single-enzyme biotransformation the various immobilization forms or reaction systems result in different effective parameters.

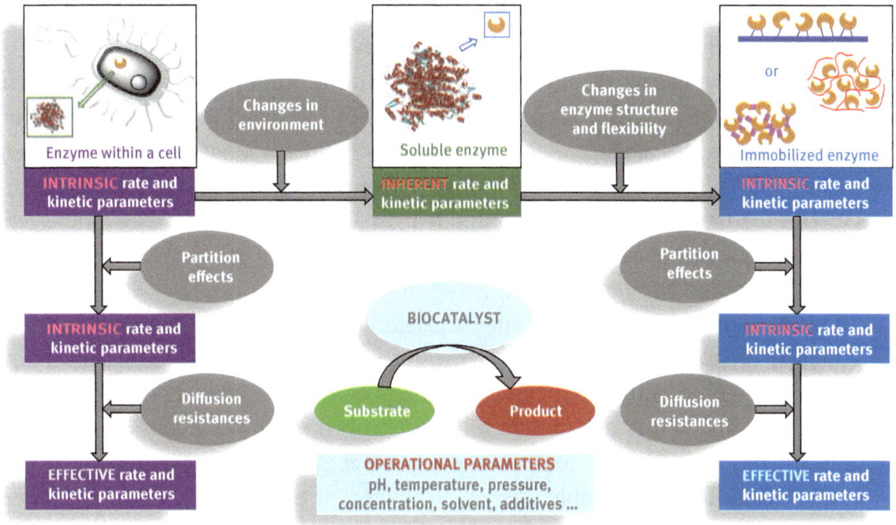

Fig. 10.5: Effect of the actual form of an enzyme applied as biocatalyst on the influence of operational parameters on an enzyme-catalyzed biotransformation.

Thus, when one changes the immobilization or reactor type of single unit in a multi-unit system or changes the usage of an enzyme from a single-enzyme unit to a multi-enzyme or chemoenzymatic unit, redetermination of the effective parameters may be necessary.

Bioprocess engineering is an essential part of biocatalytic processes in the context of sustainable industrial production of chemicals [1, 29]. In cells, enzymes rarely function in isolation, most often in reaction networks rather than in linear sequences. This allows the removal of inhibitory products and the shifting of unfavorable equilibria, and the coupled reactions of enzyme networks can also be used to regenerate cofactors or prepare substrates. Rigorous mathematical descriptions of microbial cells and consortia thereof will enable deeper biological understanding and lead to powerful in silico cellular models [30]. These levels of complexity need to be studied, understood, and modeled to decide when to use isolated enzymes or when to use whole-cell biocatalysis instead.

10.2.2.1 Operational parameters

Among the operational parameters influencing the outcome of biotransformation, pH and temperature are of particular importance. Further important parameters, especially in continuous-flow systems are the composition of solvent, the substrate concentration, biocatalyst loading, flow rate, pressure. Last but not least, catalytic efficiency, selectivity, and stability of the biocatalyst are also key issues [1–12].

In many enzyme-catalyzed reactions, pH control is an important feature because most of the biocatalytic reactions occur in aqueous solution and the biotransformation consumes or produces acidic or basic species. Additionally, a large fraction of enzymes operates within a narrow pH window and therefore the pH control is of utmost importance. This feature influences strongly the reactor selection, and therefore whole-cell- and soluble enzyme–based processes are best operated in stirred tank reactors or membrane reactors when pH control is required [1].

A study of the continuous-flow kinetic resolution of three different racemic amines by variously immobilized *Candida antarctica* lipase B biocatalysts in the 0–70 °C temperature range on enantiomer selectivity and specific reaction rate indicated, that temperature effect depended significantly both on the substrate and on the mode of immobilization [31]. Alteration of the enantiomer selectivity in the kinetic resolutions of three differently flexible amines a function of temperature was rationalized by the various flexibility of the lipase in its different forms. The results indicated that the optimal method of immobilization depended both on the nature of the substrate and on the reaction conditions.

10.2.2.2 Kinetics

There are seminal reviews on the opportunities and challenges of using microfluidic methods in biocatalyzed processes with isolated enzymes or whole cells for both analytical and chemical synthesis [1–12]. Immobilization of biocatalysts is widely used in microreactor systems because it allows continuous operation, easy separation of product and biocatalyst, and often stabilizes the biocatalyst. However, immobilization of the biocatalyst can alter the activity of the enzyme due to conformational changes, steric hindrances (e.g., overcrowded or non-oriented binding of the enzyme, which causes diffusion restriction of substrates accessing and products egressing the active site), all resulting in changes in reaction kinetics (Fig. 10.5) [32].

Enzyme kinetics: Study of the rate of chemical reactions catalyzed by enzymes (which are usually but not necessarily protein molecules that convert other molecules, called substrates). The substrate molecules bind to the active site of the enzyme and are transformed into product(s) by a series of steps corresponding to the mechanism of the enzyme. Enzyme kinetics investigates the reaction rate as a function of changing reaction conditions (pH, temperature, substrate concentration, etc.).

Enzyme inhibition: When an **inhibitor** interacts with an enzyme, it decreases the enzyme's catalytic efficiency. An irreversible inhibitor binds to the enzyme's active site – most often covalently – producing a permanent loss of catalytic ability even if the inhibitor's concentration is decreased. A reversible inhibitor – usually by forming a noncovalent complex with the enzyme – decreases the catalytic efficiency temporarily. If the inhibitor is removed, the enzyme's catalytic efficiency returns to its original level.

Substrate/product inhibition: Enzyme inhibition at elevated concentration of the substrate/product.

The immobilized enzyme is often present within porous polymer, ceramic, or silica particles [7, 16]. In such cases, the substrate must also diffuse through the porous medium to reach the enzyme. Thus, the diffusion resistance of the particles must be considered together with the external mass transfer resistance [33].

One of the main obstacles to biotechnological process design and control is the problem of measuring the most important physical, kinetic and biochemical parameters [34, 35]. There exist efficient experimental methods to determine kinetic parameters even within a microreactor [36–38]. Another option is to model and simulate processes within microreactors using state-of-the-art mathematical modeling techniques [39].

For modeling microbioreactors, where the intraparticle and external diffusion resistance are significant, multicompartment models are required to accomplish adequate model accuracy [35]. However, monocompartment models, in which the internal mass transport by diffusion and substrate conversion is considered, are still used in various applications due to the simplicity of the model [40]. In addition, substrate conversion is often studied only when enzyme kinetics approach first- or zero-order kinetics [33, 41].

The simulation approach allows to estimate certain kinetic parameters and to optimize the microfluidic reactor system with significantly decreased time and cost [35, 39, 40, 42].

10.2.2.3 Choice of reactor

Although there are variations in many implementations, size and geometry of the reactors applicable for biotransformations in continuous-flow mode, there exist two basic theoretical types of continuous-flow reactor: continuous stirred tank reactor (CSTR) and continuous plug-flow reactor (CPFR). A systematic study on reactor selection for continuous biocatalytic production of pharmaceuticals with focus on residence time distribution and on the unique mass balance affected by enzyme kinetics for each reactor type revealed that CPFR should generally be the system of choice [43].

In various microfluidics platforms mostly the CPFR-type microreactors are applied in different setups (Fig. 10.6) [44].

However, there are cases, such as the necessity of pH control as mentioned in Section 10.2.1, where they may need to be coupled with a CSTR or replaced entirely by a series of CSTRs, which can approximate plug-flow behavior. In this respect, continuous production in enzyme membrane reactors (EMR), such as in a cascade of continuously operated EMRs, can be considered theoretically as CSTRs [45]. In

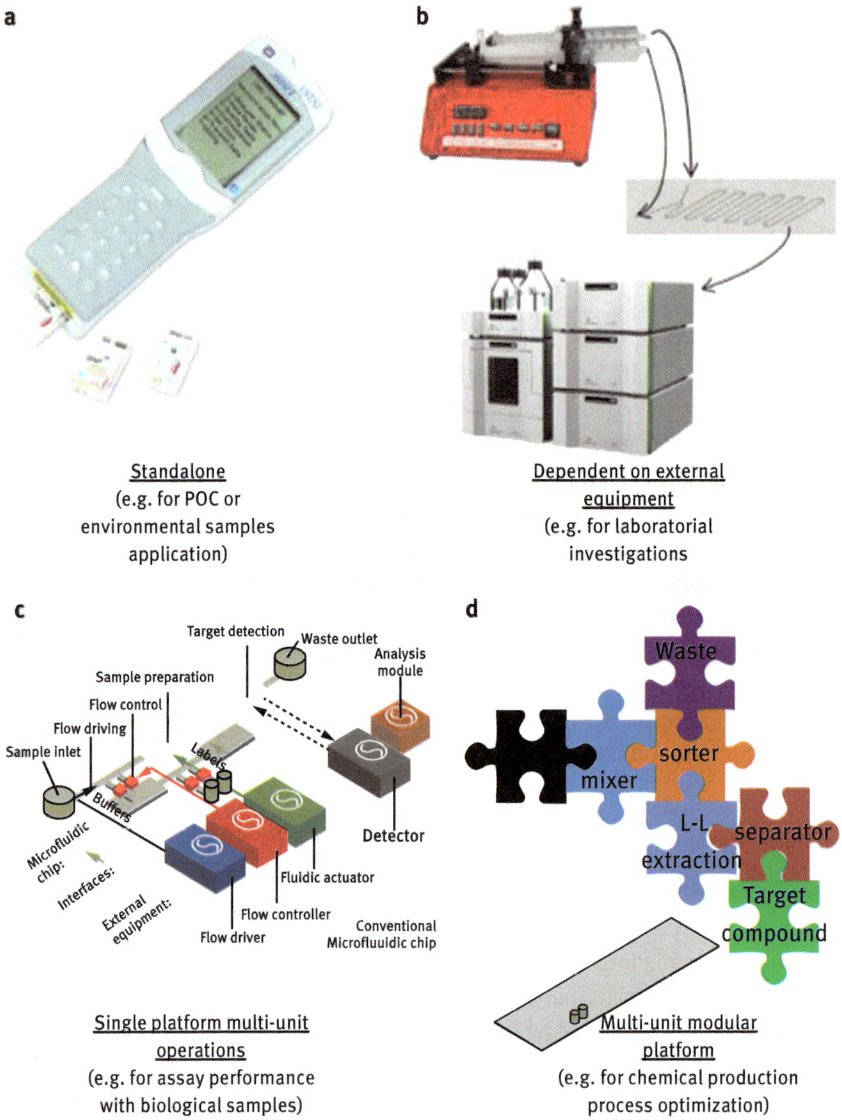

Fig. 10.6: Simplified representation of different types of microfluidic systems or platforms: (a) standalone platform (containing all the components required to carry out the target process or assay); (b) standard microfluidic system (that depends on certain external equipment such as pumps and analytical devices to carry out the target process or assay); (c) conventional approach to microfluidic platforms (miniaturization and integration of all the necessary unit operations in the same chip); and (d) modular approach to microfluidic platforms (integration of all the required unit operations as separate miniaturized unit operations that are interchangeable and replaceable by new units if needed). Adapted from Fernandes et al. [44]. *Copyright © 2018, Elsevier.*

practice, no reactor behaves ideally; instead, it falls within the mixing limits of the ideal CSTR and CPFR.

Analysis of a proposed end-to-end continuous manufacturing process for Sitagliptin, the active pharmaceutical ingredient of the leading dipeptidyl peptidase-4 inhibitor antidiabetic drug by using process modeling and optimization indicated the supremacy of a plug-flow microreactor-based process over the batch implementation [46].

(For related topics please see Volume 1, Chapter 3, Title: Technology overview/Overview of the devices)

10.2.2.4 Analytical methods and in-process control

Many optical, optofluidic [47–50], spectroscopic (such as UV [49], Vis [49], NIR [49], Raman [50], NMR [51, 52]), or MS [53, 54] and electrochemical [55] detection methods are available that allow monitoring a wide range of reaction variables (oxygen, pH, glucose, carbon dioxide, or specific reagents/products) online and real time (Fig. 10.7).

Devices exist for in-line monitoring of reaction conditions such as microfluidic flow rate [56, 57], temperature [58], pH [59], pressure [60], or oxygen [48]. Although monitoring operational parameters and reaction variables in cascade systems and individual reaction rates at different positions in the cascade system can significantly shorten process development time and provide a basis for quality-oriented design approaches, these methods have been used primarily in individual continuous-flow enzymatic reactions, and only very rarely in continuous enzymatic cascade systems [26, 34]. Since many detection methods depend on the appropriate chemical, optical, or spectroscopic properties that allow easy measurement, general detection methods may be of great interest. NMR integrated with continuous-flow systems opens the door to advanced reaction monitoring techniques that have a high level of information content in real time and may be applicable to self-optimize processes using one or more NMR methodologies [51, 52]. Integration of microreactors by chiral analysis and the detection of enantiomers in an unlabeled manner is an important step in the study of stereoselective biocatalytic transformations [61]. Regrettably, there are many methods to detect substrate and product concentrations which cannot currently be performed in-line. The existing in-line monitoring and detection tools can be integrated with various control units enabling highly automated control of the continuous microfluidic system [44].

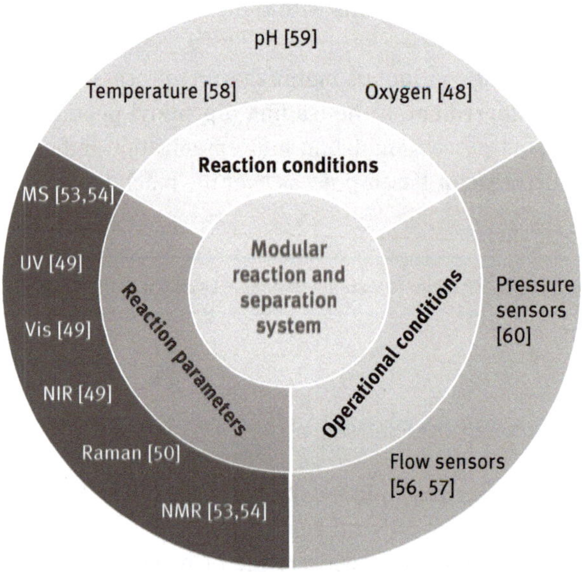

Fig. 10.7: Options for monitoring modular reaction or separation systems to improve control over reactions. The monitoring has to address reaction parameters (e.g., substrates and products), reaction conditions (e.g., pH and temperature), and operational conditions (e.g., flow rates and pressure). Several in-line spectroscopic methods and sensors have been developed and implemented in microfluidic systems which can now address this. Adapted from Gruber et al. [26]. *Copyright © 2017, CY BY 4.0.*

10.2.2.5 Optimization of CF reactors with enzymes

Continuous-flow implementation is an ideal method for efficient execution and computer-controlled optimization of chemical and biocatalyzed reactions due to its inherent advantages such as precise control of reaction time, temperature, and composition. The parameters of the various processes can be pre-programmed, each output can be analyzed automatically, and the protocol can be iteratively repeated or optimized. Selected examples from the vast number of documented optimization procedures indicate several possibilities and strategies of optimizing a single step continuous-flow biotransformation.

Optimization of operating temperature for continuous packed bed immobilized glucose isomerase reactor with pseudo linear kinetics was studied [62]. This optimization problem was based on reversible pseudo linear kinetics and the thermal deactivation of the enzyme and the substrate protection during the reactor operation was considered as well. This method involved the solution of two coupled nonlinear ordinary differential equations of the initial value type during a one-dimensional unconstrained optimization with bounds on the reactor operating temperature.

Design of experiments (DoE) was used to optimize the reaction conditions (pH value and temperature) in batch for optimization of the synthesis of a statin side chain precursor in continuous-flow continuous with immobilized deoxyribose-5-phosphate aldolase (DERA) [63]. In the DoE process the effect of the two crucial process parameters, temperature, and pH value were evaluated on intermediate and product formation in two cycles: (i) a rough, full-factorial lattice was designed for screening the process settings (temperature between 28 and 37 °C and the pH between 6.0 and 8.0); (ii) a fine full-factorial lattice was laid in the optimum of the response surface of the first cycle.

A two-step optimization was performed for continuous-flow hot water extraction and enzymatic hydrolysis using a thermostable β-glucosidase for determination of quercetin in onion raw materials [64]. First, the enzymatic hydrolysis was optimized [a three level central composite design considering temperature (75–95 °C), pH (3–6) and ethanol concentration (5–15%)]; followed by optimization of the hot water extractions from chopped yellow onions [a two-level design considering pH (2.6 and 5.5), ethanol concentration (0% and 5%) and flow rate (1 and 3 mL min^{-1}) at the optimal temperature for hydrolysis]. The optimized continuous-flow method (84 °C, 5% ethanol, pH 5.5, 3 mL min^{-1}) resulted in quercetin from onions in higher yield (e.g., 8.4 ± 0.7 μmol g^{-1} fresh onion) compared to a conventional batch extraction method.

10.3 Practical guide to CF biocatalysis

In this section some practical tips for approaching biocatalysis in flow will be discussed. It is intended as a basic guide for non-specialists to get an insight into the overall process. The fundamental steps to preparing a reactor, the initial setup of a continuous biotransformation, its optimization as well as product work-up will be also outlined.

10.3.1 Production of enzymes and cofactors

Enzyme production has been routinely done in biochemistry labs, but for a scientist with no previous knowledge in this field, even the idea handling a large biomolecule could be off putting. However, the process is relatively simple and in theory any research group, provided they can access to some basic equipment such as an incubator and a centrifuge, could obtain bespoke catalysts to work with. In fact, an ever-growing range of commercial enzymes can be purchased and trialed as any other reagent.

Most biocatalysts can be expressed recombinantly in laboratory strains of *E. coli* (the original gene coding for the target enzyme is cloned into a bacterial cell capable of translating it into the protein), alternatively, bacterial and yeast cultures spontaneously producing sufficiently high levels of one or more biocatalysts can simply be grown in suitable media. Following successful expression of the enzyme(s), if the intention is to use whole cells for the biotransformation, the cell culture is spun down, rinsed in fresh buffer, spun again and the cell paste is ready for use. The enzyme can also be extracted from the cells (assuming it is not secreted into the media) by cell lysis (mechanical or chemical), separated from the cell debris by centrifugation and purified. Enzyme purifications are generally facilitated by the presence of an affinity tag which allows selective binding to a column, elimination of any other proteins, and elution. The catalyst is now ready for the next step. Cofactors are commercially available, but they can be expensive; therefore, all cofactors that are needed stoichiometrically, for example in a redox reaction, will need to be regenerated within the reaction so that they can be used catalytically, and the overall cost of the process can be contained. Recently, the possibility of co-immobilizing the cofactor with the enzyme or enzyme-pair to create a self-sufficient resin has also been reported and could be particularly advantageous for flow systems [65].

10.3.2 In situ immobilization

In Section 10.2.1.3, a general outline of the different types of immobilization for whole cells and cell-free enzymes was given. One practical aspect of the immobilization process for cell-free enzymes is that no matter the type of immobilization, it generally involves mixing the catalyst with an activated support, rinsing off the unbound enzyme, possibly additional chemical steps to terminate further reactivity, washing again, and final storage of the immobilized catalyst till it is needed. The reactor can then be packed at any time to start the flow reaction. However, it is also possible to pack a reactor with the clean resin and perform all the immobilization steps feeding all reagents sequentially directly through the flow system. This approach offers the advantage that there is no loss of resin through filtration, which is almost inevitable when the immobilization is done in batch mode. On the other hand, true continuous in situ immobilization is not efficient for all steps; normally the time required for the enzyme to bind to the resin ranges from 2 to 16 h (depending on the chemistry); therefore, loading the resin with the enzyme containing solution and stopping the flow for the required time is advisable.

When the cell-free enzyme or whole cells are bound directly to the surface of the reactor channel (wall-coated reactors), this can also be carried out in situ. The aim is to maximize the exposed catalyst surface by creating a monolayer or a micron-thick proteinaceous layer within the microchannel [32] which will favor smooth operation of the reactor afterwards. Cells can also be immobilized in high densities and methods have been developed which ensure that the cells remain stably attached over several days of continuous microreactor operation [66].

The stability of the immobilized catalyst under flow conditions must then be tested (generally with a standard reaction) to estimate the life span of the reactor.

10.3.3 Optimization of reaction under flow conditions

Example:
Batch reaction: 10 mL total volume, 10 mM substrate, 10 mg of enzyme, achieves full conversion in 10h.

Flow starting conditions: the reactor contains 10 mg of enzyme immobilized on 1 g of resin (occupying a volume of 2 mL), reaction volume in the reactor is 0.5 mL, same 10 mM substrate. A residence time of 30 min should give full conversion (likely much shorter residence time will then be found).

When the reactor is assembled, the system is ready to go. In a multienzyme cascade, normally each reactor is optimized separately and sequentially so that the next reactor can be tested with the optimal reaction mix generated by the previous one. It is always important to compare the efficiency of the system with the equivalent batch conditions. The expectation is that in flow, the rate of the reaction is enhanced significantly but to achieve this the correct residence time must be established. A good starting point is to take the batch reaction and translate it into flow considering the fact that the volume of the reaction exposed to the total catalyst is, at any given time, normally a fraction of the batch process.

With immobilized biocatalyst it is possible to further increase the rate of the reaction by increasing the temperature by a few degrees with respect to the optimal temperature established for the free enzyme without significantly affecting its stability. In the absence of in line monitoring equipment, the collected fractions need to be manually checked for product conversion. If the product has high affinity for the resin or the reactor material (more commonly observed with hydrophobic resins and hydrophobic substrates as mentioned), it is common to observe several initial fractions with an apparent lower conversion yield which then stabilizes as the process continues. The resin is at that point saturated with the product and a steady state of production is achieved.

If the substrate is excessively retained in the resin, this could negatively affect the efficiency of the enzyme. The use of a segmented flow could then be explored. Generally, biotransformations are carried out in aqueous buffers, but a second inlet could be used to inject a different solvent at specific intervals. The solvent should be capable of stripping the resin from the trapped product and the true yield of the reaction can then be calculated. The tolerance of the catalyst to the segmented flow should of course be tested in case that the enzyme loses activity more rapidly when in contact with the solvent.

10.3.4 Downstream processing and recycling

Flow technology allows for several downstream processing modules to be integrated in the overall process. This is already well known in flow chemistry and here some of the most used systems will be discussed. As mentioned, many biotransformations are carried out in buffers but it is often possible through acidification/basification inlets to change the protonation state of a product or an unreacted starting material so that it can be extracted in a solvent.

Liquid–liquid extraction module. Several processes have been reported which include a liquid–liquid extraction system. Normally, an inlet with a non-water miscible solvent is position right after the (last) reactor. The mixing of the two phases takes place within the junction and the separation follows (Fig. 10.8).

Depending on the size of the separator, different volumes can be handled. If a segmented flow is used, with an inlet of solvent upstream of a reactor, the relative volumes of the two phases may be sufficient downstream to enable efficient separation. In such case, no additional solvent inlet is required.

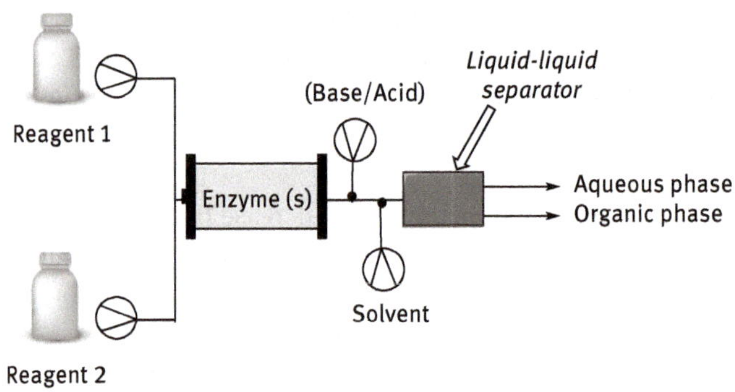

Fig. 10.8: Scheme showing the positioning of a liquid–liquid separator in a flow. apparatus.

Purification and scavenger modules. A range of columns containing differently derivatized resins (sulfonic, aminated, etc.) could be used downstream of a flow process, or even intercalated between reactors, to purify the product from unwanted byproducts. The process is very simple and again routinely used in flow chemistry, the downside is that these columns not always can be regenerated and can be quite costly, but they allow complete automation of the system.

Recycling. Under flow conditions, the amount of reagents added in excess which is then lost downstream can be significant. This is particularly true in biocatalytic processes, especially those which require cofactors that are generally added in the mobile phase. It is possible, however, to recycle the water and/or organic phases, by simply feeding them back into the system, provided they can be "cleaned up" from unwanted molecules exploiting the same technology mentioned above. Cofactors are water soluble and it is conceivable that they will remain in the aqueous phase following a liquid–liquid extraction. This phase can be fed back into the system, supplemented with fresh starting material and reused several times [67].

10.4 Examples of CF biosynthetic syntheses

Below a selection of different examples of biosynthetic reactions in flow will be presented. While this is by no means exhaustive, it should give an overview of the various approaches that have been developed at large scale in the past and at lower scale in the last few years.

10.4.1 Large-scale biocatalysis in CF

Biocatalysis has been applied in many applications as an alternative to chemical catalysis in various fields. The most striking examples of using enzymes in organic syntheses are the preparation of pure enantiomeric forms of various chiral drugs and the synthesis of important agents for the taste, food, and fragrance industries [68]. Biocatalysts are also used on a large scale to produce special and even bulk chemicals. Applicability is greatly influenced by the scalability of processes using enzymes in the production of industrial chemicals.

The very high activity of glycohydrolases (such as amylases, cellulases, and pectinases) has long been used for industrial purposes. Carbohydrate manipulating enzymes are widely used in the hydrolysis of various polysaccharides and in glucose production, now completely replacing acid catalysis. The use of glucose isomerase to produce fructose can be considered as a real biotransformation. This method using immobilized glucose isomerase (IGI) for production of high-fructose corn syrup (HFCS; 42–55% fructose, residual glucose and <4% oligosaccharides) at ~14 million t a^{-1} scale (dry weight) [69] is by far the largest scale synthetic biotransformation. The low catalyst consuming process (~0.05 g IGI kg^{-1} HFCS) is operated continuously in packed bed reactors at 60 °C with high efficiency [STY ~ 1 kg L^{-1} h^{-1} (calculated on dry HFCS)] [70, 71].

By using lipase biocatalysts in combination with proper acylating agents, kinetic resolution of racemic amines can be performed with high efficiency (Fig. 10.9). An example is the enantiomer selective acylation of racemic 1-phenyl-2-ethanamine according to the process patented by BASF [72].

The kinetic resolution of the racemic amine performed in continuous-flow mode using a column packed with immobilized lipase B from *Candida antarctica* and isopropyl 2-methoxyacetate as acylating agent resulted in one step a mixture of almost enantiomerically pure (*S*)-1-phenyl-2-ethanamine and the (*R*)-amide (Fig. 10.9). Large-scale

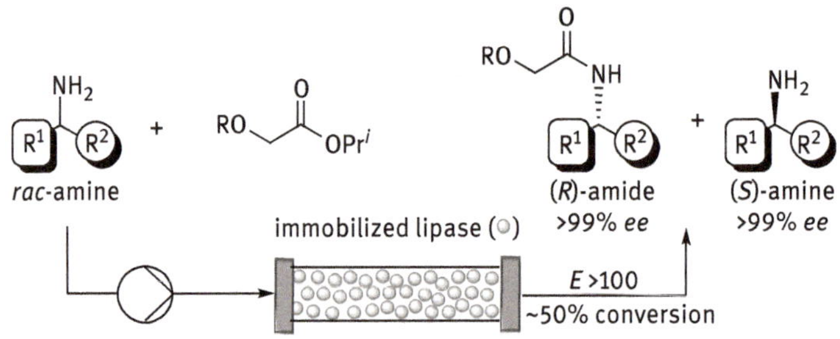

Fig. 10.9: Large-scale kinetic resolution of racemic amines with isopropyl 2-alkoxyacetates in continuous-flow mode.

production of the (*S*)-amine has been carried out in a cGMP-compliant plant (cGMP: current good manufacturing practices) of BASF on a scale of more than 1,000 t a^{-1}. In another BASF plant, 2,500 t a^{-1} (*S*)-(1-methoxy)-2-propylamine – being a synthesis intermediate of an herbicide – can be produced by this process [73]. Recently, the isopropyl 2-propoxyacetate among a series of isopropyl 2-alkoxyacetates was found to be the most efficient acylating agent in kinetic resolution of several racemic amines including racemic 1-phenyl-2-ethanamine under continuous-flow conditions [74].

Researchers of Pfizer developed a continuous enzymatic process for an efficient synthesis of (*R*)-3-(4-fluorophenyl)-2-hydroxy propionic acid at multikilogram scale using soluble D-lactate dehydrogenase from *Leuconostoc mesenteroides* (D-LDH) and formate dehydrogenase from *Candida boidinii* (FDH) (Fig. 10.10) [75]. (*R*)-3-(4-Fluorophenyl)-2-hydroxy propionic acid is a building block for the synthesis of Rupintrivir, a rhinovirus protease inhibitor. The production was performed in a 2.2 L continuously operated EMR (CEMR) using flow rate ratio of ~20 for recirculation versus filtration resulting in a residence time of 3 h leading to >90% conversion and high space–time yield (560 g L^{-1} day^{-1}). In a period of 9 days without adding fresh D-LDH and FDH, the enzymes lose their activity at a rate of only about 1% per day. The product was obtained with excellent enantiomeric excess (ee > 99.9%) and good overall yield (68–72%).

10.4.2 Examples of continuous-flow biotransformations

Whole cells in flow: early days. One of the first examples of biotransformations in flow was reported in 1990 by Lee and Chang [76] for the continuous production of acrylamide (Fig. 10.11). The immobilization of *Brevibacterium* sp. CH_2 whole cells, which displayed high nitrile hydratase activity and high tolerance to acrlylonitrile, was carried out in polyacrylamide gel and used to pack a jacked glass column. To maintain the enzymatic activity stable over time, the whole reactor and feeding reservoirs were kept at 4 °C. Due to the tolerance to adapted cell strain, the concertation of acrylonitrile starting material could be increased from 2% (used in biotransformation up until

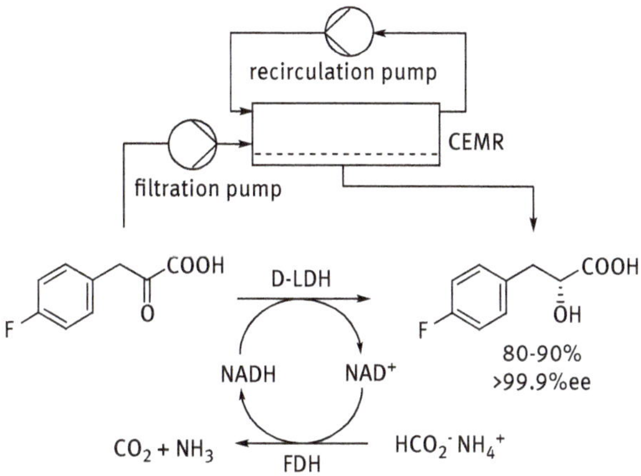

Fig. 10.10: Kilogram-scale production of (*R*)-3-(4-fluorophenyl)-2-hydroxy propionic acid by bioreduction in continuously operated enzyme membrane reactor (CEMR).

then) to 6%. To further increase the productivity of the system beyond the state of the art, the authors run two reactors in series with an additional feed of pure acrylonitrile before the second reactor. These allowed yields up to 55% with respect to the average 20% conversion achieved to date.

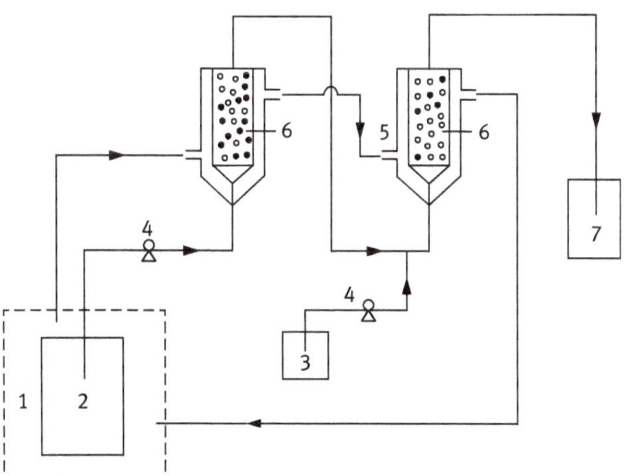

Fig. 10.11: Original scheme displaying a reactor assembly for acrylamide production in 1990. (1) Water bath kept at 4 °C, (2) 6% acrylonitrile solution, (3) pure acrylonitrile, (4) pump, (5) water jacket, (6) packed bed reactor, and (7) product collection. Adapted from Lee and Chang [76]. *Copyright © 1990, Springer.*

Many things have changed and evolved in the 30 years from this original paper with significant and unthinkable innovation at the time.

Self-sustainable flow biocatalysis with the recycling of "waste waters." Contente and Paradisi developed an original closed-loop platform of packed-bed flow reactors for the biocatalytic cascade synthesis of alcohols from amines based on enzyme covalent immobilization [67]. A transaminase for the haloadapted bacterium *Halomonas elongata* (HEWT) and a suitable reductase (horse liver alcohol dehydrogenase – HLADH, or a ketoreductase from *Pichia glucozyma* – KRED) were covalently immobilized on suitable supports using, to set up two packed-bed reactors. While the HLADH could be used also to recycle the cofactor by adding ethanol to the reaction mix, the KRED required a coupled enzyme system (a glucose dehydrogenase from *Bacillus megaterium* – *Bm*GDH) which was immobilized separately and a mix-bed reactor containing both KRED and *Bm*GDH was assembled (Fig. 10.12). A continuous-flow serial connection of these reactors was used for the synthesis of high-value primary and secondary alcohols from a variety of commercial amines via a carbonyl intermediate. A liquid–liquid extraction followed by purification of the product in the organic phase allowed the isolation of the alcohol product in excellent yield and purity.

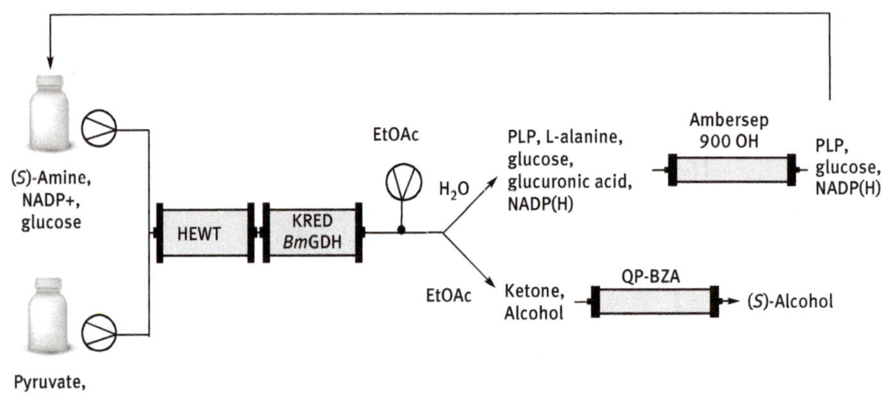

Fig. 10.12: Scheme showing multienzyme synthesis of chiral alcohols from amine with a transaminase followed by a redox reactor, with the in situ recycling of the cofactor, and reuse of the waste waters.

Of industrial relevance, the conversion of dopamine to hydroxytyrosol was achieved with 45 min residence time using a biphasic stream toluene/buffer and yielding 75% product recovery. Similarly, tryptophol and histaminol were synthesized from the corresponding biogenic amines tryptamine and histamine yielding 70% and 68% product recovery, respectively.

An additional feature of this work was the effort of recovering and reutilizing the wastewater stream which contained the cofactor. A scavenger column was introduced in line to partially purify the waters which could be recirculated for 20 times without further addition of the cofactor and virtually no loss of activity of either reactor.

Large-scale enzyme-based synthesis of melatonin. An example which shows the capacity to scale up the production of an enzyme-mediated reaction in flow was developed by Contente et al [77]. Here a highly productive biocatalytic reaction on a multigram scale for the continuous-flow synthesis of melatonin in an ultra-efficient closed-loop strategy was reported. The acetyltransferase from *Mycobacterium smegmatis* (*Ms*AcT) was covalently immobilized on glyoxyl activated agarose with a ratio of 1 mg $g^{-1}{}_{support}$ enzymatic loading, resulting in 73% recovered activity.

A packed-bed flow reactor was set up with just 1.6 g of charged resin and used initially for the continuous synthesis of *N*-(1-phenylethyl)acetamide from 1 M 1-phenylethanamine (120 g L^{-1}) in a biphasic flow system (5:15 aqueous phase/EtOAc), with 5 min residence time, allowing *N*-acetylation with 90% conversion. This covalent immobilization strategy significantly increased the enzyme stability in the presence of different esters and high temperatures. In addition, this covalent immobilized biocatalyst showed excellent operational stability, which allowed for an extensive utilization of the packed-bed flow reactor with an unprecedent amide production of 56 g day^{-1} using only 1.6 mg of MsAcT. This methodology was then successfully applied to the synthesis of melatonin from 5-methoxytryptamine (Fig. 10.13), as well as melatonin analogues, on a 0.5 M scale yielding very high molar conversions (up to 37 g day^{-1}).

As the acyl donor was EtOAc, this could be recovered following the isolation of the product and reused, once again closing minimizing the environmental impact of the process.

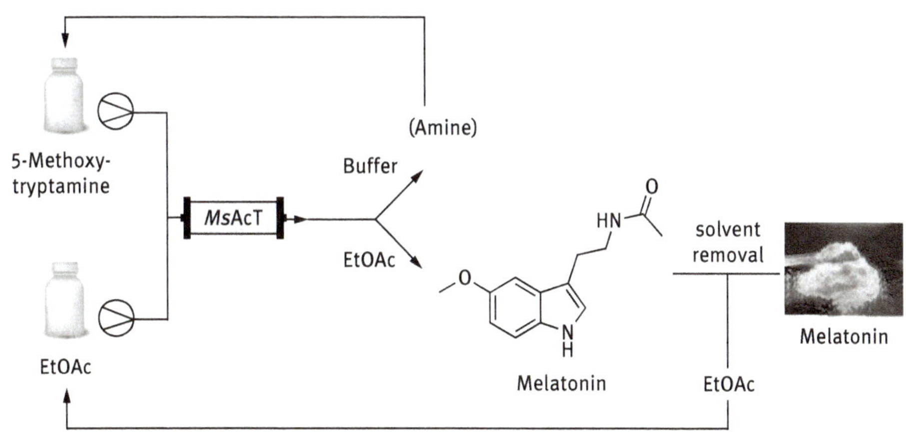

Fig. 10.13: Scheme of the flow setup for the synthesis of melatonin mediated by an acetyl transferase.

Chemoenzymatic continuous-flow dynamic kinetic resolution (DKR) *of amines by lipase catalysis.* Farkas et al. described a new chemoenzymatic system that allows the DKR of racemic benzylamines (*rac*-**1a–f**) in a completely continuous-flow manner (Fig. 10.14) [78].

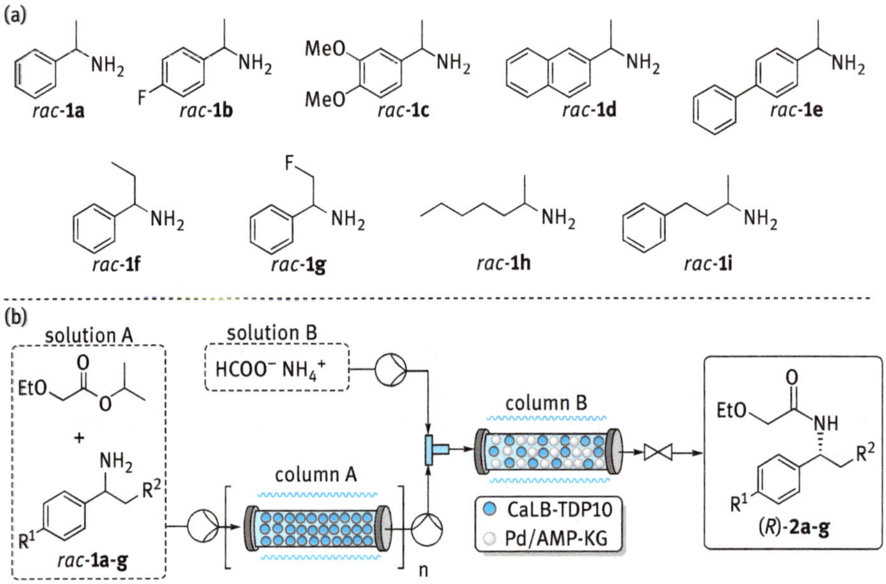

Fig. 10.14: Continuous-flow dynamic kinetic resolution of racemic amines by lipase-catalyzed acylation with isopropyl 2-ethoxyacetates. Adapted with permission from Farkas et al. [78]. *Copyright © 2018, American Chemical Society.*

For the DKR process, kinetic resolution with a robust, sol–gel matrix-immobilized form of *Candida antarctica* lipase B (CaLB-TDP10) was combined with palladium fixed on aminopropyl-functionalized silica (Pd/AMP-KG) and ammonium formate hydrogen source performing mild racemization. This DKR system, which contains ammonium formate completely dissolved in 2-methyl-2-butanol, could be operated at medium–high temperatures (60–70 °C). The study showed the racemization process only be compatible with benzylamines that do not have an easily reducible function without a reduction side reaction. The optimized DKR system consisting a purely KR unit (columns ($n = 1$ or 2) packed with CaLB-TDP10) and a mixed bed DKR unit (one Pd/AMP-KG + column packed with CaLB-TDP10) could be successfully applied to convert six valuable benzylamines (*rac*-**1a–f**) to (*R*)-amide [(*R*)-**2a–f**] in medium or high isolated yields (57–96%) with excellent enantiomeric purity (>98.8%).

Continuous-flow system comprising a capillary microreactor coupled to an extractor for kinetic resolution of acylated amino acids. An example of systems integrated with downstream units is the full kinetic resolution *N*-acetylamino acids in a capillary PFR – microextractor system containing cross-linked enzyme aggregate (CLEA) of aminoacylase co-cross-linked with polylysine (Fig. 10.15) [79].

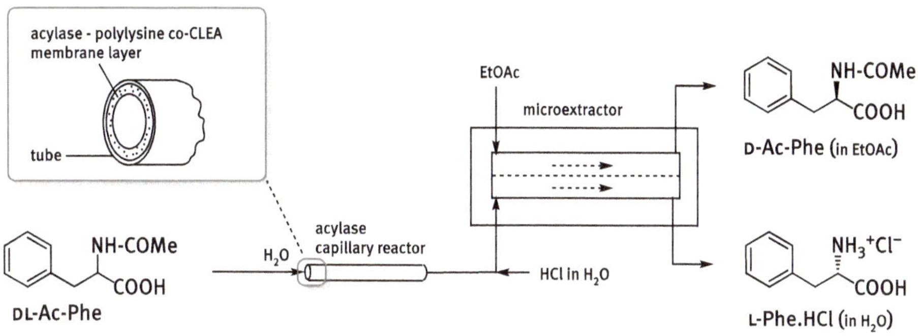

Fig. 10.15: Continuous-flow system for kinetic resolution of racemic amino acids (Honda et al. [79]). First, the L-enantiomer in the racemic substrate (4 mM) is converted by enantiomer selective hydrolysis into an amino acid through the tubing acylase-reactor. After acidifying with 0.2 M HCl the acetyl-D-Phe in the acidic aqueous phase is extracted selectively to the ethyl acetate phase flowing along the silicon-side in a microextractor.

The performance of the acylase-capillary PFR was evaluated with racemic *N*-acetyl phenylalanine (Ac- D,L-Phe, 1 mM) at a flow rate of 1.0 μL min^{-1} yielding the L-isomer in high enantiomeric purity (99.2–99.9% ee). The full KR of Ac-D,L-Phe and six further *N*-acetylamino acids were performed in this integrated system at a flow rate of 0.5 μL min^{-1} in the capillary PFR giving the L-isomers in 92.9–99.7% ee and 38–48.6% yield.

Biotransformation of propargylglycine by PAL in continuous-flow magnetic nanoparticle-based system with in-line UV detection. In microfluidic systems, protein-coated magnetic nanoparticles (MNPs) can optionally either co-flow with the fluid or even anchor with a magnet at defined site(s) letting to pass the free-flowing fluid through the regions of the anchored MNPs. Enzyme-coated MNPs can be fixed at certain positions in a lab-on-a-chip system using a permanent magnet (Fig. 10.16). In this microfluidic system called "Magne-Chip," multiple magnetic cells can be used for biotransformation, analysis, or protein biochemistry studies. Weiser et al. applied this microfluidic reactor equipped with in-line UV detection to study the mechanism of phenylalanine ammonia-lyase (PAL) immobilized on MNPs [80]. With this "Magne-Chip" system, it was first demonstrated that the acyclic, nonaromatic L-propargylglycine can be converted to (*E*)-pent-2-en-4-ynoate by PAL catalysis. This reaction strongly questioned the idea of a mechanism for the PAL reaction assuming a Friedel–Crafts-type attack on the aromatic ring of the substrate.

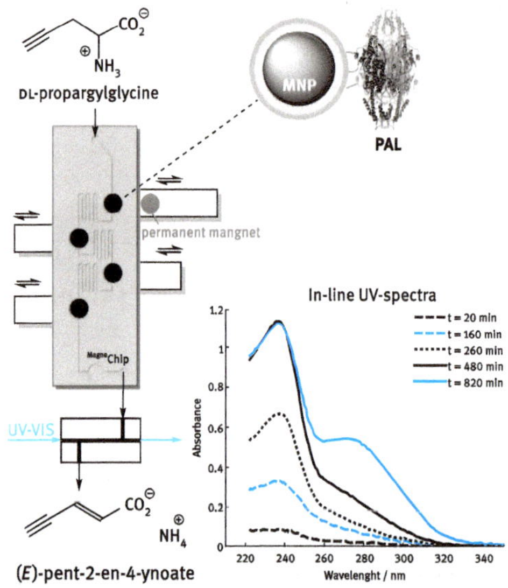

Fig. 10.16: Ammonia elimination from DL-propargylglycine in a light-protected microfluidic reactor with multiple magnetic cells filled with PAL immobilized on MNPs and equipped with an in-line UV/vis detector (reaction in D_2O at pD 8.8, 37 °C). Adapted with permission from Weiser et al. [80]. *Copyright © 2015, Wiley-VCH Verlag GmbH&Co. KGaA.*

10.4.3 More illustrative examples

Huffman and coworkers (Merck & Co.) reported in 2019 an outstanding approach to the synthesis of Islatravir, a nucleoside analogue used in the treatment of HIV [81]. which traditionally required between 12 and 18 steps, and can now be achieved exclusively enzymatically in fewer than half the steps with just five key enzymes (plus an additional four auxiliary ones) (Fig. 10.17). The enzymes were heavily engineered to ensure a compatibility window throughout the process, and some had been also immobilized to facilitate filtration from the reaction. The steps were run sequentially which maximized the yield.

This cascade would be an ideal candidate to be moved into flow, many of the challenges have been already addressed and flow could offer a further level of automation.

The possibility of using a multienzyme system of comparable complexity under flow conditions is indicated by an early report on cascade biocatalysis. Seven enzymes involved in the biosynthetic pathway were immobilized via His-tags to Ni-NTA agarose resin for the continuous production of uridine diphosphate galactose (UDP-galactose) from galactose and uracil monophosphate [82]. The multienzyme

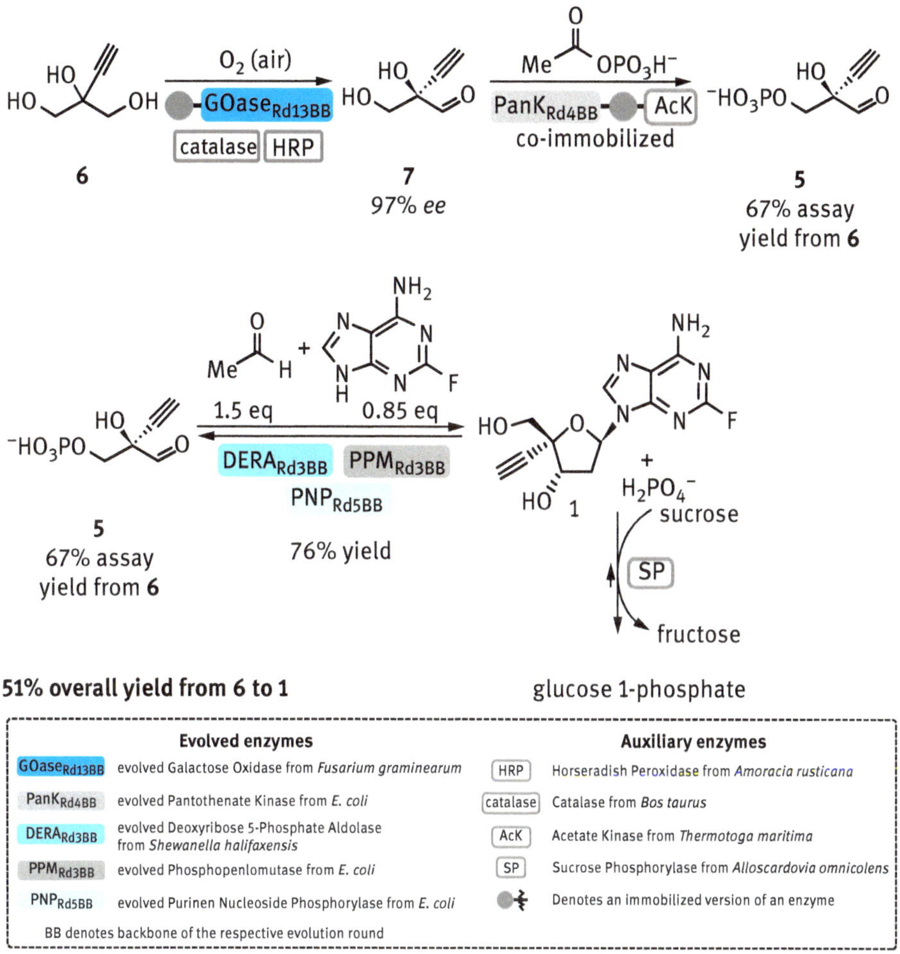

Fig. 10.17: Fully assembled biocatalytic pathway of Islatravir. *Copyright © 2019 The Authors, some rights reserved; exclusive licensee American Association for the Advancement of Science. No claim to original U.S. Government Works* [81].

resin was packed into a column to perform the bioconversion from galactose to UDP-galactose with an overall conversion of 50% by recirculation of the reaction mixture through the column over a period of 48 h.

10.5 Challenges and future opportunities

The challenges in continuous-flow biocatalysis and future opportunities of microfluidics-based enzyme-catalyzed applications are well reviewed [4–6, 11, 83, 84].

Performing biocatalytic transformations in continuous flow has the potential to assist in bioprocess intensification in many areas from academia to industry. Biocatalytic process development comprising enzyme/cell engineering and screening, substrate, medium, and operational parameter optimization, as well as reactor engineering and further process integration significantly benefit from the implementation of microflow processing. However, the scope and efficiency of current examples leave room for significant improvements, particularly with respect to catalytic efficiency and stability of the biocatalysts, ISPR, and implementation of multistep reactions.

Increasingly efficient and smart enzyme immobilization strategies continue to be developed, along with the expanding number and kind of available biocatalysts. Eventually, the realization of biocatalysts in flow with immobilized enzymes needs to surpass the trial-and-error approaches and evolve toward immobilization engineering [85] applying rational designs that fulfill the process requirements [16]. Importantly, the use of immobilized enzymes in microfluidic bioreactors further extends the applicability of biocatalysis by enabling creation of "plug-and-play" biocatalyst systems. The multiple "off-the-shelf" enzymic units can be assembled into cascade systems and be integrated with other enabling technologies [26].

Implementation of multienzyme cascade systems benefit from automatization of microfluidic devices, integration of in-line analytical tools, and emerging use of mathematical tools such as DoE, data analysis, and process simulations. This is expected to significantly improve the impact of biocatalysis on industry and medical applications. Creation of compartmentalized cascade reactions in continuous flow enabling end-to-end processing will have a significant impact on sustainable manufacturing of biobased products.

10.5.1 Upscaling and integration

Instrumentation of the continuous-flow processes depends on the number of molecules to be handled/synthesized and on the amount of the required compound(s). The examples in Section 10.4.1 and the literature survey for biocatalytic processes implemented so far indicated that instrumentation pattern of the biocatalytic processes implemented in continuous flow still follows the general trend for continuous-flow syntheses in pharma industry as summarized by Wheeler et al. in 2007 (Fig. 10.18) [86].

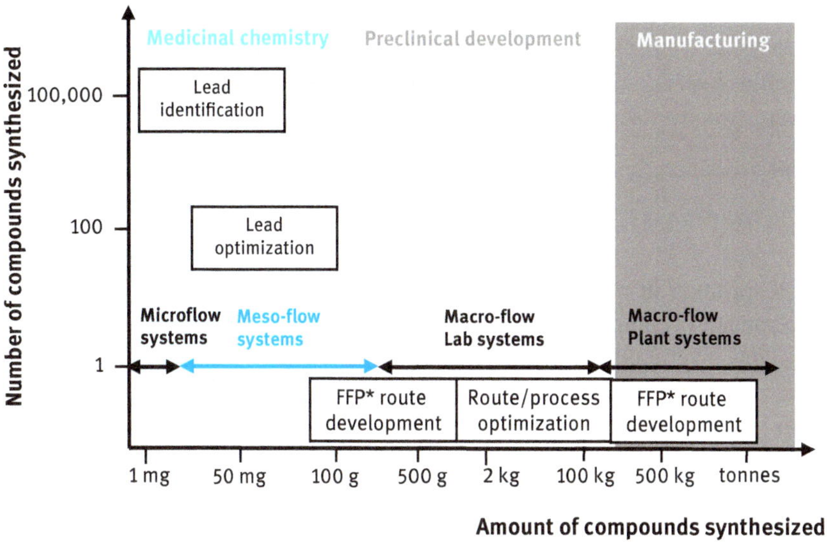

Fig. 10.18: Instrumentation of continuous-flow processes involved in the development of a drug (*FFP: Fit For Purpose). Adapted with permission from Wheeler et al. [86]. *Copyright © 2007, American Chemical Society.*

The typical tasks for biosensing, high throughput screening of enzymes or substrates, and bioprocess development experiments (information driven applications) can be best addressed in small-scale/high compound number cases by using microfluidic systems or in smaller number of compounds cases by applying larger-scale meso-flow systems. When the task is to develop an industrially relevant technology for production of a single target compound (product-driven applications) one of the two possible upscaling strategies should be selected [87]. Most often the *scale-out* concept (the extension of product collection time or increase of the length or size of the micro-/meso-flow reactor) is applied for scale-up. Alternatively, the *numbering up* principle (parallel operation of many identical micro-/meso-flow reactor units) can be envisaged. However, from a process engineering point of view this strategy is far from ideal because identical flow properties should be provided for each individual flow system, the system requires highly complex online monitoring and replacement of the exhausted biocatalyst is more complicated.

This is especially obvious, when one considers integration of a single enzyme/single step unit with other units, such as with a premixing unit, an in situ product removal unit, an in-line detection unit, or other chemical or enzymic reactor units [26]. In the case of the parallelized solution multiple connections should be made between the numerous small reactors and the additional units which rapidly and unnecessarily increase the complexity of the system. All examples of continuous-flow implementation of biocatalyzed processes at large scale (Section 10.4.1) utilized the *scale-out* concept and were implemented in macroflow plant systems.

The challenges and solutions of integration to chemoenzymatic or multienzyme systems were already discussed in the preceding sections.

10.5.2 On-demand fabrication of bioreactors

Recently, 3D printing in microfluidics has fascinated substantial interest due to the rapid development of commercial 3D printers [88–90]. A possible use of 3D printing in continuous-flow microfluidics is to create on-demand reactors, often mentioned as "reactionware." While 3D printing became common for continuous flow synthesis, the first 3D-printed continuous-flow biocatalytic reactor manufactured from Nylon Taulman 618 appeared only in 2017 [91]. The 3D-printed reactor channels were activated with glutaraldehyde for immobilization of an (*R*)-selective ω-transaminase (TA). The TA reactor was applied for kinetic resolution of racemic methylbenzylamine yielding 49% of the (*S*)-amine with >99% ee. Importantly, reactor stability was tested, and little to no effect on performance was observed over 48 h storage.

3D-printed "labware" has been used for high-throughput immobilization of enzymes [92], for cell immobilization applicable for propionic acid fermentation [93], or for enzyme immobilization after proper chemical modifications [94].

Another possibility for 3D-printing-based biocatalytic applications when a "bio-ink"- containing thermostable enzymes are used to create hydrogel-based scaffolds for various continuous-flow applications [95–97].

10.5.3 Machine learning in continuous monitoring and control

In recent years, there has been a growing interest in computer-aided syntheses. Achieving this is a serious challenge due to the large dimensions of the chemical and reaction space, even if we consider only processes involving purely chemical steps. It is an even greater challenge if the syntheses are to be performed in a continuous-flow mode [98, 99]. Although continuous flow offers a number of potential benefits, not all reactions are capable of operating continuously. Much of the consideration raised for computer-aided syntheses by purely chemical methods are also valid for biocatalyzed processes, and these computational methods are starting to enter the field of general and flow biocatalysis [1, 84, 100].

Further readings

- Heckmann CM, Paradisi F. Looking Back: A Short History of the Discovery of Enzymes and How They Became Powerful Chemical Tools. ChemCatChem 2020, 12(24), 6082–6102.
- Noël T, Luque R, editors. Accounts on Sustainable Flow Chemistry. Cham: Springer International Publishing; 2020.
- Poppe L, Nógrádi M, editors. Stereochemistry and Stereoselective Synthesis: An Introduction. Weinheim: Wiley-VCH Verlag GmbH & Co. KGaA; 2016.
- Grunwald P., editor. Industrial Biocatalysis. Singapore: Jenny Stanford Publishing; 2015.
- Grunwald P., editor. Pharmaceutical Biocatalysis: Chemoenzymatic Synthesis of Active Pharmaceutical Ingredients. Singapore: Jenny Stanford Publishing; 2019.

Study questions

10.1 What are the advantages and disadvantages of using cell-free enzymes or whole cells in flow systems?

10.2 What are the challenges of a multienzyme cascade in flow?

10.3 In redox system requiring a cofactor, how can we ensure that the cost of the process is contained?

10.4 What is the use of scavenger columns in a flow setup?

10.5 What are the differences between inherent rate and kinetics of an enzyme and effective rate and kinetics of a biocatalyst using this enzyme?

10.6 Which are the most important operational parameters of an enzyme-based continuous-flow reactor?

10.7 Which are the most important kinetic considerations for the biocatalyst in a continuous-flow reactor?

10.8 Which are the mostly used reactor types for biocatalytic reaction in a continuous-flow mode?

References

[1] Woodley JM, Advances in biological conversion technologies: new opportunities for reaction engineering, React Chem Eng, 2020, 5, 632–640.

[2] Paradisi F, Flow Biocatalysis, Catalysts, 2020, 10, 645.

[3] Bolivar JM, Valikhani D, Nidetzky B, Demystifying the flow: biocatalytic reaction intensification in microstructured enzyme reactors, Biotechnol J, 2019, 14, 1800244.

[4] De Santis P, Meyer L-E, Kara S, The rise of continuous flow biocatalysis – fundamentals, very recent developments and future perspectives, React Chem Eng, 2020, 5, 2155–2184.

[5] Thompson MP, Peñafiel I, Cosgrove SC, Turner NJ, Biocatalysis using immobilized enzymes in continuous flow for the synthesis of fine chemicals, Org Process Res Dev, 2019, 23, 9–18.

[6] Žnidaršič-Plazl P, The promises and the challenges of biotransformations in microflow, Biotechnol J, 2019, 14, 1800580.

[7] Bolivar JM, López-Gallego F, Characterization and evaluation of immobilized enzymes for applications in flow reactors, Curr Opin Green Sustain Chem, 2020, 25, 100349.

[8] Wohlgemuth R, Plazl I, Žnidaršič-Plazl P, Gernaey KV, Woodley JM, Microscale technology and biocatalytic processes: opportunities and challenges for synthesis, Trends Biotechnol, 2015, 33, 302–314.

[9] Tamborini L, Fernandes P, Paradisi F, Molinari F, Flow bioreactors as complementary tools for biocatalytic process intensification, Trends Biotechnol, 2018, 36, 73–88.

[10] Hajba L, Guttman A, Continuous-flow biochemical reactors: biocatalysis, bioconversion, and bioanalytical applications utilizing immobilized microfluidic enzyme reactors, J Flow Chem, 2016, 6, 8–12.

[11] Zhu Y, Chen Q, Shao L, Jia Y, Zhang X, Microfluidic immobilized enzyme reactors for continuous biocatalysis, React Chem Eng, 2020, 5, 9–32.

[12] Britton J, Majumdar S, Weiss GA, Continuous flow biocatalysis, Chem Soc Rev, 2018, 47, 5891–5918.

[13] Wang X, Saba T, Yiu HHP, Howe RF, Anderson JA et al., Cofactor NAD(P)H regeneration inspired by heterogeneous pathways, Chem, 2017, 2, 621–654.

[14] Mutti FG, Knaus T, Scrutton NS, Breuer M, Turner NJ, Conversion of alcohols to enantiopure amines through dual-enzyme hydrogen-borrowing cascades, Science (80-), 2015, 349, 1525–1529.

[15] Pinto A, Contente ML, Tamborini L, Advances on whole-cell biocatalysis in flow, Curr Opin Green Sustain Chem, 2020, 25, 100343.

[16] Romero-Fernández M, Paradisi F, Protein immobilization technology for flow biocatalysis, Curr Opin Chem Biol, 2020, 55, 1–8.

[17] Polakovič M, Švitel J, Bučko M, Filip J, Neděla V et al., Progress in biocatalysis with immobilized viable whole cells: systems development, reaction engineering and applications, Biotechnol Lett, 2017, 39, 667–683.

[18] Mateo C, Palomo JM, Fernandez-Lorente G, Guisan JM, Fernandez-Lafuente R, Improvement of enzyme activity, stability and selectivity via immobilization techniques, Enzyme Microb Technol, 2007, 40, 1451–1463.

[19] Jiao S, Li F, Yu H, Shen Z, Advances in acrylamide bioproduction catalyzed with Rhodococcus cells harboring nitrile hydratase, Appl Microbiol Biotechnol, 2020, 104, 1001–1012.

[20] Xu K, Chen X, Zheng R, Zheng Y, Immobilization of multi-enzymes on support materials for efficient biocatalysis, Front Bioeng Biotechnol, 2020, 8, 1–17.

[21] Hartley CJ, Williams CC, Scoble JA, Churches QI, North A et al., Engineered enzymes that retain and regenerate their cofactors enable continuous-flow biocatalysis, Nat Catal, 2019, 2, 1006–1015.

[22] Hanefeld U, Gardossi L, Magner E, Understanding enzyme immobilisation, Chem Soc Rev, 2009, 38, 453–468.

[23] Cantone S, Ferrario V, Corici L, Ebert C, Fattor D et al., Efficient immobilisation of industrial biocatalysts: criteria and constraints for the selection of organic polymeric carriers and immobilisation methods, Chem Soc Rev, 2013, 42, 6262.

[24] Finnigan W, Citoler J, Cosgrove SC, Turner NJ, Rapid model-based optimization of a two-enzyme system for continuous reductive amination in flow, Org Process Res Dev, 2020, 24, 1969–1977.

[25] Xue R, Woodley JM, Process technology for multi-enzymatic reaction systems, Bioresour Technol, 2012, 115, 183–195.

[26] Gruber P, Marques MPC, O'Sullivan B, Baganz F, Wohlgemuth R et al., Conscious coupling: the challenges and opportunities of cascading enzymatic microreactors, Biotechnol J, 2017, 12, 1700030.

[27] Ringborg RH, Woodley JM, The application of reaction engineering to biocatalysis, React Chem Eng, 2016, 1, 10–22.

[28] Finnigan W, Cutlan R, Snajdrova R, Adams JP, Littlechild JA et al., Engineering a seven enzyme biotransformation using mathematical modelling and characterized enzyme parts, ChemCatChem, 2019, 11, 3474–3489.

[29] Noorman HJ, Heijnen JJ, Biochemical engineering's grand adventure, Chem Eng Sci, 2017, 170, 677–693.
[30] Straathof AJJ, Wahl SA, Benjamin KR, Takors R, Wierckx N et al., Grand research challenges for sustainable industrial biotechnology, Trends Biotechnol, 2019, 37, 1042–1050.
[31] Boros Z, Falus P, Márkus M, Weiser D, Oláh M et al., How the mode of Candida antarctica lipase B immobilization affects the continuous-flow kinetic resolution of racemic amines at various temperatures, J Mol Catal B Enzym, 2013, 85–86, 119–125.
[32] Bolivar JM, Nidetzky B, Smart enzyme immobilization in microstructured reactors, Chim Oggi/ Chemistry Today, 2013, 31, 50–54.
[33] Doran MP, Engineering Principles Second Edition, Second ed., Waltham, Academic Press, 2013.
[34] Semenova D, Fernandes AC, Bolivar JM, Rosinha Grundtvig IP, Vadot B et al., Model-based analysis of biocatalytic processes and performance of microbioreactors with integrated optical sensors, N Biotechnol, 2020, 56, 27–37.
[35] Baronas R, Kulys J, Petkevičius L, Modelling the enzyme catalysed substrate conversion in a microbioreactor acting in continuous flow mode, Nonlinear Anal Model Control, 2018, 23, 437–458.
[36] Hess D, Yang T, Stavrakis S, Droplet-based optofluidic systems for measuring enzyme kinetics, Anal Bioanal Chem, 2020, 412, 3265–3283.
[37] Hess D, Rane A, DeMello AJ, Stavrakis S, High-throughput, quantitative enzyme kinetic analysis in microdroplets using stroboscopic epifluorescence imaging, Anal Chem, 2015, 87, 4965–4972.
[38] Neun S, Zurek PJ, Kaminski TS, Hollfelder F, Ultrahigh throughput screening for enzyme function in droplets, In: Methods in Enzymology. Elsevier Inc., 2020, 317–343.
[39] Di Maggio J, Paulo C, Estrada V, Perotti N, Diaz Ricci JC et al., Parameter estimation in kinetic models for large scale biotechnological systems with advanced mathematical programming techniques, Biochem Eng J, 2014, 83, 104–115.
[40] Bailey R, Jones F, Fisher B, Elmore B, Enhancing design of immobilized enzymatic microbioreactors using computational simulation, Appl Biochem Biotechnol – Part A Enzym Eng Biotechnol, 2005, 122, 639–652.
[41] Konti A, Mamma D, Hatzinikolaou DG, Kekos D, 3-Chloro-1,2-propanediol biodegradation by Ca-alginate immobilized Pseudomonas putida DSM 437 cells applying different processes: mass transfer effects, Bioprocess Biosyst Eng, 2016, 39, 1597–1609.
[42] Miložič N, Lubej M, Lakner M, Žnidaršič-Plazl P, Plazl I, Theoretical and experimental study of enzyme kinetics in a microreactor system with surface-immobilized biocatalyst, Chem Eng J, 2017, 313, 374–381.
[43] Lindeque R, Woodley J, Reactor selection for effective continuous biocatalytic production of pharmaceuticals, Catalysts, 2019, 9, 262.
[44] Fernandes AC, Gernaey KV, Krühne U, Connecting worlds – a view on microfluidics for a wider application, Biotechnol Adv, 2018, 36, 1341–1366.
[45] Kragl U, Vasic-Racki D, Wandrey C, Continuous production of L-tert-leucine in series of two enzyme membrane reactors, Bioprocess Eng, 1996, 14, 291–297.
[46] Ho C-H, Yi J, Wang X, Biocatalytic continuous manufacturing of diabetes drug: plantwide process modeling, optimization, and environmental and economic analysis, ACS Sustain Chem Eng, 2019, 7, 1038–1051.
[47] Pfeiffer SA, Nagl S, Microfluidic platforms employing integrated fluorescent or luminescent chemical sensors: a review of methods, scope and applications, Methods Appl Fluoresc, 2015, 3, 034003.

[48] Sun S, Ungerböck B, Mayr T, Imaging of oxygen in microreactors and microfluidic systems, Methods Appl Fluoresc, 2015, 3, 034002.
[49] Yue J, Schouten JC, Nijhuis TA, Integration of microreactors with spectroscopic detection for online reaction monitoring and catalyst characterization, Ind Eng Chem Res, 2012, 51, 14583–14609.
[50] Chrimes AF, Khoshmanesh K, Stoddart PR, Mitchell A, Kalantar-zadeh K, Microfluidics and Raman microscopy: current applications and future challenges, Chem Soc Rev, 2013, 42, 5880.
[51] Sans V, Porwol L, Dragone V, Cronin L, A self optimizing synthetic organic reactor system using real-time in-line NMR spectroscopy, Chem Sci, 2015, 6, 1258–1264.
[52] Lee WG, Zell MT, Ouchi T, Milton MJ, NMR spectroscopy goes mobile: using NMR as process analytical technology at the fume hood, Magn Reson Chem, 2020, 58, 1193–1202.
[53] Roberts KM, Fitzpatrick PF, Measurement of Kinetic Isotope Effects in an Enzyme-Catalyzed Reaction by Continuous-Flow Mass Spectrometry, In: Methods in Enzymology. Elsevier Inc., 2017, 149–161.
[54] Burkhardt T, Kaufmann CM, Letzel T, Grassmann J, Enzymatic assays coupled with mass spectrometry with or without embedded liquid chromatography, ChemBioChem, 2015, 16, 1985–1992.
[55] Rackus DG, Shamsi MH, Wheeler AR, Electrochemistry, biosensors and microfluidics: a convergence of fields, Chem Soc Rev, 2015, 44, 5320–5340.
[56] Li X, Mao Y, Zhu Z, Zhang Y, Fang Z et al., Digital flow rate sensor based on isovolumetric droplet discretization effect by a three-supersurface structure, Microfluid Nanofluidics, 2019, 23, 102.
[57] Wu T, Shen J, Li Z, Xing F, Xin W et al., Microfluidic-integrated graphene optical sensors for real-time and ultra-low flow velocity detection, Appl Surf Sci, 2021, 539, 148232.
[58] Kopparthy VL, Guilbeau EJ, Highly sensitive microfluidic chip sensor for biochemical detection, IEEE Sens J, 2017, 17, 6510–6514.
[59] Gruber P, Marques MPC, Sulzer P, Wohlgemuth R, Mayr T et al., Real-time pH monitoring of industrially relevant enzymatic reactions in a microfluidic side-entry reactor (μSER) shows potential for pH control, Biotechnol J, 2017, 12, 1600475.
[60] Strniša F, Bajić M, Panjan P, Plazl I, Sesay AM et al., Characterization of an enzymatic packed-bed microreactor: experiments and modeling, Chem Eng J, 2018, 350, 541–550.
[61] Krone KM, Warias R, Ritter C, Li A, Acevedo-Rocha CG et al., Analysis of enantioselective biotransformations using a few hundred cells on an integrated microfluidic chip, J Am Chem Soc, 2016, 138, 2102–2105.
[62] Faqir NM, Optimization of operating temperature for continuous immobilized glucose isomerase reactor with pseudo linear kinetics, Eng Life Sci, 2004, 4, 450–459.
[63] Grabner B, Pokhilchuk Y, Gruber-Woelfler H, DERA in flow: synthesis of a statin side chain precursor in continuous flow employing deoxyribose-5-phosphate aldolase immobilized in alginate-luffa matrix, Catalysts, 2020, 10, 137.
[64] Lindahl S, Liu J, Khan S, Nordberg Karlsson E, Turner C, An on-line method for pressurized hot water extraction and enzymatic hydrolysis of quercetin glucosides from onions, Anal Chim Acta, 2013, 785, 50–59.
[65] Velasco-Lozano S, Benítez-Mateos AI, López-Gallego F, Co-immobilized phosphorylated cofactors and enzymes as self-sufficient heterogeneous biocatalysts for chemical processes, Angew Chemie Int Ed, 2017, 56, 771–775.
[66] Stojkovič G, Žnidaršič-Plazl P, Covalent Immobilization of Microbial Cells on Microchannel Surfaces, In: Methods in Molecular Biology., 2020, 417–426.

[67] Contente ML, Paradisi F, Self-sustaining closed-loop multienzyme-mediated conversion of amines into alcohols in continuous reactions, Nat Catal, 2018, 1, 452–459.
[68] Wu S, Snajdrova R, Moore JC, Baldenius K, Bornscheuer UT, Biocatalysis: enzymatic synthesis for industrial applications, Angew Chemie Int Ed, 2021, 60, 88–119.
[69] Børge Poulsen P, Buchholz K, History of Enzymology with Emphasis on Food Production, In: Handbook of Food Enzymology, Whitaker JR, Voragen AGJ, Wong DWS, editors, Basel, Marcel Dekker AG, 2003, 1–10.
[70] OECD-FAO Agricultural Outlook (Edition 2020). 2021. doi:10.1787/4919645f-en
[71] DiCosimo R, McAuliffe J, Poulose AJ, Bohlmann G, Industrial use of immobilized enzymes, Chem Soc Rev, 2013, 42, 6437.
[72] Ditrich K, Balkenhohl F, Ladner W Separation of optically active amines. United States Pat. 19991–19995.
[73] Breuer M, Ditrich K, Habicher T, Hauer B, Keßeler M et al., Industrial methods for the production of optically active intermediates, Angew Chemie Int Ed, 2004, 43, 788–824.
[74] Oláh M, Kovács D, Katona G, Hornyánszky G, Poppe L, Optimization of 2-alkoxyacetates as acylating agent for enzymatic kinetic resolution of chiral amines, Tetrahedron, 2018, 74, 3663–3670.
[75] Tao J, McGee K, Development of a continuous enzymatic process for the preparation of (R)-3-(4-Fluorophenyl)-2-hydroxy propionic acid, Org Process Res Dev, 2002, 6, 520–524.
[76] Lee CY, Chang HN, Continuous production of acrylamide using immobilized Brevibacterium sp. CH2 in a two-stage packed bed reactor, Biotechnol Lett, 1990, 12, 23–28.
[77] Contente ML, Farris S, Tamborini L, Molinari F, Paradisi F, Flow-based enzymatic synthesis of melatonin and other high value tryptamine derivatives: a five-minute intensified process, Green Chem, 2019, 21, 3263–3266.
[78] Farkas E, Oláh M, Földi A, Kóti J, Éles J et al., Chemoenzymatic dynamic kinetic resolution of amines in fully continuous-flow mode, Org Lett, 2018, 20, 8052–8056.
[79] Honda T, Miyazaki M, Yamaguchi Y, Nakamura H, Maeda H, Integrated microreaction system for optical resolution of racemic amino acids, Lab Chip, 2007, 7, 366.
[80] Weiser D, Bencze LC, Bánóczi G, Ender F, Kiss R et al., Phenylalanine ammonia-lyase-catalyzed deamination of an acyclic amino acid: enzyme mechanistic studies aided by a novel microreactor filled with magnetic nanoparticles, ChemBioChem, 2015, 16, 2283–2288.
[81] Huffman MA, Fryszkowska A, Alvizo O, Borra-Garske M, Campos KR et al., Design of an in vitro biocatalytic cascade for the manufacture of islatravir, Science (80-), 2019, 366, 1255–1259.
[82] Liu Z, Zhang J, Chen X, Wang PG, Combined biosynthetic pathway for de novo production of UDP-galactose: catalysis with multiple enzymes immobilized on agarose beads, ChemBioChem, 2002, 3, 348–355.
[83] Jensen KF, Flow chemistry-Microreaction technology comes of age, AIChE J, 2017, 63, 858–869.
[84] Woodley JM, Accelerating the implementation of biocatalysis in industry, Appl Microbiol Biotechnol, 2019, 103, 4733–4739.
[85] Weiser D, Nagy F, Bánóczi G, Oláh M, Farkas A et al., Immobilization engineering – How to design advanced sol–gel systems for biocatalysis?, Green Chem, 2017, 19, 3927–3937.
[86] Wheeler RC, Benali O, Deal M, Farrant E, MacDonald SJF et al., Mesoscale flow chemistry: a plug-flow approach to reaction optimisation, Org Process Res Dev, 2007, 11, 704–710.
[87] Wegner J, Ceylan S, Kirschning A, Ten key issues in modern flow chemistry, Chem Commun, 2011, 47, 4583.
[88] Au AK, Huynh W, Horowitz LF, Folch A, 3D-printed microfluidics, Angew Chemie Int Ed, 2016, 55, 3862–3881.

[89] Ngo TD, Kashani A, Imbalzano G, Nguyen KTQ, Hui D, Additive manufacturing (3D printing): a review of materials, methods, applications and challenges, Compos Part B Eng, 2018, 143, 172–196.
[90] Bhattacharjee N, Urrios A, Kang S, Folch A, The upcoming 3D-printing revolution in microfluidics, Lab Chip, 2016, 16, 1720–1742.
[91] Peris E, Okafor O, Kulcinskaja E, Goodridge R, Luis SV et al., Tuneable 3D printed bioreactors for transaminations under continuous-flow, Green Chem, 2017, 19, 5345–5349.
[92] Spano MB, Tran BH, Majumdar S, Weiss GA, 3D-printed labware for high-throughput immobilization of enzymes, J Org Chem, 2020, 85, 8480–8488.
[93] Belgrano F, Dos S, Diegel O, Pereira N, Hatti-Kaul R, Cell immobilization on 3D-printed matrices: a model study on propionic acid fermentation, Bioresour Technol, 2018, 249, 777–782.
[94] Ye J, Chu T, Chu J, Gao B, He B, Versatile A, Approach for enzyme immobilization using chemically modified 3D-printed scaffolds, ACS Sustain Chem Eng, 2019, 7, 18048–18054.
[95] Chimene D, Lennox KK, Kaunas RR, Gaharwar AK, Advanced bioinks for 3D printing: a materials science perspective, Ann Biomed Eng, 2016, 44, 2090–2102.
[96] Peng M, Mittmann E, Wenger L, Hubbuch J, Engqvist MKM et al., 3D-printed phenacrylate decarboxylase flow reactors for the chemoenzymatic synthesis of 4-hydroxystilbene, Chem – A Eur J, 2019, 25, 15998–16001.
[97] Steier A, Schmieg B, Irtel Von Brenndorff Y, Meier M, Nirschl H et al, Enzyme scaffolds with hierarchically defined properties via 3D jet writing, Macromol Biosci, 2020, 20, 2000154.
[98] Plehiers PP, Coley CW, Gao H, Vermeire FH, Dobbelaere MR et al, Artificial intelligence for computer-aided synthesis in flow: analysis and selection of reaction components, Front Chem Eng, 2020, 2. Doi: 10.3389/fceng.2020.00005.
[99] Mateos C, Nieves-Remacha MJ, Rincón JA, Automated platforms for reaction self-optimization in flow, React Chem Eng, 2019, 4, 1536–1544.
[100] Lashkaripour A, Rodriguez C, Mehdipour N, Mardian R, McIntyre D et al., Machine learning enables design automation of microfluidic flow-focusing droplet generation, Nat Commun, 2021, 12, 25.

Steven V. Ley, Oliver S. May, Oliver M. Griffiths and Karin Sowa

11 Outlook, future directions, and emerging applications

11.1 Introduction: past, present, and future

The preceding chapters in this second edition make a compelling case for the application of flow chemistry and its related technologies. Its many innovations over the last 20 years have directly impacted the user community by providing new tools and methodologies to overcome numerous repetitive, scale-up, and time-consuming tasks, sadly many of which are still present in modern synthesis programs. On a higher level it has inspired new thinking by responding to the challenges of the Green Agenda [1, 2]. The technology, benefiting from the advances in data acquisition, machine learning, and artificial intelligence, is being used in the discovery of new science and even as a tool in the quest to understand the origins of life [3], or in other extraterrestrial mining and exploration ventures [4, 5]. As more people acquire the necessary skill sets, we anticipate the area will continue to see further innovations and advancements. An additional requirement going forward will be for the various equipment to work in harmony and synchronicity with each other. Electronic laboratory management systems will play a vital part in providing new opportunities for better data collection, data storing and data mining. This highlights a clear case for better data standardization, since this generated data will undoubtedly be mined in the future to feed more complex algorithms and neural networks.

Academic curiosity along with increasing industrial interest within the field has led to many new enabling technologies. In a didactic sense this is expected in any growing area as one usually sees an evolutionary progression. The continued collaboration between the fields of chemistry, engineering, informatics, and computer science has helped to cultivate this exciting new frontier. This final chapter seeks to provide an outlook and perspective on this area and offers some potential directions for the future [6].

11.2 General considerations

Commercially available flow equipment [7], its application within the field [8–10] along with a discussion of the fundamentals such as mixing efficiencies [11, 12] have already been extensively reviewed and discussed in previous chapters. Here only a brief outline of the planning, chemical sustainability, and safety decisions that might influence future adoption are discussed.

A highly advantageous aspect of converting to a continuous-flow process is the reduction in manual effort, particularly for routine tasks. As will be discussed later,

https://doi.org/10.1515/9783110693690-011

scientists will continue to dominate innovation and creativity. However, robotics will almost certainly remove labor-intensive activities that currently waste the human resource. This is often a point of much contention and philosophical debate within the scientific community, but sadly is beyond the scope of this review. In the future, laboratories will have to address these issues and how we will better integrate machinery and robotics in our daily workflows [13, 14].

11.2.1 Planning to succeed

Prior to embarking on any new continuous-flow program, benefits over conventional batch mode processes should be easily recognizable and quantifiable. The criteria should include an evaluation of process time, costs, scale-up, safety, sustainability, and appropriate chemistry. Most suitable processes lead to completion of the reaction within 2 h – preferably in less than 30 min. While slurries and solids can be accommodated these often necessitate the use of additional technologies and devices [15]. Single solvents are preferred especially when pursuing multistep transformations or telescoping [16]. Owing to their pyrophoric properties and water sensitivity, organometallic reagents are well suited to flow chemistry thanks to the sealed nature of the platforms [17]. The current interest in photo-redox induced transformations along with other novel process windows [18, 19] are well adapted to using continuous flow and becomes apparent particularly for scale-up [20]. Traditional benefits of flow chemistry, namely handling of exotherms and the use of hazardous or undesirable reagents, continue to dominate the literature. To date most of these flow applications are based on previous batch reaction where there is a desire to improve upon some undesirable aspect of the process.

11.2.2 The right chemistry

Good synthesis design is a key requirement for any reaction process, regardless of whether it is in flow or batch. A common misconception is that a poor reaction can be made good simply by converting it to a flow process – this is not the case. A poor reaction in any mode is still a bad reaction. Any gains made through translating it into a continuous process merely masks this fact and undermines the true benefits of flow chemistry.

Another process stumbling block, be it batch or flow, is in the downstream. By avoiding solvent manipulations, purifications and premature isolations, substantial savings in both time and expense can be achieved. On a larger scale this is realized in the reduction in plant size and complexity, leading to overall process intensification.

While on the topic of chemical design it is pertinent to comment on the dominant role played by catalysts in an increasing number of chemical transformations being applied to continuous processes. A component of the 12 principles of green chemistry and a major contributor to any innovative synthesis, all types of catalysts are of primary importance [21]. To date, most catalysts, apart from those used in the petroleum industry, have predominantly been developed for use in the solution phase within batch mode applications. This situation is changing to reflect the move towards continuous processing. As a result, there has already been a refocusing of ideas towards the use of immobilized catalytic systems [22] on various support materials or wall-coated materials. These will become of greater significance in the future as route planners begin to challenge their preconceptions and move away from the confines of batch mode chemistry. Increasingly we can expect to see more use of multiple catalyst combinations in flow, possibly combined with biotransformations to build greater molecular complexity. This concept will be discussed further in Section 11.4.2. Looking ahead, more emphasis will be placed on catalyst regeneration and recovery. This is likely to involve the renewed use of scavengers and immobilized reagents. These tools will play a decisive role in the chemists' attempts to avoid unnecessary purification or premature isolation and will also support greater telescoping opportunities in the future.

11.2.3 Sustainability: cleaner and greener

There are many synergies between the principles of green chemistry and flow chemistry [21, 23, 24]. A key imperative to improving sustainability is to leverage the principles of Process Intensification [25]. This aspiration is readily achieved through the incorporation of continuous-flow technology. One particular intensification technique that has been gaining considerable traction lately is the use of solvent-free reactions [26]. Employing an atom-efficient hydrogen borrowing protocol, it has been demonstrated that simple amines can be coupled with alcohols using a Ruthenium catalyst [27]. These coupling partners were premixed without the use of solvents and were directly introduced into a steel coil reactor operating at 250 °C. In this particular example (Fig. 11.1), morpholine was reacted with neat benzyl alcohol to give the coupled product in near quantitative yield.

The pure product stream from the reactor was transformed from liquid to the corresponding hydrochloride salt upon mixing with the hydrogen chloride flow stream. The resulting hydrochloride salt, which formed as a slurry, was pumped continuously through an open flow tube. The final product was then collected by filtration. Remarkably, this small footprint reactor generated 1.2 kg of material over 9 h.

Increasingly the stockpiling of large chemical inventories is becoming unacceptable. There is now a requirement to provide a duty of care when planning a new chemical process and to better manage our chemical resources. It is no longer acceptable to use excessive stoichiometries or ignore the proliferation of unwanted by-products in waste streams. To overcome these concerns there is likely to be a renewed focus on

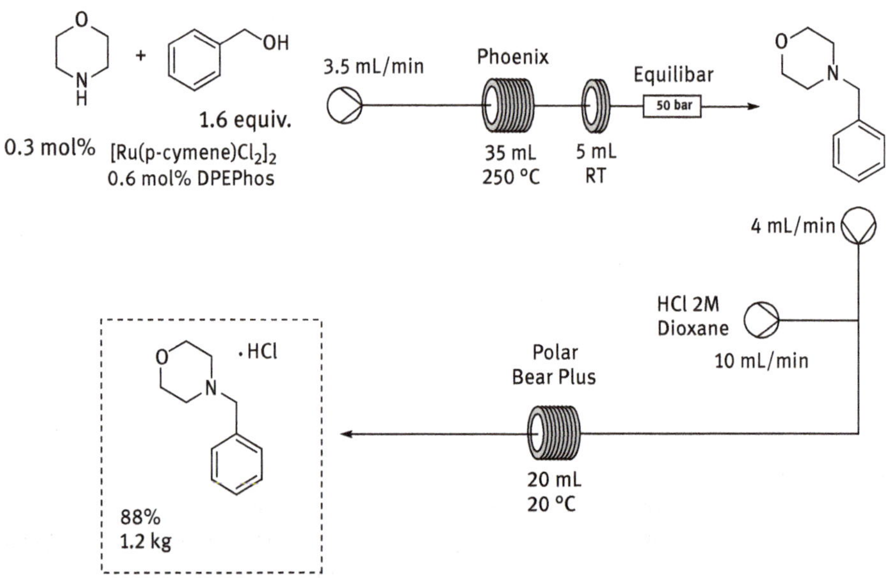

Fig. 11.1: Preparation of a tertiary hetrocyclic amine using solvent-free conditions.

improving downstream processing. For example, industry has started to apply the use of immobilized reagents and scavengers [28]. These systems are particularly suited to continuous-flow processes with much already written on the subject. Indeed our group's original aspirations for flow chemistry arose from a need to make better use of the polymer-supported reagents we had conceived for batch mode processes [29]. In fact some of our first published work (2006) on flow was based around the use of these immobilized reagents [30]. A key exponent for their continued use in this area is down to the ease with which these polymer-supported reagents and scavengers can be loaded into disposable or reusable columns. Spent reagents are fully retained on the support material and are easily recovered, allowing them to be recycled multiple times. An attractive application for these immobilized materials is in producing multiple sequential transformations. These can often achieve a greater number of telescoped steps than would otherwise be possible in batch. For example, a series of packed reagent and scavenger columns was used to prepare various triazoles [31]. Following initial mixing of all the substrates consisting of the Ohira-Bestmann reagent, a benzylic alcohol and an azide, the resulting solution was first passed through a polymer-supported 2,2,6,6-tetramethylpiperidin-1-yl)oxyl (TEMPO) column at 60 °C. This reagent caused selective conversion of the alcohol to the corresponding aldehyde. Next, the flow stream continues to a T-piece where it is mixed with a potassium butoxide solution. Here the aldehyde is converted to the corresponding homologated acetylene through engagement of the Ohira-Bestmann reagent. By flowing the resulting mixture through a copper-supported catalyst, cycloaddition of the azide with the acetylene occurs to give the corresponding triazole product. At this stage the flow stream contains multiple by-products. Chromatography would normally be required at this point to purify the product. This problem was solved by passing the reaction mixture through a series of commercially available scavenger columns to give the final pure crystalline triazole (Fig. 11.2).

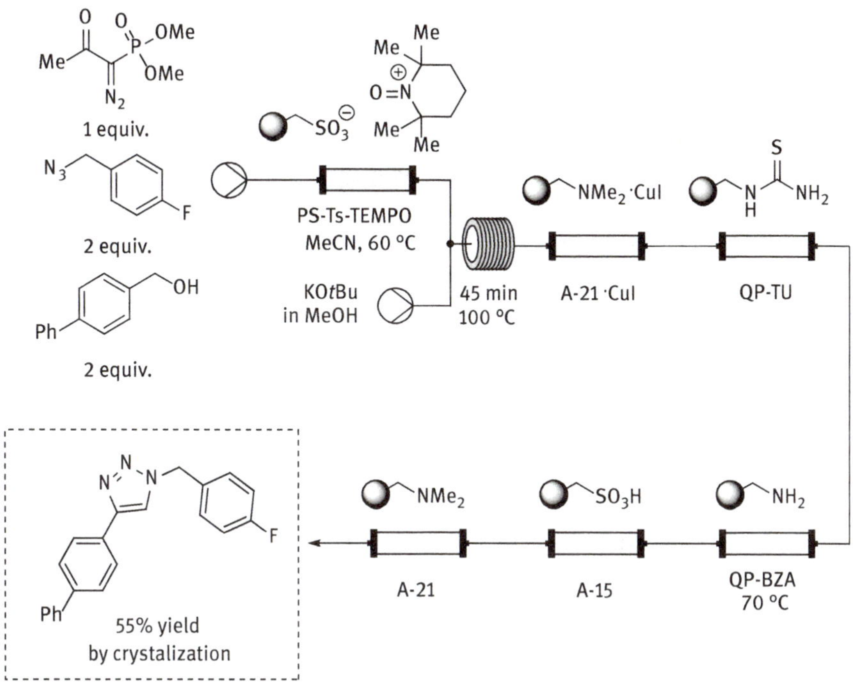

Fig. 11.2: Preparation of various triazoles using polymer-supported reagents and scavengers.

Although this example demonstrates an early use of multiple immobilized catalysts and scavengers in the preparation of a simple heterocycle, the overall concept is general and can be applied to make many active pharmaceutical ingredients (APIs) and natural products [32]. The technique is also suitable for recovering expensive asymmetric catalysts. This feature is particularly attractive and is now becoming a consideration in synthesis design. Alternative techniques employ flow electrodeposition methods for selective removal and recycling of precious metals; however, these have yet to see wider adoption.

Given that many important chemical reactions used in synthesis generate unwanted waste materials, there is considerable interest in creating value from these by-products. Revalorization and product repurposing are becoming more popular, particularly within industry, with substantial efforts currently under way to incorporate these processes in new and existing synthesis programs.

i *(For related topics please see Volume 2, Chapter 6, Title:* From green chemistry principles to sustainable flow chemistry*)*

11.2.4 Safety: protecting the user

As discussed in other chapters, continuous-flow processes offer substantial and tangible safety advantages to their equivalent batch mode counterparts [9]. Often following the development of a method, it is left to technicians to reproduce the formulated routes and perform larger scale-ups. Many of the lessons and observations, however well documented, can often be lost. This represents a potential safety concern for contract research organizations (CROs) and other chemical plant operating facilities. Having a technology that removes many of the uncertainties associated with either long production or scale-up has immeasurable advantages for safety.

11.2.5 Data, data, data

As will be seen through this final chapter, data plays a vital role in scientific discovery. The characterization and fundamental understanding of a chemical process along with its conditions is often divisive in the development of a safe, robust, and efficient process. Early discussion of what is known and, more importantly, what remains to be determined about a reaction is a very important consideration. Early identification of potential hazards such as isotherms, reaction kinetics, poor mixing, and heat transfer will ultimately lead to more successful campaigns. This is true whether in batch or continuous mode. A major benefit of a continuous process is the rapid exploration of the chemical space and the *in situ* capture of information. We will see later how this generated knowledge is used to continuously feedback in self-optimizations. Also discussed are the challenges, along with some of the solutions, that have been applied to data collection, data storage, and data mining.

11.3 Current progress

Broadly speaking, the steady increase and development of commercial flow equipment over the past decade has been well adopted and embraced by industry and academia. A diverse range of applications and examples can be found throughout the literature. Equipment vendors through collaborations with numerous chemical research groups have subsequently furnished the area with a myriad of platforms and devices that have proven to be useful for a range of applications. They have also provided investigators with a means to enter the field by removing many of the engineering challenges associated with the area. However, it is clear that no single piece of equipment will satisfy all possible reaction conditions and requirements [33]. As a consequence, it has become increasingly common to find investigators developing in-house systems [34]. A casualty of this approach is the interoperability of these platforms due to the plethora of bespoke hardware and software. This makes integration between systems challenging. With no open standard, or indeed any standardization in hardware or software, the many potential advantages of continuous processing for data collection and particularly future data mining are lost.

11.3.1 Upstream

The ability to rapidly assemble or to safely reconfigure a reactor is a critical factor – although this is often overlooked. Likewise, the choice of other components such as the pumps can directly influence the reaction through turbulence and cavitations. At the heart of any flow chemistry platform is the reactor. Its design directly influences the outcome of a chemical transformation, more so than the universal round-bottom flask. Even for a conventional single coil flow reactor, the material of manufacture, volume, length, surface properties, and chemical resistance all have an impact on the reaction.

To bring about a successful chemical transformation, the understanding of thermal parameters, mixing characteristics, bulk transfer properties, and kinetics are key elements in any reactor design. The ability to investigate larger chemical space by exploiting high temperatures and pressures can substantially expand the range of chemical reactions. Similarly, being able to control low temperatures accurately, preferably without the use of cryogens, from room temperature to below −90 °C is now possible with off-the-shelf equipment. This opportunity gives access to chemistries that are not presently popular within many industrial settings. The generation of exotherms and use of reactions prone to thermal runaway are problematic within batch mode operation: continuous-flow technology can readily control these phenomena and even use the excess energy from exothermic reactions to drive additional downstream chemical transformations.

Other classically difficult reactions to control are those involving toxic, flammable and pressurized gases. Engineering advances in recent years have seen the development of new reactor designs for performing liquid–gas reactions [35–38]. Systems where the reactive gas is generated within the device and subsequently consumed in a chemical transformation are now popular. Pioneering work in this field led to the commercial development of the H-Cube® reactor for performing hydrogenation reactions. Here hydrogen gas is generated through electrolysis of water. The generated gas is mixed with the target substrate in a T-piece and subsequently flowed through a cartridge containing a catalyst [39].

The development of tube-in-tube reactor technology provided a safer and more convenient on-board generative method for toxic gases such as ozone, diazomethane, carbon monoxide, phosgene, and hydrogen cyanide. These specially designed reactors work by gas diffusion through semipermeable Teflon® AF2400 polymer membranes [40]. They have been used to successfully deliver gases for a range of chemical transformations which would have otherwise been difficult.

Without trivializing either the engineering complexity or the associated chemical hazards, multiple gas reactors can be combined to deliver efficient multistep chemical transformations without the need for isolation or purification. A notable example of this approach involved a triple gas combination using ethylene, hydrogen, and carbon monoxide to convert an aromatic halide to the corresponding branched aldehyde [41]. Here the reaction proceeded through a tandem Heck-Hydroformylation sequence (Fig. 11.3) to produce the desired aldehyde.

Of additional note was the use of an enantioselective preparation of O-acetylcyanohydrins by a three-step telescoped continuous process [42]. Here the modular components enabled an accurate control of two sequential biotransformations, safe handling of an *in situ*–generated hazardous gas, and in-line stabilization of products. This method proved to be advantageous over the existing batch protocols.

A final comment on reactor design relates to safety. Chemical research laboratories are hazardous environments. The potential for chemical exposure is an ever-present risk. This necessitates very strict policies and operating procedures. All future reactor designs operating within these environments will need to align with these demanding safety standards. Increasingly as flow chemistry machinery gets smarter we anticipate it will become self-monitoring and able to autonomously respond to potential problems. Given the future desire for the equipment to conduct long self-optimization campaigns safely, the stability and robustness of flow reactors will need to improve. Better real-time monitoring combined with intelligent safety algorithms could enable earlier detection and identification of failure or loss of containment. This would allow systems to perform safe shutdown procedures before a reaction gets out of control and material is lost. Flow platforms of the future are likely to offer more opportunities for remote access and for improved laboratory data management. Although serial communication remains the norm, some commercial systems already offer interfaces for local area connections through TCP/IP and basic http protocols.

11.3.2 Downstream

There is a perception that the upstream aspects of a multistep synthesis are the most important. To that end, a disproportionate amount of attention has focused on the upstream. While undoubtedly important, if the downstream workup is not properly

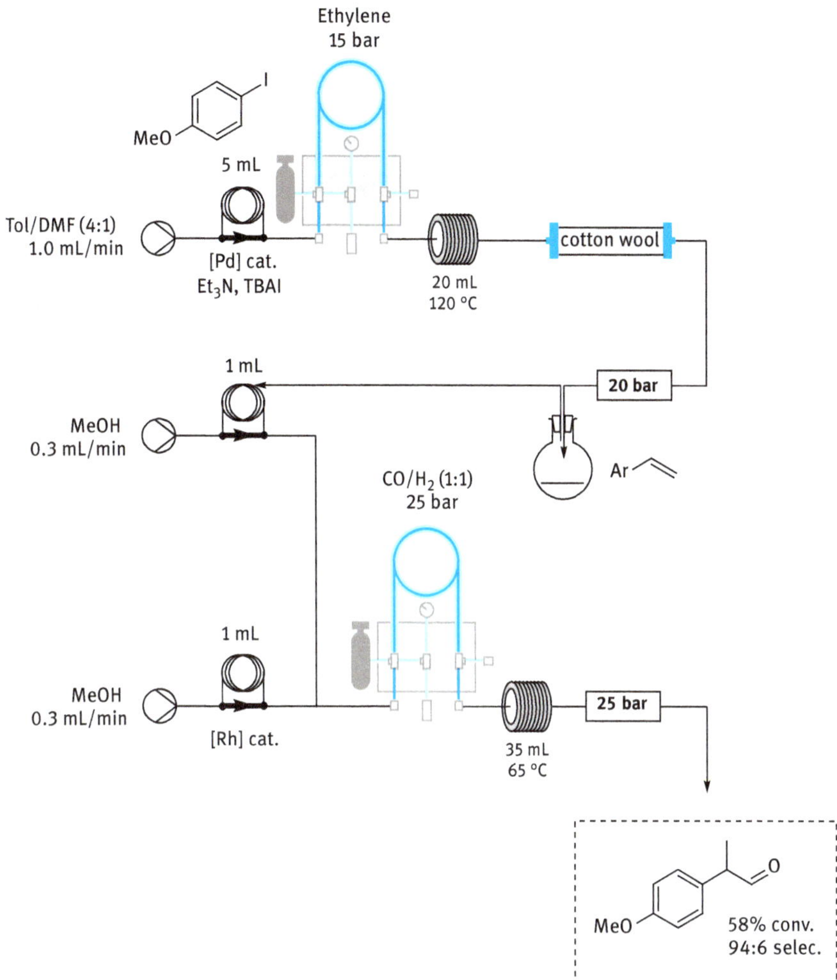

Fig. 11.3: An example of a triple gas combination using ethylene, hydrogen and carbon monoxide to convert an aromatic halide to the corresponding branched aldehyde.

factored into the design process, product purification and isolation becomes increasingly difficult. This is all too often the case as can be seen by the numerous pumping technologies and reactors compared to the noticeably smaller number of in-line workup technologies.

At the initial discovery stage workup is usually relegated to offline purification methods in the form of high-performance liquid chromatography (HPLC) and increasingly ultra-performance liquid chromatography (UPLC). During scale-up, the downstream operations become significant components and can determine the difference between success or failure of a campaign. Many good synthetic methods require a range of workup steps, most of which are considered to be routine in the hands of an experienced and skilled chemist – but here lies the problem. What is relatively easy for a human can be a phenomenally complicated task for a machine and is thus a bottleneck to many existing continuous-flow processes. New equipment and innovative ideas are needed if the area is to progress. To take a simple example of liquid–liquid extraction: while readily accomplished with a separating funnel for lab scale batches up to several grams, this relatively trivial workup becomes remarkably challenging under continuous-flow conditions. A partial solution to this issue led to the development of various membrane based separation technologies, some of which are now commercialized [43, 44]. A more creative and humanistic approach saw the use of camera monitored gravity separation devices. These took advantage of the availability of open-source machine vision algorithms [45]. In spite of these developments there remains a need for further innovation and for a greater variety of off-the-shelf solutions, particularly those which offer improved robustness over long periods of continuous processing.

In a similar vein, in-line solvent switching and solvent evaporation are still problematic, although some basic solutions do exist [46]. Should in-line filtration be necessary, this often requires the development of bespoke systems [47]. A challenge for the upstream as well as the downstream is the handling of slurries [15]. These requirements nearly always add additional complexity and limit planning options. Interestingly deliberate generation of crystalline products in flow is becoming more widespread. This is especially true within the field of polymorphism. Here precise crystal structures and sizes are key requirements since they directly impact drug formulations and pharmacokinetics. This requirement has led to the development of several ingenious technical innovations, allowing specific polymorphic structures to be isolated [48].

Downstream processes often involve the manipulation of significant volumes of solvents. Accordingly in-line recovery systems for continuous solvent reuse are increasingly relevant in the delivery of the Green Agenda. Owing to the convenience of many routine workup methods such as chromatography, crystallization, and distillation we have become particularly reliant on these techniques. These manual skills do not easily transfer into continuous processes – certainly at the types of scales most modern campaigns require. Indeed at larger scale (>100 ton year^{-1}) these manipulations, on account of improved engineering, become easier as can be seen by their routine use in the petroleum and fine chemical industries. For APIs, where the quantities are substantially less (<100 kg year^{-1}), the miniaturization of these techniques becomes substantially more challenging but is achievable [49–51]. At the lab scale (>100 mg day^{-1}), these engineering challenges are magnified further since they become prone to blockages and foulage, however. As a consequence immobilized reagents and scavengers are becoming more prevalent. They are also more likely to be used to minimize in-line waste products and allow process designers to effectively circumvent many downstream workups. However, it is unlikely they will fully solve the problem and continued research into in-line downstream processing will still be required in order to furnish the area with new technologies and solutions.

11.3.3 Feedback and control

There is a growing appetite for faster data collection as kinetic, structural and quantitative data are often used during process optimization [52]. Increasingly this data is also being used to feed machine learning algorithms as we move away from existing statistical methodologies, as we will see later. Real-time system monitoring and integrated feedback mechanisms are fast becoming the norm [53]. Previous chapters have discussed in detail the many improvements in process analytical technologies (PAT) and its application in flow chemistry. Here the focus of our discussion will be on emerging technologies and pointing out where further ingenuity is needed.

Building on the advances in analytical science over the past decade, PAT has furnished the area with a myriad of spectrographic, electronic and optical sensors and detectors. These have been deployed with great effect to monitor a range of chemical transformations and reaction conditions. PAT has become a cornerstone for the laboratory of the future [54, 55] finding substantial use within the agrochemical, petroleum and fine-chemical industries. However, it is in the pharmaceutical industry that PAT has been deployed during continuous API synthesis campaigns for compliance and is forming a cornerstone for current good manufacturing practices (cGMPs) [56]. Continuous-flow processes lend themselves to feedback and control. Unlike traditional glassware and batch mode processes, fine reaction control and process granularity are achievable. This becomes particularly desirable in the pursuit of high commodity materials [57] and the manufacture of APIs [58–60], whereby process control and continual monitoring is required for compliance. Too often in batch mode processes entire batches have to be written off due to a deviation outside the defined process parameters – this is both costly and time consuming. Continuous processes on the other hand have many advantages in this area. Using continuous feedback and control, a deviation can now be detected and appropriate action triggered to either stabilize the conditions or divert that segment, thus saving the production run.

In their basic form, feedback and control systems are based on simple analogue triggers. These are usually either voltages or currents. The ability to control and trigger equipment was a substantial innovation, along with the microcontroller, which led to the establishment of Industry 3.0 in the 1970s. Several industrial standards exist and process logic controllers (PLCs) have been utilizing these signals to perform simple logic operations, controlling air conditioning units to car manufacturing plants. Industry 4.0 takes this a step further by incorporating more humanistic logic and decision making. This, coupled with the greater interconnectivity made possible by the emergence of the Internet of Things (IoT), allows machines to communicate directly with each other without the need for human oversight.

In 2011, Ley *et al.* demonstrated that a secondary pump could be triggered by monitoring a target peak from a Fourier transform infrared (FTIR) spectrum using a Mettler-Toledo® FlowIR® (Fig. 11.4). This direct feedback control was used during the synthesis of selected pyrazoles and crotylated intermediates [61].

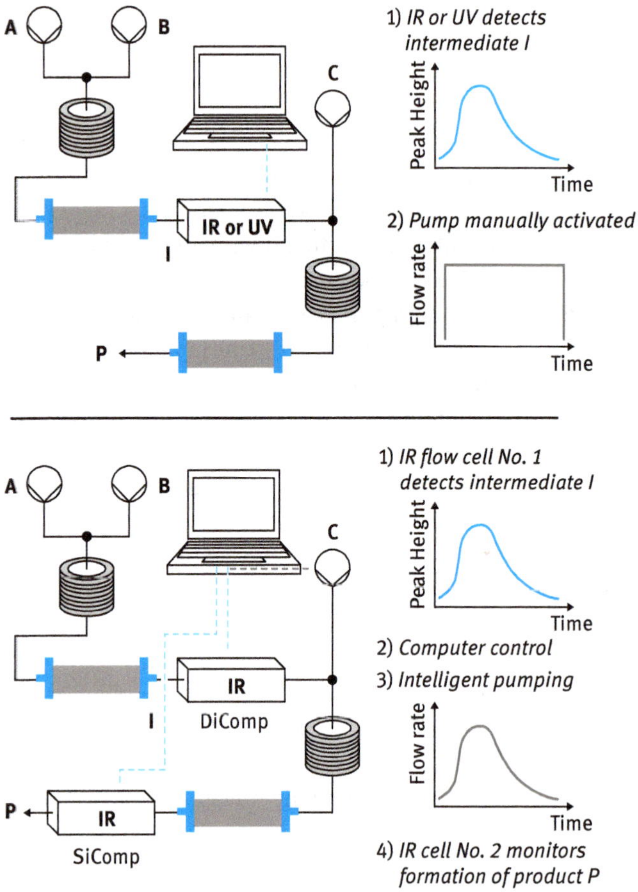

Fig. 11.4: Top: IR/UV detection with manual pump control. **Bottom:** IR-detector coupled to a computer to control a pump.

Here a piston pump was able to be triggered once a steady state had been achieved. To our knowledge this was one of the first examples where a lab-scale continuous-flow process dynamically changed to match varying flow profiles each with unique steady state characteristics. During the course of the experiment the connected pump was able to synchronize with these process changes, with no user input, cutting down on the volume of reagents used.

Fitzpatrick and Ley later used a similar approach for an Appel reaction. This time the FTIR device was used to monitor the production of triphenylphosphine oxide. Additionally in the same paper in-line mass spectrometry was used to monitor the conversion of 3-cyanopyridine to the corresponding nicotinamide [62]. Here using an Advion® CMS mass spectrometer, the target ion was converted to an analogue voltage signal connected to an Arduino® microcontroller. Using an analogue digital converter, the voltage signal was passed to a Raspberry Pi® microprocessor running a modified simplex optimization algorithm. The target ion peak intensity was monitored and the reaction continually cycled until an optimum conversion was achieved. The ability to take complex spectrographic information and reduce it to a simple voltage trace or Boolean is a powerful tool and has helped spawn a new generation of self-optimization platforms based on a range of single and multiple objective function optimization algorithms.

(For related topics please see Volume 1, Chapter 1, Title: Fundamentals of Flow Chemistry *and Chapter 6,* Fundamentals of Continuous Manufacturing, PAT: Sampling, Analysis, and Automation*)*

i

11.3.4 Self-optimization

Optimization remains a key component to any chemical process. Over the years statisticians and mathematicians have developed a range of techniques to tackle the problems of time, cost, and availability of data. At its core an optimization generally seeks to maximize or minimize one component. In chemistry this is often maximizing either yield or conversion, while minimizing other factors such as materials and time.

By far the most ubiquitous method for both batch mode and continuous flow is the statistical design of experiments (DoE) approach [63, 64]. Here the desired space is defined by the user and is then systematically probed to produce empirical data to develop a model. This is then used to plot a surface to allow an estimation of the maxima or minima within the defined space. Although robust and offering many advantages to the often used, but seldom recommended, one-variable-at-a-time technique, these models require a large amount of empirical data and consume a lot of time and resource. This is problematic when performing multiple batch experiments where the time required to stabilize reaction conditions and to set up iterative reactions can be particularly time consuming. Early continuous systems with liquid handler frontends were able to perform material manipulations and sequential DoE runs. A major benefit of using continuous-flow systems for this type of optimization is the faster mass and heat transfer characteristics of the reactors, which reduce the time taken between sequential runs since conditions stabilize quickly. Additionally, the smaller reactor volumes reduce the amount of materials required.

Over the years, alternatives to full and partial factorial designs have been postulated; however, these often result in a compromise in interactions between multiple variables (alienizing) while attempting to cut down on the amount of data required. More recently, linear ramps have been used to reduce material consumption and obtain more data points per run [65]. These have been used to great effect to glean kinetic data for various in silico optimization models, but still require significant volumes of materials.

A review of chemical optimization algorithms shows that Simplex and SNOBFIT have so far been the focus of most chemical optimization research (Fig. 11.5) [66]. The development and application of multiobjective function algorithms, as can be seen in preceding chapters, has arguably been the main focus of many chemical research groups investigating continuous-flow optimizations in recent years. Interestingly, there is still sparse literature and precedence for using optimization algorithms for multistep processes.

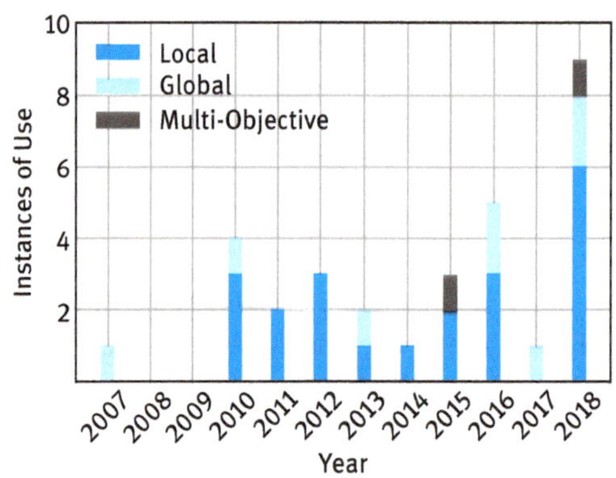

Fig. 11.5: A summary of algorithms used in self-optimizing chemical platforms.

Throughout the 1960s many so-called "black box" algorithms were developed – most notably the simplex algorithm by Dantzig. This single objective function was developed further by Nelder and Mead and consists of a user-defined convex polyhedra formed of $n + 1$ vertices or simplex. The feasible design space is then explored. This approach is routinely used in engineering applications to maximize or minimize a target variable – the objective function. In chemical applications it is still used to optimize magnetic fields (shimming) in nuclear magnetic resonance instruments.

Stable Noisy Optimization by Branch and Fit (SNOBFIT) is another optimization algorithm, this time developed by Huyer and Neumaier. It was applied to an automated chemical platform for the optimization of an AstraZeneca API [67]. It has since been used by multiple groups and is a popular choice for solving chemical optimization problems [66]. De Mello and co-workers developed a wrapper for the standard MATLAB® SNOBFIT algorithm [68]. The wrapper provides high-level access to the optimization algorithm and is intended for chemical optimizations. Interestingly it was also one of the first such algorithms to be hosted on GitHub as a free and open-source chemical optimizer. Several variations have appeared in the literature in recent years to try and incorporate multi-objective functions. Thompson sampling efficient multiobjective optimization (TSEMO) [69] is one such other "black box" optimization algorithm. Developed in the 1970s, unlike Simplex and SNOBFit this comprises a multiobjective function. Bourne and coworkers applied the TSEMO algorithm to a continuous-flow platform for two case studies. First, an aromatic substitution reaction was optimized between 2,4-difluoronitrobenzene and morpholine. Here the objective was to form the desired ortho product and minimize the undesired *para*- and *bis*-adducts. The second study consisted of an *N*-benzylation of α-methylbenzylamine with a benzyl bromide to form a secondary amine and minimize the formation of the tertiary amine. Data was collected and fed back to the algorithm using in-line HPLC data (Fig. 11.6).

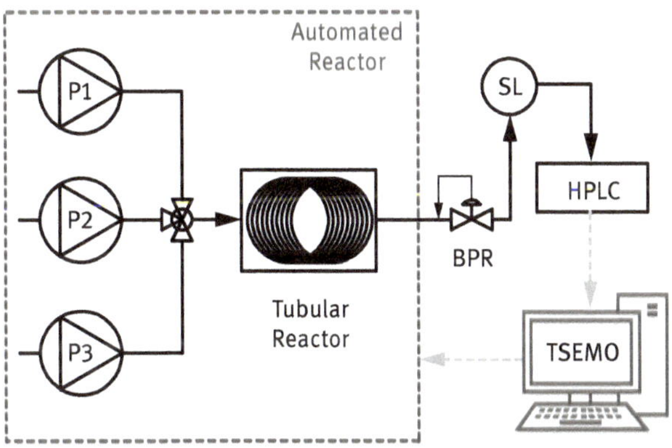

Fig. 11.6: Schematic of the feedback control system used during a TSEMO multiobjective optimization.

The algorithm was able to successfully identify a set of optimal conditions between E-Factor and space–time yield by use of a trade-off curve known as the Pareto front. The algorithm also required fewer experiments compared to sequential single-objective optimizations.

A key advantage of continuous-flow processes is the ability to telescope reactions, allowing multistep synthesis without the need for isolation. This would seem to be an obvious area to deploy these multi-objective algorithms. Rather than optimizing individual steps, the entire process could be optimized taking into account the impact of each upstream stage along with any downstream processes. Bourne and Lapkin demonstrated this by identifying key conditions based on the Pareto fronts for a multistep optimization [69]. Applying this to more complicated, multistep processes will undoubtedly have a wide appeal within industry. The wider availability and use of open-source sharing platforms such as GitHub® will likely lead to increased adoption in the future.

11.4 The future

11.4.1 Integrating batch and flow: A hybridized approach

Up to now most flow processes have been compared to their corresponding batch contemporaries, many of which were first discovered over a century ago. Despite this unfair comparison there are now many successful examples of continuous processes that have replaced existing batch methods.

A major inhibitor to wider adoption is in no small part due to education and end-user skill sets. Chemists, for the most part, still overwhelmingly gravitate toward and foster a batch mode mentality. Time constraints and business pressures also conspire to favor a conservative approach, which disproportionately favors established methods over newer ideas. Whether it be batch or flow, the best method should be employed to fully exploit the chemistry; better still would be a platforms approach. Indeed, there are now many such systems in operation that are fully integrated at larger scales.

It seems remarkable that basic equipment for organic synthesis has largely remained unchanged for the past 150 years. The use of the ubiquitous round-bottom flask, distillation kit, separating funnels, filtration devices, and evaporation techniques have graced our laboratories for years. It is time that a more concerted effort be made to design new systems capable of switching from batch to flow, and perhaps, more importantly, from flow to batch – even the design of a fume cupboard constrains our ability to be more innovative. Many flow chemistry systems are poorly suited to existing laboratory infrastructure, suggesting more attention should be spent on new laboratory designs. The automation of synthesis is not trivial, nor should it be, but it is not unreasonable to want to harness the best methods and modern technologies. Synthesis planning usually employs known chemistries for the majority of the anticipated steps. Although creative ideas often feature, the primary goal is defined as the final target structure. Many of the tools of synthesis are taken for granted and new methodologies and techniques are rarely considered when there are already existing processes. Too often the quickest, easiest, or most convenient path is taken rather than the most efficient, cleanest, or scalable. This situation needs to change. Whether to use batch or flow or a hybrid approach, this consideration should become part of the initial route design.

11.4.2 Mimicking nature: the role of biotransformations

Often seen as an independent topic, biotransformations can provide both interesting and strategically unique chemical processes for novel applications in synthesis. Being able to achieve asymmetric conversions through reductive and oxidative transformations or the formation of new carbon–carbon bonds makes this an attractive option. Although there is considerable literature on using biotransformations during flow processes, synthesis chemists tend to disproportionately focus on the negatives such as solvent incompatibility, stability of the enzyme system, limited selectivity, and downstream processing problems to name a few. This is unfortunate since the opportunities far outweigh the problems.

The use of directed evolution methods to improve enzyme selectivity and function together with advances in immobilization techniques that facilitate recycling has reenergized interest in the area. Romeo-Fernândez and Paradisi, in an informative review, make a strong case for immobilization technologies in multistep transformations in flow [70]. Although there are many reported single pot multienzyme processes, most involve elaborate natural product-like compounds as substrates. An attractive goal and application would be the late-stage functionalization and tailoring of unnatural precursors.

The development of multiple flow cartridges containing specific immobilized enzymes and cofactor systems has obvious applications for generating complex molecular arrays. Indeed many of the functions and physical operations of flow equipment resemble how cells assemble molecules. Compartmentalization of chemical steps, surface reactivity, and transport mechanisms are common phenomena in cells, which can be mimicked in flow. Many lessons can be learnt through mimicking these bioprocessing features and through biomimetics that can usefully be integrated into future flow reactor designs [71].

11.4.3 Seeing the future: machine vision

Chemistry is an empirical and practical science based on observation and experimentation. During a synthesis a chemist will perform a range of manual manipulations, transfers, and observations. These are often seen as trivial to the trained chemist but dissecting any one of these skills reveals just how challenging translating these into the digital world can be. While working at the bench multiple visual cues such as color changes, phase boundaries, meniscuses, and liquid levels are used to make decisions. These are subtle tasks that machines currently find difficult to mimic.

To a limited extent a basic liquid–liquid separation is able to be automated using remote camera monitoring. This is largely thanks to the availability of opensource computer vision libraries such as OpenCV. Other physical properties and phenomena such as refractive index, light absorption, or color differences outside the visible spectrum can also be utilized using specialized cameras and detectors. The addition of cameras also removes a level of user subjectivity leading to improved reproducibility.

Augmented reality (AR) is a technology and field of research that sits at the interfaces of virtual reality and computer graphics. In AR applications, data is overlaid onto the real-world environment, endowing it with additional information in a new interactive manner. The widespread adoption of increasingly powerful and compact computing devices provides great potential for the development of new applications in this area [72]. AR has already impacted chemistry for educational purposes, and we believe the technology will become a valuable tool for presenting and conveying chemical information as well as reaction monitoring. An obvious disadvantage often encountered when using continuous-flow systems is the lack of visibility since reactors are sealed systems. Unlike round-bottom flasks, which are made from glass, it is often not feasible to use this as a material for flow reactors. Although photoredox reactors do utilize such materials these are often encased to protect the user from the light source. This lack of transparency can limit the user's ability to visualize the reaction. One solution is to use AR systems to build a human-interpretable image of the reaction inside the reactor based on in-line process analytical technologies. Translating the vast amount of data generated from these sensors and visualizing it in real time as a suitable format for human interpretation is no easy task and will require continued research and refinement in the coming years.

Furthermore, many observations are subjective. Greater proliferation of cameras will undoubtedly allow more objective observation and easier documentation [73]. Such observations are particularly valuable when describing precipitation, foaming or gas production. These can be invaluable components for advanced reaction optimization programs. For example still images taken every 60 s were used during the scale-up of a re-crystallization. From these frames the exact temperature and time that crystallization occurred was determined. This enabled an improved crystallization process to be developed over 12 h that afforded enhanced enantioselectivity of the desired product. Similarly, the successive screening of crystallization conditions commonly employed in the pharmaceutical industry is both time- and resource-heavy. Camera-based optimization would be particularly beneficial as an alternative to existing techniques such as RAMAN and powder X-ray diffraction.

Another obvious area where the use of remote monitoring could benefit continuous-flow systems is for the detection of leaks or failures. Having the ability to observe an environment beyond the visible range such as ultraviolet or infrared allows for tiny defects, temperature spikes, or toxic gases to be detected. Increasingly sophisticated robotic platforms are already a reality within the manufacturing sector. These are likely to become more prominent in laboratories in the future and will be far more flexible and capable than existing X Y Z-based machines which currently occupy our benches. These are currently very inflexible platforms, performing only singular tasks. They are not designed to be connected to multiple systems or be routinely reconfigured. Coupling humanistic robotic platforms with machine vision will allow these to work more collaboratively with chemists during the running of experiments. This would be beneficial for both batch and flow or a combination of the two as discussed earlier.

This has already been realized to a certain extent by the integration of a robotic arm for selecting flow modules. In a similar vein to our discussion about early identification and intervention by continuous-flow systems to detect potential faults, the incorporation of robotic arms will enable these systems to be corrected and reset remotely. This will help further to remove the user from potential chemical hazards but also greatly reduce any downtime, which is particularly undesirable during the running of long API campaigns.

11.4.4 Machine learning

Traditional chemical self-optimization algorithms are not examples of machine learning since they are based on established statistical methodologies [74]. Machine learning algorithms utilize trained or reinforced neural networks to make decisions and predictions or, in our example, to optimize a reaction. Where such machine learning algorithms have been deployed reductions of almost 70 experimental runs over SNOBFIT and DoE have been achieved. This demonstrates the potential advantages of using machine learning over the established statistical "black box" methodologies. Similarly the use of machine learning has been demonstrated as a retrosynthetic tool alongside its function as a chemical reaction optimizer [75]. Coupling this approach to a robotic platform, Jenson and coworkers show that the system could derive several pathways to a target, then using a robotic arm fabricate an appropriate flow system from modular parts to test and optimize each route [76].

Moore's law has continued to be robust since it was first postulated in 1965 due to the increased availability, lower cost, and improved processing power of microprocessors. Many new technologies we currently take for granted in smart devices, such as facial recognition and biometric security, are made possible through this increased processing power. The Ley group was one of the first to take advantage of such advancements, most notably using the Raspberry Pi® platform, for chemical synthesis [77]. Within science, many disciplines are leveraging this processing power to teach more complex algorithms or to process big data. Cheminformatics was an early adopter and has used it to fill our knowledge gaps by predicting chemical properties. These have been used to feed more complex models and provide more complete data sets.

Machines are not limited by the often conservative nature of chemists. Although a scientist's intuition will undoubtedly always out perform a machine's, conversely a machine is able to see past prejudices, having the strategic advantage of being able to access the combined stored chemical knowledge. This has its limitations –

most obviously the availability of data and more importantly its structure. All too often this invaluable knowledge and experience is consigned to the pages of a researcher's lab book, whereas in the open literature these could prove invaluable training data sets for future algorithms. Going forward, it is anticipated that there will be a tighter coupling of machine learning and continuous processing since the two complement one another well.

There are four principle areas where artificial intelligence is being applied within chemistry; computer-aided synthesis planning, process optimization, chemical reaction and property prediction, and high-throughput reaction discovery. These all rely on having access to large amounts of data from both successful and "failed" reactions. Even though our amassed chemical knowledge seems large we are still nowhere near the levels of big data. The amount of categorized data used to train face along with voice and optical character recognition algorithms is many orders of magnitude larger than our own. These have been helped by the Internet and especially social media. Unfortunately, chemistry doesn't have this luxury and is still heavily reliant on correctly characterized empirical data generated from experiments.

In the future, to help augment machine vision other computer senses are likely to develop and evolve. Voice recognition is already widely adopted among natural language interface devices such as Alexa, Siri, and Google Home. These could easily be incorporated into chemical control software and laboratory hardware in the future. There is already precedent for voice control in industrial processes with examples of PLCs being controlled through voice commands. This is becoming easier thanks to cloud services such as Microsoft Azure, Google Cloud, IBM's Watson, and Amazon Web Services, to name but a few.

While on the subject of senses we have already covered vision and voice, another obvious area, although not often used these days by chemists, is scent. Optical, electrochemical, catalytic bed, and metal oxide sensors have been employed in hazardous and safety critical environments with great success for the past 50 years, many of which were developed for the mining industry for the detection of toxic or flammable gases. Following their miniaturization and lower cost these are now increasingly used to monitor emissions and are a key enabler for future smart cities. Such devices are already used as oxygen depletion sensors in equipment rooms or where large volumes of toxic, flammable, or asphyxiant gases are used and stored. It is conceivable that these too could be utilized more in laboratory environments. Coupled with continuous-flow systems these could provide another level of safety. Better still would be the incorporation of these sensors into flow systems allowing them to monitor the potential buildup or release of hazardous gases and perform preemptive safe shutdown procedures.

11.4.5 AI: Artificial intelligence

Artificial intelligence is still a relatively untapped resource for synthetic chemistry outside the realms of cheminformatics. However, owing to the parallel developments of flow chemistry and PAT this coming together of technologies has led to an explosion of interest in the area and is opening up new areas of research.

The term "artificial intelligence" was first coined by John McCarthy when he held the first academic conference on the subject in 1956. It is often used interchangeably to describe one or more subfields and is a term that has become synonymous with the future. It is often used unsparingly to describe any processes or techniques involving computation, which in fact are not themselves true artificial intelligence. Before diving in to this complex area, it is worth broadly defining some of the key concepts.

Machine learning is about the automation of analytical model building. It uses methods from neural networks, statistics, and empirical data to find hidden insights in data without being explicitly programmed where to look or what to conclude.

Neural network is a type of machine learning inspired by the workings of the human brain. It is a computing system made up of interconnected units, similar in concept to neurons, which process information by responding to external inputs, relaying information between units. The process requires multiple passes of the data to find connections and derive meaning from undefined data.

Deep learning uses huge neural networks with many layers of processing units, taking advantage of advances in computing power and improved training techniques to learn complex patterns in large amounts of data. Common applications include image and speech recognition.

Computer or machine vision uses pattern recognition and deep learning to recognize what is in a picture or video. When machines can process, analyze, and understand images, they can capture images or videos in real time and interpret their surroundings.

Natural Language Processing (NLP) is the ability of computers to analyze, understand and generate human language, including speech. The next stage of NLP is natural language interaction, which allows humans to communicate with computers using normal, everyday language to perform tasks.

While machine learning is based on the idea that machines should be able to learn and adapt through experience, artificial intelligence refers to a broader idea where machines can execute tasks humanistically to a particular problem or decision.

Despite its many virtues and applications one must be aware that occasionally artificial intelligence might not be the most suitable solution. Indeed, there are examples where artificial intelligence has been applied and has not been able to outperform a human. This is particularly apparent within the creative and innovative spheres, not only in chemistry but other disciplines. There is often a tendency for research to get caught up in the moment and try and apply the latest and greatest breakthroughs to everything: in essence a solution looking for a problem. This type of approach should be avoided. As stated in the beginning of this final chapter, both a clear and quantifiable criteria should first be applied – read Marcus Aurelius and ask, “what is it you seek?”

11.4.6 Data collection and storage

Flow platforms by virtue of their continuous nature produce large amounts of process and chemical data. Advances in process analytical technologies, as discussed earlier, means access to real-time spectrographic data is now a reality. Unfortunately, data standardization has yet to be realized and there is still little consensus within the chemical community on the best way to collect and store this valuable data.

Typically during batch mode, off-line samples are taken and analyzed. Depending on the analysis this can take several minutes or even hours, particularly if additional sample preparation is required. This has obvious limitations when attempting to optimize or monitor a process. Continuous flow has largely removed this barrier. It has allowed this data to be used for real-time system control as discussed earlier. It has also presented an opportunity for experimentalists to provide more complete data sets that can be used to feed future algorithms.

Historically scientists, philosophers and researchers have used journals, manuscripts and notebooks to capture and share their observations and conclusions. The idea of electronic laboratory notebooks (ELNs) is far from novel. In fact its origins can be traced back to the early 1970s. As computers have become a permanent presence within laboratories attempts have been made to digitized and archive this data. This is often stored as flat data in the form of text files. Increasingly researchers are taking advantage of open-source repositories and sharing platforms such as GitHub®. Previously text (.txt) and comma-separated value (.csv) files have and to some extent remain the *de facto* standard for sharing and publishing data sets. These have ostensibly been replaced by the more scalable and flexible extensible markup language (XML) and increasingly Java Script Object Notation (JSON) formats. These have the advantage of being able to be read by web based technologies such as Java Script and PHP. Additionally these formats are better suited to modern databases like MongoDB® with architectures based on NoSQL. This offers better scalability over relational databases, for example Microsoft SQL Server®, Oracle SQL Server, and their open-source alternatives MariaDB® and MySQL®.

Having easy access to raw data and particularly in a digitalized form is vital for easy searching and mining. Along with reading, writing data is also a key component to any data storage system. The failure of existing laboratory data capture systems lies in their over-complication, which often manifests itself as confusing user interfaces. These are usually a result of system vendor’s attempts to produce a one-size-fits-all package (Fig. 11.7) [78].

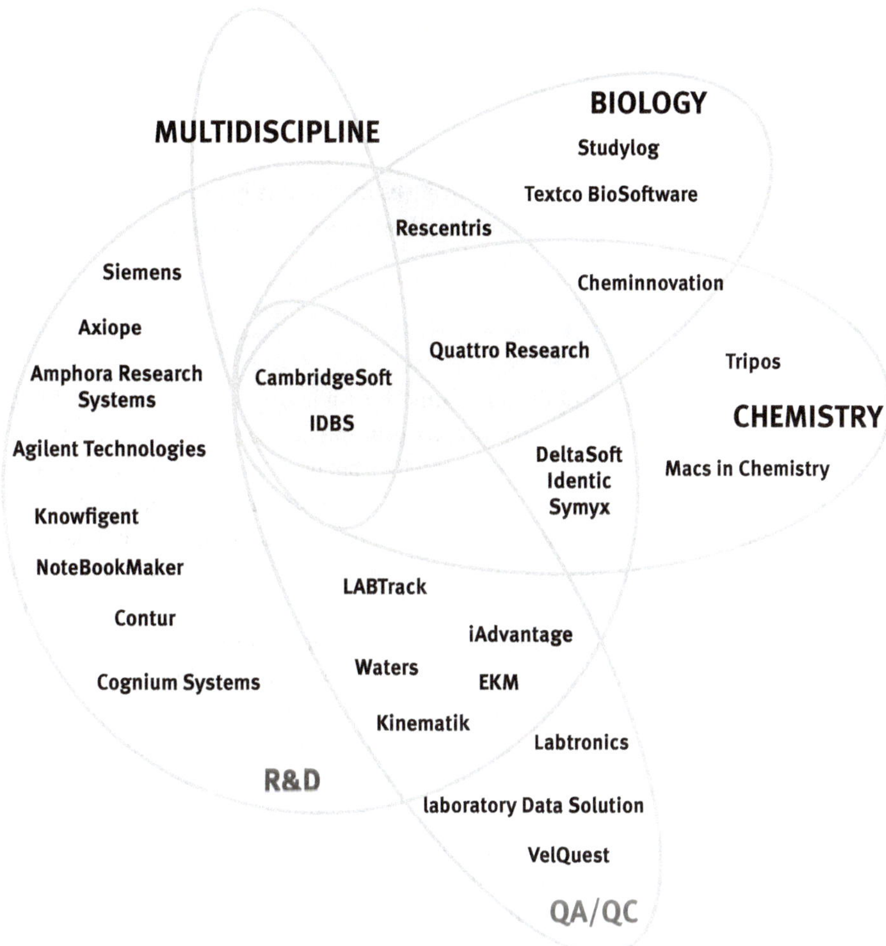

Fig. 11.7: Venn diagram demonstrating the multiple intersecting requirements from multiple scientific disciplines.

The ability to directly save generated data in real time during a chemical process offers many advantages. Several equipment vendors already take advantage of open data standards such as OPC UA. Unfortunately, there is still no consensus among continuous-flow equipment suppliers on what standard to adopt. Given that many laboratories have acquired and invested in multivendor equipment, especially in research groups, this creates further problems in the recording of data in conventional ELNs. Although beyond the scope of this review additional considerations towards data security and integrity will need to be addressed.

Self-optimizing reactor systems, along with high-throughput screening and reaction discovery platforms are out-pacing existing data storage solutions. As additional technologies such as machine vision take hold, managing and storing this data will require new innovations to prevent the data from becoming unusable due to its volume. Big data is one particular area within the scientific community that has been gaining traction over the past decade. As more digitized chemical knowledge becomes available, it is likely that the discipline will utilize the advances and research made within this field. Couple this with advances in machine learning and automated synthesis platforms the future of continuous-flow chemistry is bright.

11.4.7 Printing the future: 3D printing

Like so many innovations it often takes time for their true value to be identified. Additive manufacturing or more simply 3D printing has been a mainstay within engineering and manufacturing disciplines for the past 20 years. Its advantages become clear when combining it with modern computer-aided design (CAD) and computational fluid dynamics software. Researchers are able to rapidly visualize, validate and realize their designs. It is pleasing to see that researchers in flow chemistry are making use of these 3D-printing techniques. This is particularly apparent for developing reactor prototypes. What is more interesting is the development of the technology to print not only more complicated designs but also in a range of materials. An example where both of these innovations have been applied was in the use of a commercial selective laser melting technique to print a stainless steel reactor (Fig. 11.8). The printed reactor was used to perform a continuous low-temperature (−60 °C) deprotonation and difluoromethylation of diphenylacetonitrile with fluoroform. Of additional note was the ability of the printed component to be seamlessly integrated into a continuous-flow system by connecting to existing syringe pumps and laboratory equipment (Fig. 11.8) [79].

The use of 3D-printed parts is particularly attractive not only for the development of new reactors but also for existing components such as T-pieces, unions and adapters [80]. Existing laboratory equipment can quickly be customized to be used for one-off applications. Improved printer resolutions, even for hobbyist printers (<0.1 mm), can reliably print complex features such as screw threads. This eliminates the need for thread tapping or manual lathes. Equipment that would otherwise be relegated to single operations can be quickly integrated with other equipment simply by printing a new part.

An example of using 3D technology to exploit existing flow equipment was published by researchers at University College London [81]. Here a low-cost polypropylene continuous-flow reactor was fabricated using a hobbyist printer (Ultimaker 2+) for use in a Uniqsis® FlowSyn® continuous-flow reactor for a range of SNAr reactions. Several heterocyclic compounds were successfully synthesized, including the core structure of the natural product (±)-γ-lycorane and structurally complex compounds based on the tetracyclic core of erythrina alkaloids (Fig. 11.9).

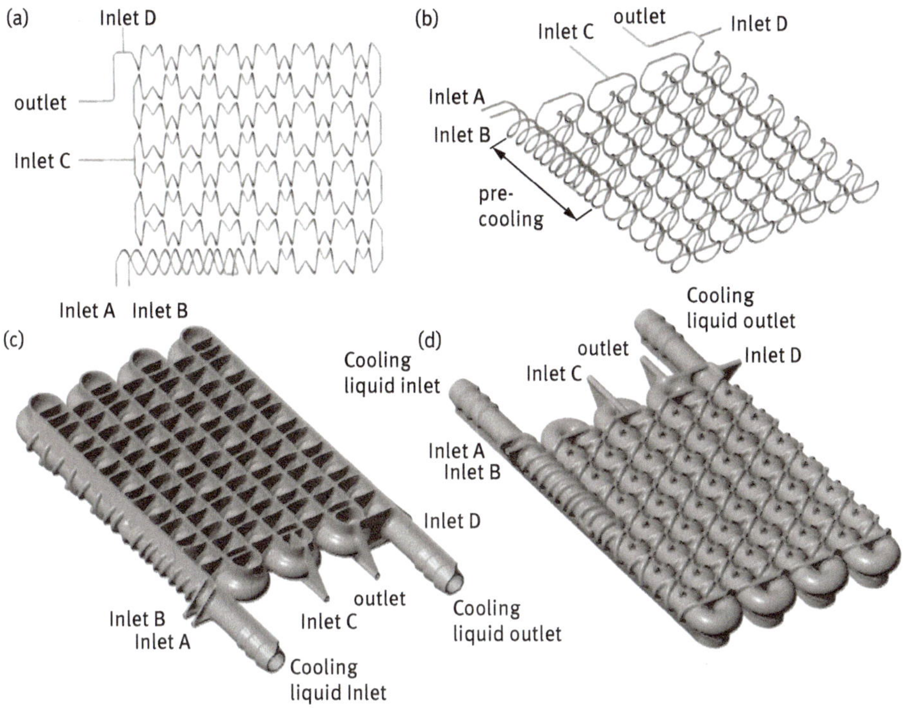

Fig. 11.8: CAD drawings of 3D printed flow reactor. **(A)** Top view of reaction channels; total length 4 m (1.89 mL reaction volume). **(B)** Perspective view. **(C and D)** How channels are directly incorporated onto the cooling coil is demonstrated.

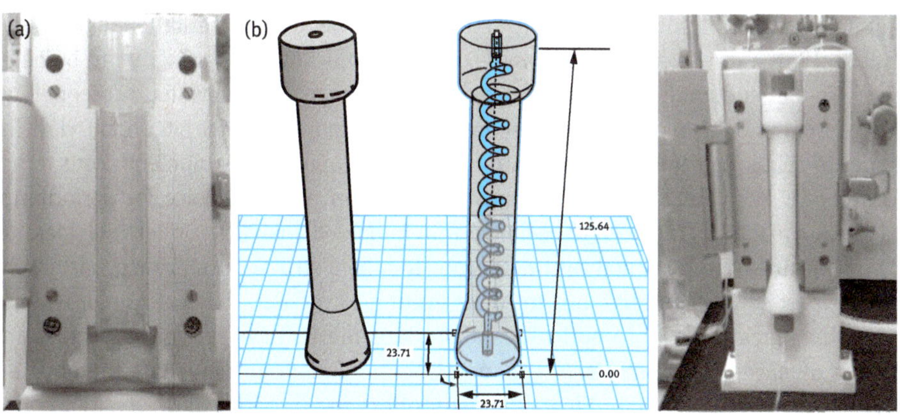

Fig. 11.9: **(A)** Heating block for glass reactor. **(B)** Design of the reactor showing internal coil in CAD software. **(C)** 3D-printed PP reactor fitted in the reactor space on the Uniqsis® FlowSyn®.

The low cost, wide availability, and chemical inertness of polypropylene demonstrated this as an attractive alternative to more expensive PTFE and 316L-printed reactors, both of which require expensive equipment and filaments. These printing technologies currently sit outside the reach of most research laboratories; however, the successful use of a hobbyist 3D printer and available off-the-shelf filaments demonstrates exciting possibilities for the technology and how it can be exploited by research groups in the near feature.

A shared advantage with continuous-flow platforms is the ability to reproduce printed components and propagate designs through open source STL file formats. Of special interest is the ability to print larger monolithic components. This has recently become a reality through so-called continuous 3D printers. Here the static heated bed has been replaced with a moving conveyor, adding an additional dimension without any increase in platform size. It is likely these systems and the process of designing components will become more common place within chemical laboratories. Already, research groups are exploring the technology and developing the skill set necessary to fully utilize these benefits.

i *(For related topics please see Volume 2, Chapter 9, Title:* Advanced Manufacturing Techniques in Reactor Engineering: 3DP *Applications in Flow Chemistry)*

11.4.8 High-throughput synthesis: faster is better

High-throughput reaction screening in batch is a tested strategy for reaction discovery and methodology. There is currently no concerted effort to develop a similar approach in flow. Highlighted earlier, there needs to be a fundamental change in the mindset if the true benefits and rewards of continuous flow processing are to be realized. To that end, initial reaction discoveries should be made in flow and subsequently transferred to batch mode rather than the other way round as is the situation today. In order to derisk this approach, innovative high-throughput flow equipment needs to be developed. Microliter plug-flow systems would be an obvious choice for performing screening reactions, but need to be linked with improved analytics and technologies.

Molecules are designed and made not for their structural elegance, although this is perfectly acceptable, but for their eventual function. This can range widely from biological activity to optical, catalytic, polymeric, or electrical properties. In nearly all cases the synthesis is separated from the functional screening thereby delaying the discovery cycle. Flow chemistry offers the opportunity to make a compound and immediately screen it for activity within the same platform. Being able to capture the functional readout in real time provides useful information that can be used immediately and fed back into the next round of the design cycle leading to improved targets. This closed-loop approach, in principle, offers rapid serial processing toward a defined functional goal. While not yet validated as a general concept in flow chemistry, the presence of published studies determining linked synthesis and biological activity are encouraging.

A notable example of this closed-loop approach described the preparation of a series of Abl kinase inhibitors. This fully integrated platform synthesized 21 compounds through an automated process, affording a novel hinge binding motif with $pIC_{50} > 8$ against Abl kinase. The system consisted of synthesis on a microfluidic chip coupled to and integrated with liquid chromatography mass spectrometry (LC-MS) and a reformatting device prior to in-line biological assay (Fig. 11.10). Once generated the data re-enters the cycle where a computed compound that is judged to be an improvement from the previous structural motif is prepared for a further round of synthesis and screening. This is an iterative closed-loop process and is stopped once a level of biological activity has been achieved [82].

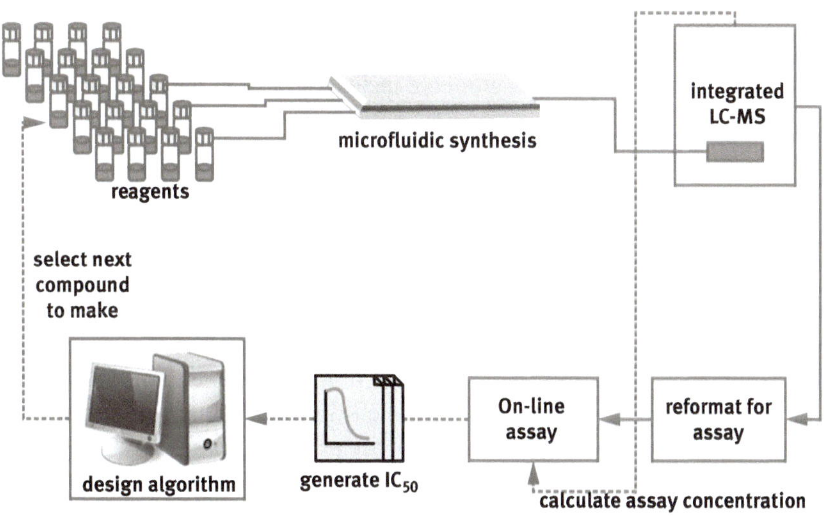

Fig. 11.10: Schematic of integrated flow platform with online assay and feedback.

Unfortunately, this is one of the few examples of flow-based synthesis that has been directly coupled to a biological activity assay screen. It is anticipated that this area will grow and begin to incorporate additional emerging technologies and tools. As flow chemistry expands into new domains, particularly for material production, linked evaluation of functional properties are likely to follow by exploiting rapid information feedback to generate new self-optimizing discovery cycles.

11.4.9 Education: teaching the next generation

A key enabler for the area is the ability of process designers to think and plan in flow and to move beyond existing batch mode thinking. Education and early adoption in the design phase will be crucial. This is likely to be a slow process as new generations come online and current generations learn to adapt. To ease this transition and to power the next generation, machine learning will play a huge part in the discovery and optimization of chemical processes in the future.

It is fair to say that the area has not expanded quite as rapidly as the early indications may have suggested – why is this? It is mostly likely due to the conservative character of synthetic chemists who over many years have developed skills based around batch processing and traditional glassware. Further to this, educational programs pointing out the opportunities arising from continuous processing have not entered mainstream teaching curriculums. Although flow chemistry is starting to appear as a taught course in several undergraduate syllabuses, the situation is changing and is being driven by our data-rich society. New generations are less fearful of this integration of technology and are more inclined to live in harmony with their machines.

11.5 Final thoughts

The machine-enabled approach delineated in this second edition of *Flow Chemistry* has undoubtedly impacted the way we view advances in robotics, machine learning, and continuous processing. There is a need to rapidly exploit these technologies and to sustainably provide wider access to chemical data. Harnessing the best of human endeavor and of the machines, these are exciting times whereby engagement with other disciplines is opening up unforeseen opportunities. Flow chemistry and continuous processing methods can afford high levels of reproducibility and consistency during the production of QDs, nanomaterials, polymers, and other advanced functional materials, such as dyes, beyond conventional methods [57]. Over the years, it has become clear that early education and training in the area are necessary to provide a justifiable comparison with more traditional methods. In synthesis it is all too easy to be trapped into using only the conventional skills and mechanistic understanding of our training. This degree of conservatism expresses itself when one notices that the fundamental tools of synthesis have not changed significantly over time.

Additionally, several data obstacles and engineering challenges remain and will be the focus of much future research. A substantial stumbling block remains the lack of empirical data to feed and train increasingly complex algorithms and neural networks. Current work around this problem has seen much ingenuity and innovation in the use of models to predict chemical properties in an attempt to fill this data gap. A second equally challenging hurdle is the miniaturization of flow technology to perform more advanced workups in the downstream and at a scale that allows deployment of these systems in laboratories.

Finally, there is the challenge of better cross-discipline co-operation. As can be seen by the preceding chapters many challenges have been solved within their respective disciplines. As this text attests, these challenges take the combined efforts of chemical engineers, systems developers, equipment vendors, statisticians, and chemists. The lack of a universal open-source platform, defined software ecosystem, and common modular architecture remains a big challenge for the area and is likely to be the focus of much future research and debate. Going forward, cooperation and collaboration will become even more important along with the role of robotics and machine learning. These will undoubtedly propel the field of continuous processing to greater heights. Continuous flow chemistry has arguably been the single biggest paradigm shift in recent years in the development of the synthesis chemists' tool box.

References

[1] Horváth IT, Anastas PT, Introduction: green chemistry, Chem Rev, 2007, 2167–2168.
[2] Anastas P, Eghbali N, Green chemistry: principles and practice, Chem Soc Rev, 2010, 39, 301–312.
[3] Ritson DJ, Battilocchio C, Ley SV, Sutherland JD, Mimicking the surface and prebiotic chemistry of early earth using flow chemistry, Nat Commun, 2018, 9, 1821.
[4] Sipos G, Bihari T, Milánkovich D, Darvas F, Flow chemistry in space–a unique opportunity to perform extraterrestrial research, J Flow Chem, 2017, 7, 151–156.
[5] Hessel V, Sarafraz MM, Tran NN, The resource gateway: microfluidics and requirements engineering for sustainable space systems, Chem Eng Sci, 2020, 225, 115774.
[6] Ley SV, Chen Y, Fitzpatrick DE, May O, A new world for chemical synthesis?, Chimia, 2019, 73, 792–802.
[7] Britton J, Jamison TF, The assembly and use of continuous flow systems for chemical synthesis, Nat Protoc, 2017, 12, 2423–2446.
[8] Pastre JC, Browne DL, Ley SV, Flow chemistry syntheses of natural products, Chemical Society Reviews, Royal Society of Chemistry, 8849–8869, Dec. 07, 2013.
[9] Plutschack MB, Pieber B, Gilmore K, Seeberger PH, The Hitchhiker's guide to flow chemistry, Chem Rev, 2017, 117, 11796–11893.
[10] Britton J, Raston CL, Multi-Step Continuous-Flow Synthesis, Chemical Society Reviews, Royal Society of Chemistry, 1250–1271, Mar. 07, 2017.
[11] Rahimi M, Azimi N, Parsamogadam MA, Rahimi A, Masahy MM, Mixing performance of T, Y, and oriented Y-micromixers with spatially arranged outlet channel: evaluation with Villermaux/Dushman test reaction, Microsys Technol, 2017, 23, 3117–3130.
[12] Reckamp JM, Bindels A, Duffield S, Liu YC, Bradford E, Ricci E, Susanne F, Rutter A, Mixing performance evaluation for commercially available micromixers using villermaux–dushman

reaction scheme with the interaction by exchange with the mean model, Org Process Res Dev, 2017, 21, 816–820.
[13] Ley SV, Fitzpatrick DE, Ingham RJ, Myers RM, Organic synthesis: march of the machines, Angew Chem Int Ed, 2015, 54, 3449–3464.
[14] Ley SV, Fitzpatrick DE, Myers RM, Battilocchio C, Ingham RJ, Machine-assisted organic synthesis, Angew Chem Int Ed, 2015, 54, 10122–10136.
[15] Browne DL, Deadman BJ, Ashe R, Baxendale IR, Ley SV, Continuous flow processing of slurries: evaluation of an agitated cell reactor, Org Process Res Dev, 2011, 15, 693–697.
[16] Ouchi T, Mutton RJ, Rojas V, Fitzpatrick DE, Cork DG, Battilocchio C, Ley SV, Solvent-free continuous operations using small footprint reactors: a key approach for process intensification, ACS Sustain Chem Eng, 2016, 4, 1912–1916.
[17] Movsisyan M, Delbeke EIPP, Berton JKET, Battilocchio C, Ley SV, Stevens CV, Taming hazardous chemistry by continuous flow technology, Chem Soc Rev, 2016, 45, 4892–4928.
[18] Hessel V, Novel process windows – gates to maximizing process intensification via flow chemistry, Chem Eng & Technol, 2009, 32, 1641–1641.
[19] Hessel V, Kralisch D, Kockmann N, Noël T, Wang Q, Novel process windows for enabling, accelerating, and uplifting flow chemistry, ChemSusChem, 2013, 6, 746–789.
[20] Chaudhuri A, Kuijpers KPL, Hendrix R, Hacking J, Shivaprasad P, Emanuelsson EAC, Noel T, Van Der Schaaf J, Process Intensification of Photochemical Oxidations using a High Throughput Rotor-Stator Spinning Disk Reactor: A Strategy for Scale Up, 2020.
[21] Newman SG, Jensen KF, The role of flow in green chemistry and engineering, Green Chem, 2013, 15, 1456.
[22] Oliveira PHR, Santos BMS, Leão RAC, Miranda LSM, San Gil RAS, Souza ROMA, Finelli FG, From immobilization to catalyst use: a complete continuous-flow approach towards the use of immobilized organocatalysts, ChemCatChem, 2019, 11, 5553–5561.
[23] Ley SV, On being green: can flow chemistry help?, Chem Rec, 2012, 12, 378–390.
[24] Dallinger D, Kappe CO, Why Flow Means Green – Evaluating the Merits of Continuous Processing in the Context of Sustainability, Elsevier B.V., 2017, 6–12.
[25] Rogers L, Jensen KF, Continuous manufacturing – the green chemistry promise?, Green Chem, 2019, 21, 3481–3498.
[26] Tanaka K, Toda F, Solvent-Free Organic Synthesis, Chem Rev, 2000, 100, 1025–1074.
[27] Labes R, Mateos C, Battilocchio C, Chen Y, Dingwall P, Cumming GR, Rincón JA, Nieves-Remacha MJ, Ley SV, Fast continuous alcohol amination employing a hydrogen borrowing protocol, Green Chem, 2019, 21, 59–63.
[28] Ley SV, Baxendale IR, Bream RN, Jackson PS, Leach AG, Longbottom DA, Nesi M, Scott JS, Storer RI, Taylor SJ, Multi-step organic synthesis using solid-supported reagents and scavengers: a new paradigm in chemical library generation, J Chem Soc, Perkin Trans, 1, 2000, 3815–4195.
[29] Ley SV, Baxendale IR, New tools and concepts for modern organic synthesis, Nat Rev Drug Discov, 2002, 1, 573–586.
[30] Baxendale IR, Deeley J, Griffiths-Jones CM, Ley SV, Saaby S, Tranmer GK, A flow process for the multi-step synthesis of the alkaloid natural product oxomaritidine: a new paradigm for molecular assembly, Chem Commun, 2006, 2566–2568.
[31] Baxendale IR, Ley SV, Mansfield AC, Smith CD, Multistep synthesis using modular flow reactors: Bestmann-Ohira reagent for the formation of alkynes and triazoles, Angew Chem Int Ed, 2009, 48, 4017–4021.
[32] Cossy J, Arseniyadis S (Eds), Modern Tools for the Synthesis of Complex Bioactive Molecules, Hoboken, NJ, USA, John Wiley & Sons, Inc., ISBN 978-0-470-61618-5, 2012. Myers RM, Roper KA, Baxendale IR, and Ley SV, The evolution of immobilised reagents and their application in flow

chemistry for the synthesis of natural products and pharmaceutical compounds. Chapter 11, p. 359–394.
[33] Coley CW, Eyke NS, Jensen KF, Autonomous discovery in the chemical sciences part i: progress, Angew Chem Int Ed, 2020, 59, 22858–22893.
[34] O'Brien M, Hall A, Schrauwen J, Van Der Made J, An open-source approach to automation in organic synthesis: the flow chemical formation of benzamides using an inline liquid-liquid extraction system and a homemade 3-axis autosampling/product-collection device, Tetrahedron, 2018, 74, 3152–3157.
[35] Brzozowski M, O'Brien M, Ley SV, Polyzos A, Flow chemistry: intelligent processing of gas–liquid transformations using a tube-in-tube reactor, Acc Chem Res, 2015, 48, 349–362.
[36] Mallia CJ, Baxendale IR, The use of gases in flow synthesis, Org Process Res Dev, 2016, 20, 327–360.
[37] Mo Y, Imbrogno J, Zhang H, Jensen KF, Scalable thin-layer membrane reactor for heterogeneous and homogeneous catalytic gas–liquid reactions, Green Chem, 2018, 20, 3867–3874.
[38] Hone CA, Kappe CO, Membrane microreactors for the on-demand generation, separation, and reaction of gases, Chemi – Eur J, 2020, 26, 13108–13117.
[39] Jones R, Gödörházy L, Szalay D, Gerencsér J, Dormán G, Ürge L, Darvas F, Novel A, Method for high-throughput reduction of compounds through automated sequential injection into a continuous-flow microfluidic reactor, QSAR Comb Sci, 2005, 24, 722–727.
[40] O'Brien M, Baxendale IR, Ley SV, Flow ozonolysis using a semipermeable teflon af-2400 membrane to effect gas–liquid contact, Org Lett, 2010, 12, 1596–1598.
[41] Kasinathan S, Bourne S, Tolstoy P, Koos P, O'Brien M, Bates R, Baxendale I, Ley SV, Syngas-Mediated C-C Bond Formation in Flow: selective Rhodium-Catalysed Hydroformylation of Styrenes, Synlett, 2011, 2011, 2648–2651.
[42] Brahma A, Musio B, Ismayilova U, Nikbin N, Kamptmann SB, Siegert P, Jeromin GE, Ley SV, Pohl M, An orthogonal biocatalytic approach for the safe generation and use of HCN in a multistep continuous preparation of chiral O-acetylcyanohydrins, Synlett, 2016, 27, 262–266.
[43] Hornung CH, Mackley MR, Baxendale IR, Ley SV, A microcapillary flow disc reactor for organic synthesis, Org Process Res Dev, 2007, 11, 399–405.
[44] Adamo A, Heider PL, Weeranoppanant N, Jensen KF, Membrane-based, liquid–liquid separator with integrated pressure control, Ind Eng Chem Res, 2013, 52, 10802–10808.
[45] O'Brien M, Koos P, Browne DL, Ley SV, A prototype continuous-flow liquid–liquid extraction system using open-source technology, Org Biomol Chem, 2012, 10, 7031.
[46] Deadman BJ, Battilocchio C, Sliwinski E, Ley SV, A prototype device for evaporation in batch and flow chemical processes, Green Chem, 2013, 15, 2050.
[47] Zhang H, Lakerveld R, Heider PL, Tao M, Su M, Testa CJ, D'Antonio AN, Barton PI, Braatz RD, Trout BL, Myerson AS, Jensen KF, Evans JMB, Application of continuous crystallization in an integrated continuous pharmaceutical pilot plant, Cryst Growth Des, 2014, 14, 2148–2157.
[48] Scott CD, Labes R, Depardieu M, Battilocchio C, Davidson MG, Ley SV, Wilson CC, Robertson K, Integrated plug flow synthesis and crystallisation of pyrazinamide, React Chem & Eng, 2018, 3, 631–634.
[49] Adamo A, Beingessner RL, Behnam M, Chen J, Jamison TF, Jensen KF, Monbaliu J-CM, Myerson AS, Revalor EM, Snead DR, Stelzer T, Weeranoppanant N, Wong SY, Zhang P, On-demand continuous-flow production of pharmaceuticals in a compact, reconfigurable system, Science, 2016, 352, 61–67.
[50] Berton M, De Souza JM, Abdiaj I, McQuade DT, Snead DR, Scaling continuous API synthesis from milligram to kilogram: extending the enabling benefits of micro to the plant, J Flow Chem, 2020.

[51] Heider PL, Born SC, Basak S, Benyahia B, Lakerveld R, Zhang H, Hogan R, Buchbinder L, Wolfe A, Mascia S, Evans JMBB, Jamison TF, Jensen KF, Development of a Multi-Step Synthesis and Workup Sequence for an Integrated, Continuous Manufacturing Process of a Pharmaceutical, Org Process Res Dev, 2014, 18, 402–409.

[52] Reizman BJ, Jensen KF, Feedback in flow for accelerated reaction development, Acc Chem Res, 2016, 49, 1786–1796.

[53] Struble TJ, Alvarez JC, Brown SP, Chytil M, Cisar J, DesJarlais RL, Engkvist O, Frank SA, Greve DR, Griffin DJ, Hou X, Johannes JW, Kreatsoulas C, Lahue B, Mathea M, Mogk G, Nicolaou CA, Palmer AD, Price DJ, Robinson RI, Salentin S, Xing L, Jaakkola T, Green WH, Barzilay R, Coley CW, Jensen KF, Current and future roles of artificial intelligence in medicinal chemistry synthesis, J Med Chem, 2020, 63, 8667–8682.

[54] Sagmeister P, Williams JD, Hone CA, Kappe CO, Laboratory of the future: a modular flow platform with multiple integrated PAT tools for multistep reactions, React Chem & Eng, 2019, 4, 1571–1578.

[55] Hopkin MD, Baxendale IR, Ley SV, The lab of the future, Chem Today, 2011, 29, 28–34.

[56] Gouveia FF, Rahbek JP, Mortensen AR, Pedersen MT, Felizardo PM, Bro R, Mealy MJ, Using PAT to accelerate the transition to continuous API manufacturing, Anal Bioanal Chem, 2017, 409, 821–832.

[57] Myers RM, Fitzpatrick DE, Turner RM, Ley SV, Flow chemistry meets advanced functional materials, Chem – Eur J, 2014, 20, 12348–12366.

[58] Baumann M, Baxendale IR, The synthesis of active pharmaceutical ingredients (APIs) using continuous flow chemistry, Beilstein J Org Chem, 2015, 11, 1194–1219.

[59] Gérardy R, Emmanuel N, Toupy T, Kassin V-E, Tshibalonza NN, Schmitz M, Monbaliu J-CM, Continuous flow organic chemistry: successes and pitfalls at the interface with current societal challenges, Eur J Org Chem, 2018, 2018, 2301–2351.

[60] Gutmann B, Cantillo D, Kappe CO, Continuous-flow technology-a tool for the safe manufacturing of active pharmaceutical ingredients, Angew Chem Int Ed, 2015, 54, 6688–6728.

[61] Lange H, Carter CF, Hopkin MD, Burke A, Goode JG, Baxendale IR, Ley SV, A breakthrough method for the accurate addition of reagents in multi-step segmented flow processing, Chem Sci, 2011, 2, 765.

[62] Fitzpatrick DE, Battilocchio C, Ley SV, A novel internet-based reaction monitoring, control and autonomous self-optimization platform for chemical synthesis, Org Process Res Dev, 2016, 20, 386–394.

[63] Nunn C, DiPietro A, Hodnett N, Sun P, Wells KM, High-throughput automated design of experiment (DoE) and kinetic modeling to aid in process development of an API, Org Process Res Dev, 2018, 22, 54–61.

[64] Basavaraju G, Lydia SD, Rajanna R, Flow process development and optimization of halo-amine coupling through customized flow processing equipment using DoE approach, J Flow Chem, 2020, 10, 571–582.

[65] Aroh KC, Jensen KF, Efficient kinetic experiments in continuous flow microreactors, React Chem & Eng, 2018, 3, 94–101.

[66] Clayton AD, Manson JA, Taylor CJ, Chamberlain TW, Taylor BA, Clemens G, Bourne RA, Algorithms for the self-optimisation of chemical reactions, React Chem & Eng, 2019, 4, 1545–1554.

[67] Holmes N, Akien GR, Blacker AJ, Woodward RL, Meadows RE, Bourne RA, Self-optimisation of the final stage in the synthesis of EGFR kinase inhibitor AZD9291 using an automated flow reactor, React Chem & Eng, 2016, 1, 366–371.

[68] Walker BE, Bannock JH, Nightingale AM, Demello JC, Tuning reaction products by constrained optimisation, React Chem & Eng, 2017, 2, 785–798.
[69] Schweidtmann AM, Clayton AD, Holmes N, Bradford E, Bourne RA, Lapkin AA, Machine learning meets continuous flow chemistry: automated optimization towards the Pareto front of multiple objectives, Chem Eng J, 2018, 352, 277–282.
[70] Romero-Fernández M, Paradisi F, Protein immobilization technology for flow biocatalysis, Curr Opin Chem Biol, 2020, 55, 1–8.
[71] Kamptmann SB, Ley SV, Facilitating biomimetic syntheses of borrerine derived alkaloids by means of flow-chemical methods, Aust J Chem, 2015, 68, 693.
[72] Musio B, Mariani F, Śliwiński EP, Kabeshov MA, Odajima H, Ley SV, Combination of Enabling Technologies to Improve and Describe the Stereoselectivity of Wolff-Staudinger Cascade Reaction, Synthesis (Germany), 2016, 48, 3515–3526.
[73] Ley SV, Ingham RJ, O'Brien M, Browne DL, Camera-enabled techniques for organic synthesis, Beilstein J Org Chem, 2013, 9, 1051–1072.
[74] Clayton AD, Schweidtmann AM, Clemens G, Manson JA, Taylor CJ, Niño CG, Chamberlain TW, Kapur N, Blacker AJ, Lapkin AA, Bourne RA, Automated self-optimisation of multi-step reaction and separation processes using machine learning, Chem Eng J, 2020, 384, 123340.
[75] Baylon JL, Cilfone NA, Gulcher JR, Chittenden TW, Enhancing retrosynthetic reaction prediction with deep learning using multiscale reaction classification, J Chem Inf Model, 2019, 59, 673–688.
[76] Coley CW, Thomas DA, Lummiss JAM, Jaworski JN, Breen CP, Schultz V, Hart T, Fishman JS, Rogers L, Gao H, Hicklin RW, Plehiers PP, Byington J, Piotti JS, Green WH, Hart AJ, Jamison TF, Jensen KF, A robotic platform for flow synthesis of organic compounds informed by AI planning, Science, 2019, 365, eaax1566.
[77] Fitzpatrick DE, O'Brien M, Ley SV, A tutored discourse on microcontrollers, single board computers and their applications to monitor and control chemical reactions, React Chem & Eng, 2020, 5, 201–220.
[78] Bird CL, Willoughby C, Frey JG, Laboratory notebooks in the digital era: the role of ELNs in record keeping for chemistry and other sciences, Chem Soc Rev, 2013, 42, 8157–8175.
[79] Gutmann B, Köckinger M, Glotz G, Ciaglia T, Slama E, Zadravec M, Pfanner S, Maier MC, Gruber-Wölfler H, Oliver Kappe C, Design and 3D printing of a stainless steel reactor for continuous difluoromethylations using fluoroform, React Chem & Eng, 2017, 2, 919–927.
[80] Alimi OA, Akinnawo CA, Onisuru OR, Meijboom R, 3-D printed microreactor for continuous flow oxidation of a flavonoid, J Flow Chem, 2020, 10, 517–531.
[81] Rao ZX, Patel B, Monaco A, Cao ZJ, Barniol-Xicota M, Pichon E, Ladlow M, Hilton ST, 3D-printed polypropylene continuous-flow column reactors: exploration of reactor utility in S N Ar reactions and the synthesis of bicyclic and tetracyclic heterocycles, Eur J Org Chem, 2017, 2017, 6499–6504.
[82] Desai B, Dixon K, Farrant E, Feng Q, Gibson KR, van Hoorn WP, Mills J, Morgan T, Parry DM, Ramjee MK, Selway CN, Tarver GJ, Whitlock G, Wright AG. Rapid discovery of a novel series of Abl kinase inhibitors by application of an integrated microfluidic synthesis and screening platform. J Med Chem. 2013, 56, 3033–3047.

Answers to the study questions

1.1
$$k = \alpha \cdot I^{\beta}$$

where

k is the rate constant,

α is a constant depending on the type of photochemistry.

I is the light intensity

β is a constant depending on the photon flux.

First regime: For lower light intensities, β is 1.0. This means that the rate constant increases linearly with increasing photon fluxes.

Second Regime: For intermediate light intensities, β is around 0.5 (i.e. square root behavior).

Third Regime: For high light intensities, β becomes 0 and, consequently, the reaction rate becomes independent of the light intensity. In the latter situation, all catalyst molecules are permanently active and not all photons are being absorbed.

1.2 1. Since photocatalysis is chromoselective, it is important to match the emission wavelength of the light source with the absorption characteristics of the photochemical transformation. E.g., use LEDs with narrow emission wavelengths.

2. To avoid lost irradiation and to maximize radiation transfer to the reactor, the distance between the light source and the reactor needs to be optimized (see also Figure 4A-C). In addition, to prevent dilution of energy, light can be focused to the reactor by using suitable mirrors and/or waveguides.

3. Due to the attenuation effect and the correlation between reaction rate and light intensity, microreactors are preferred as reactors. They ensure a more uniform irradiation of the reaction medium, resulting in equal reaction conditions across the diameter of the microchannel.

1.3 The advantages brought by flow chemistry:

1. Microreactors provide a better penetration of light resulting in a homogeneous energy distribution over the entire reaction medium (Lambert-Beer law). This allows to reduce the residence times (only 10 min reaction time) and to avoid byproduct formation (in this case decarboxylation could occur at longer operation times).

2. The second advantage is the use of a gas liquid flow regime (segmented or Taylor flow), which provides a reproducible interfacial area and enhances mixing (through enhanced mixing patterns) and reduces gas-liquid mass transfer.

3. The pressure is elevated (0.34 MPa corresponds to almost 3.5 Atm), which increases the solubility of the gaseous reagent into the organic phase. Using a back pressure regulator, one can easy use elevated pressures in flow.

https://doi.org/10.1515/9783110693690-012

1.4 Solids can be handled by
1. applying ultrasound which breaks up particle agglomerates (the smaller the particles, the easier to handle) and prevents particle setting at the reactor walls. See also: Chemical Engineering Journal 2021, 428, 130968.
2. packing the solids in packed-bed reactor or coating them on the reactor walls.
3. by encapsulating the solids in a non-wetting liquid bubble (Taylor-flow enabled suspension). This prevents the solids from touching the reactor walls and due to the flow-induced recirculation patterns in the bubble the solids are kept in suspension.

1.5 The volume of a microreactor can be expressed using the following equation:

$$V = \frac{\pi}{4} \cdot N \cdot L \cdot D^2$$

where N is the number of the channels connected in parallel, L is the length of the channel and D is the hydrodynamic channel diameter.
The first option is to increase the total number of channels (N) which are placed in parallel (numbering up).
The second option is called sizing up and is about increasing the dimensions of the reactor. This can be done either by increasing the length of the reactor (L) or its diameter (D). However, bear in mind that increasing the diameter (D) can become an issue due to the observed attenuation effect of photon transport (Bouguer-Lambert-Beer law). However, by combining high photon flux light sources and increased mass transfer, these hurdles can be partially overcome and the diameter can be increased (up to one-to-several mm).
As each scale up strategy has its merits, there is not a single strategy that can reach the scale up factors required for industrial implementation. It is clear that different approaches need to be combined to reach a satisfying production capacity. The total scale up factor of the reactor volume can be expressed as the product of individual scale up factors of the channel number (S_N), channel length (S_L) and the channel diameter (S_D):

$$S = S_N \cdot S_L \cdot S_D^2$$

1.6 1. to reduce some repetitive laboratory work
2. to gather more data with greater accuracy in a shorter amount of time
3. to carry out some hazardous processes in a more secure environment using automation protocols and robotics, hence not exposing the researcher to potentially hazardous situations.

2.1 Electrochemical processes are useful for storage and generation of energy, for example, in rechargeable batteries (metal hydride, Li ion, NiCd) and fuel cells. Electrochemical refinery (Cu, Al, Mg, etc.), electroplating (Ni, Cu, Ag, Au, etc.), surface treatment (oxidation of aluminum), and electrolysis (Cl_2/NaOH) are also common industrial processes.

2.2 An electrochemical cell consists of two electrodes connected to an external DC power supply (current and voltage). Both electrodes are immersed in an electrolyte, which contains chargeable anions and cations. By applying an electrical current, an oxidation reaction takes place on one electrode surface (anode) and a reduction reaction on the opposite one (cathode).
Overall reaction: $2H_2O \rightarrow 2H_2 + O_2$
Anode reaction: $6H_2O \rightarrow O_2 + 4H_3O^+ + 4e^-$ (acidic solution)
$4OH^- \rightarrow O_2 + 2H_2O + 4e^-$ (alkaline solution)
Cathode reaction: $2H_3O^+ + 2e^- \rightarrow H_2 + 2H_2O$ (acidic solution)
$2H_2O + 2e^- \rightarrow H_2 + 2OH^-$ (alkaline solution)

2.3 It is necessary to add a supporting electrolyte, which dissociates into anions and cations, to ensure the electrical transfer of charges (current) through the electrolyte. Without adding a supporting electrolyte, the ohmic resistances between the electrodes raises and the electrode potential increases. Therefore, the electrochemical reaction becomes uncontrolled and the selectivity decreases. A close proximity between the anode and the cathode, which is the case in a micro- or narrow-gap cell setup, reduces the ohmic resistance, a priori, and the concentration of the supporting electrolyte can thus be remarkably decreased.

2.4 Performing an electrochemical reaction in a potentiostatic mode (electrode potential) instead of a galvanostatic one (current/current density) allows to select the chemical reaction by the applied potential and to suppress unwanted side reactions. The potential defines the current and not vice versa. In this case, the current decreases in proportion (at least to 0) by the progress of the electrochemical reaction.

2.5 Gas evolution can be handled in several ways: The most convenient way is to operate the cell in a cycling or cascading mode. The applicable high flow rates ensure efficient removal of the gas from the flow electrolyzer, which is in turn removed from the system via the headspace of the reservoir. In the single-pass flow electrolysis, gas bubbles may be (partially) compensated by pressurizing the whole system. However, conversion, yield, and current efficiency may decrease due to reoxidation/reduction of the dissolved gas at the catalytically active electrodes.

2.6 Anodic processes are conducted at electrodes with sufficient high oxidation potentials to avoid corrosion of the electrode. These include graphite, glassy carbon (GC), and boron-doped diamond (BDD), as well as platinum and other noble metals, for example, gold. Lesser noble metals such as nickel or molybdenum can be used under certain circumstances if the electrolyte composition ensures the formation of an insoluble conductive layer on the electrode surface (→ active electrodes). In a few cases, easily oxidized metals such as magnesium, aluminum, or zinc may be employed as sacrificial anodes if the generated metal ions exhibit an auxiliary effect.

2.7 The most popular carbon-based electrode materials are graphite, GC, and BDD. Graphite is the working horse in technical electroorganic synthesis. It is inexpensive, available in a number of specifications (highly dense, felt, etc.), and easy to handle. The overpotential for O_2 evolution offers access to various anodic reactions. However, anodes might be consumed if high concentrations of oxygen or chlorine are involved and especially porous graphite cathodes are prone to cathode fouling. GC shows similar electrochemical properties as platinum and is easy in handling and purification. With reticulated vitreous carbon, a porous variety is available. GC also shows an overpotential for oxygen evolution. On the other hand, GC is very hard and brittle and, thus, difficult to machine. Additionally, it is quite costly. BDD is a sustainable high-performance electrode material. In principle, it can be produced at relatively low costs by plasma coating on a conductive support material (mostly Si or Nb). However, the production of large electrodes is expensive. The electrochemical properties are similar to or even better than platinum. In particular, the high overpotentials for hydrogen and oxygen evolution enable an extraordinary broad potential window in aqueous solution as well as in other solvents. In contrast, it is a rather novel and emerging material with which many chemists still have less experience, and its mechanical sensitivity due to the thin layer of actual diamond makes it difficult to handle, for example, machining.

2.8 When platinum is employed, for example, as an anode material, the chlorine within the electrolyte is able to form the soluble platinum tetrachloride complex, leading to a tremendous corrosion of the anode.

2.9 In anodic oxidation reactions, the counterreaction at the cathode is often the hydrogen evolution reaction. The same accounts for the reductive processes, whereby, at the anode, the oxygen evolution reaction (OER) takes place. By substituting the by-reaction with a more valuable transformation, the efficiency of this paired electrolysis is doubled. For the OER, glycerol oxidation is a valuable alternative. In some cases, both electrodes are used, for example, in domino oxidation–reduction sequences. In BASF lysmeral process, both anode and cathode reactions are coupled in an undivided cell, since both electrolysis products are inert toward the corresponding counter electrode.

2.10 The required amount of applied charge is calculated using the Faraday's law. The stoichiometry of the reaction is given by the ratio of the coefficients n_p/n_s. z is the number of electrons passed to or taken from the substrate and F is the Faraday constant. For example, for the formation of a new C,C-bond, z equals 2. In the case of homocoupling, n_p is 1 and n_s is 2:

$$Q = n^* z^* F^* \frac{n_p}{n_s}$$

In a cross-coupling reaction, the amount of charge is related to the substrate that is primarily oxidized, generally, the substrate that has the lower oxidation potential.

In a flow electrolyzer, the residence time depends on the active volume of the electrolyzer and the flow rate. The overall reaction time is given by $t_{reaction} = Q/I$. For a given volume, $V_{electrolyte}$, to be electrolyzed in a single pass, the flow rate, $\dot{V} = (V_{electrolyte} \cdot Q)/I$.

4.1
- Product intensification (broader temperature window can be used)
- Scalability
- Increase in synthesis precision with respect to molecular weight and molecular weight distributions
- All other typically observed advantages seen in flow chemistry

4.2 Only for very fast reactions such as anionic polymerizations. There, the molecular weight dispersity can be influenced, depending on the mixing efficiency. Otherwise, reaction times are typically much higher than the time required for the mixing of components.

4.3 No. In photopolymerizations, only chain activation/initiation is photoinduced. Propagation and, hence, the overall polymerization rate is determined by the temperature. Best results are typically observed when photopolymerizations is carried out at slightly elevated temperatures.

4.4 Typically, polymerizations need to proceed to full conversion in order to have no disruptive influence of the residual monomer on the next reaction stage. This requires a thorough monitoring of the reaction conditions.

4.5
- FT-IR: monomer conversion, presence of specific chemical functionalities
- NMR: structural information, monomer conversion
- MS: end group patters, end group transformations
- SEC: molecular weight distributions and average molecular weight

5.1 Nanomaterials are chemical substances or materials having diameters in the range of 1–100 nm that are manufactured and used at the nanometer scale.

5.2 Flow chemistry–based approaches offer inexpensive, scalable, reproducible, and safe routes for the development and production of uniform nanosized materials. Flow chemistry–based approaches have numerous advantages, such as, good mixing, improved heat and mass transfer, and high space–time yields, due to short diffusion lengths and high surface-to-volume ratios of the reactors, green and sustainable production (increased safety (toxic materials) and reduced costs (expensive materials)), very homogeneous conditions (reduced turbulence, low Reynolds numbers), and fast and precise adjustment and modification of process parameters (T, p, x), over conventional batch methods.

5.3 The particle formation and growth can be described by the classical nucleation theory. The theory describes the nucleation process in terms of the change in Gibbs free energy of the system, upon transfer of molecules from the liquid phase to a solid cluster, with the following equation:

$$\Delta G = -\frac{4}{V}\pi r^3 k_B T \ln(S) + 4\pi r^2 \gamma$$

where V is the molecular volume of the precipitated species, r is the radius of the nuclei, k_B is the Boltzmann constant, T is the absolute temperature, S is the saturation ratio, and y is the surface free energy per surface area.

The particles can only be formed if $S > 1$, the system is supersaturated, and the free energy term is negative, favoring generation of solid particles and their growth.

The maximum value of ΔG corresponds to the nucleus with critical radius (r^*). A thermodynamically stable nucleus exists when the radius of the nucleus reaches r^*. Therefore, the slope of ΔG at the critical radius of nucleus will be zero:

$$\frac{d\Delta G}{dr} = 0$$

In this situation, the critical radius (r^*) of the spherical nucleus can be obtained by

$$r^* = \frac{2V\gamma}{3k_B T \ln(S)}$$

The behavior of formed solid particles in the supersaturation solution depends on their size. Particles with radius smaller than r^* will dissolve because this is the only way that leads to a reduction of the particle's free energy. Similarly, if $r > r^*$, particle growth will occur.

The colloid stability is the balance of various interaction forces such as van der Waals attraction, double-layer repulsion, and steric interaction. These interaction forces have been described at a fundamental level by Deryaguin and Landau and Verwey and Overbeek (DLVO theory). In this theory, the van der Waals attraction is combined with the double-layer repulsion and an energy–distance curve can be established to describe the conditions of stability/instability. The electrical forces increase exponentially as particles approach one another and the attractive forces increase as the inverse power of separation. As a consequence, these additive forces may be expressed as a potential energy versus separation curve. A positive resultant corresponds to an energy barrier and repulsion, while a negative resultant corresponds to attraction and, hence, aggregation. It is generally considered that the basic theory and its subsequent modifications provide a sound basis for understanding colloid stability.

5.4 There are two main approaches to make nanomaterials: "top-down" and "bottom-up" technologies. Top-down approach basically relies on mechanical attrition to render large components into nano-sized substances, for example, nanomilling. The bottom-up approach relies on the arrangement of a smaller component into more complex assemblies at molecular level, for example, continuous flow metal nanocrystal production in microreactors.

5.5 Gold, silver, platinum, palladium, iron/gold, iron/copper, cobalt/platinum, and so on.

5.6 Organic nanoparticles consist of active pharmaceutical, agrochemical ingredients, nutraceuticals or food supplements, active ingredients of skin care products, etc. Examples are drugs encapsulated in liposomes or dendrimers, nanocrystals of pesticides, fungicide, and encapsulated vitamins.

6.1 Biomass- and waste-derived solvents are the preferred choice for several aspects that are linked to sustainability, economy, and bioeconomy. They can be briefly summarized as follows:

Closed or very efficient CO_2 cycle; low environmental impact compared to petrol-based solvents; generally feature lower toxicity compared to their classic counterparts; are generally associated with a simpler and more efficient disposal procedure.

6.2 By using hydrogen carriers such as formic acid or urea or liquid organic hydrogen carrier (LOHC).

6.3 The solvents have the ability to coordinate the metals present in a heterogeneous catalyst, especially, during the possible change of oxidation state during the process. Based on this state, the choice of solvent used as medium largely influences the metal leaching of the heterogeneous catalyst and, of course, its possible reuse and durability.

6.4 The solvent selection plays a crucial role in assessing the waste associated with the process and the strategy for its minimization. Solvents account for the major part of the mass of a chemical process, and, therefore, the selection should consider the toxicity, environmental impact, safety hazard, and recyclability; all factors that will eventually impact the waste associated with chemical transformation.

6.5 Open-ended question

6.6 Open-ended question

7.1 With flow chemistry, it is possible to combine reactions and methodologies that are otherwise incompatible. This makes the development of telescoped synthesis simpler.

7.2 One of the key advantages of flow is the possibility to make and use hazardous and unstable reagents. They can be prepared in a controlled manner by monitoring the residence time and temperature. Next, they are used in the subsequent reaction, avoiding their stockpiling; they are used as they are formed.

7.3 Clogging can be prevented by a careful choice of reagents and solvents, the use of wide-bore channels or tubing, agitation, sonication, and control of the fluid velocity.

7.4 Flow chemistry can be considered as an enabling technology for the pharmaceutical industry since this methodology allows the use of forbidden or forgotten chemistries, facilitating the chemical space that is inaccessible by batch chemistry.

7.5 Because, its higher volatility, fragrances conversion and yields work much better under pressurized systems, such as those in flow. Also, production can be scaled up without difficulties or risk of producing highly volatile pure compounds. Similar reasoning as with hazardous substrates applies in this case. Through continuous flow chemistry, synthesis, isolation, and dilution of the pure components can be isolated from the labor force.

7.6 It is important since the agrochemical market is estimated to be growing to some 5 billion worldwide by 2020 and it is expected to continue growing in the future. Thus, applying flow chemistry to the synthesis of agrochemicals will improve the sustainability in their production.

8.1 Nowadays, microstructure devices are used in many separation processes such as liquid–liquid extraction and gas–liquid absorption. However, it is still a challenge to carry out the process with numerous theoretical stages. Limited stages can be carried out by a combination of current flow devices but it is still difficult to implement countercurrent flow in small and confined flow paths. The high-gravity method or membrane separation processes might break the restriction of countercurrent flow in microchannels or mini channels. However, it still needs systematic innovation for the entire flow chemistry process.

8.2 In the fine chemical industry, one company usually has several chemical products. These products are often prepared from similar batch reactors. However, with the application of the flow synthesis method, the use of a universal equipment for reaction and separation will decrease for differences in reaction requirements such as different temperatures, times, solvents, and phase states. This limitation can be addressed by expanding the operating range of the flow equipment, but it is best solved via system module replacement. For example, we can change the micromixer from a microsieve dispersion mixer to a membrane dispersion mixer for switching from liquid–liquid reaction to gas–liquid reaction, but the shared equipment such as tanks and the pumps remain unchanged. This method is especially useful for the synthesis of homologues, which have few differences in molecules.

8.3 In the petrochemical industry, a continuous chemical engineering system should work continuously for at least one year without maintenance. For the fine chemical industry, the working period is sometimes dominated by the order cycles. For a flow chemistry system in industrial applications, it is best to make it possible to run for a long time to reduce the failure rate, but sometimes short-term maintenance is necessary, such as in the preparation of $CaCO_3$, as shown above. Thus, convenient and online cleaning methods and programs are the application guarantees of flow chemistry technologies.

10.1 Cell-free enzymes, especially if in purified form, offer a very clean system where the intended catalyst is not impacted by the presence of other enzymes belonging to the cell itself. The catalyst loading can be easily modified and, generally, the substrate concentration that can be achieved is very high. However, if the enzyme requires a cofactor, it needs to be added exogenously and often a second system to recycle such cofactor must be included in the process. Purification, if needed, can be time consuming and the stability of the catalyst, once purified, may be limited. On the other hand, whole cells are rapidly prepared, can house more than one catalyst, and are generally capable of recycling cofactors through their own metabolism. Disadvantages of whole cells include possible side reactions, sequestering of the substrate or product, and permeability issues of the cell wall. Probably one of the major differences is the flow rate that can be achieved; whole cell systems can be efficient if the flow rate is slow, while cell-free catalysts can be operated at significantly higher flow rates.

10.2 Given the different enzymes that are generally involved in cascade reactions, one of the main challenges is the compatibility of the reaction conditions (pH, temperature, buffer composition, etc.) with each catalyst. In addition, the rate of each catalyzed reaction should ideally be the same if multiple reactors are in use. These challenges can be overcome with careful optimization of each reaction. Of course, in-line adjustments of pH between reactors or keeping reactors at different temperatures is also a possible solution.

10.3 Cofactors must be recycled to contain the process costs. While this is also true in batch, in flow, it becomes essential as the volume lost downstream can be significant. Several recycling systems have been developed, following the same strategies used in batch, with the additional benefit that, in flow, one could also consider the recycling of the wastewaters to further reduce the cofactor requirements.

10.4 Scavenger columns are very useful downstream in a flow process to purify the product from contaminants and/or unreacted reagents. The packing material of these columns can be highly selective for functional groups that can then be exploited for targeted trapping. It is important to mention that scavenger columns could also be used between reactors, if required, to increase the purity of the incoming flow stream to the next reactor.

10.5 The inherent property of the enzyme reaction is the rate that would occur without diffusion constraints, that is, under conditions where the transport of the substrate and the product between the micro- and macroenvironment of the enzyme would be infinitely fast. Inherent enzyme activity can be observed in practice at low enzyme activity in the case of well-distributed homogeneous enzyme solutions. The intrinsic velocity and its kinetic parameters characteristics may differ from the inherent velocity and parameters when, due to electrostatic and other interactions between the matrix and the different soluble components, the distribution results in different concentrations in the micro- and macroenvironment.

The actual rate of the enzymatic reaction and the effective kinetic parameters can be observed when diffusion constraints occur in the presence or absence of distributional effects. Values can be determined from the total rate of reaction measured under standard experimental conditions.

10.6 Among the operational parameters influencing the outcome of biotransformation, pH and temperature are of particular importance. Further important parameters, especially in continuous flow systems, are the composition of solvent, the substrate concentration, biocatalyst loading, flow rate, and pressure. Catalytic efficiency, selectivity and stability of the biocatalyst are also key issues.

10.7 Immobilization of biocatalysts is widely used in microreactor systems because it allows continuous operation, easy separation of product and biocatalyst, and often stabilizes the biocatalyst. However, immobilization of the biocatalyst can alter the activity of the enzyme due to conformational changes, steric hindrances (e.g., overcrowded, or nonoriented binding of the enzyme, which causes diffusion restriction of substrates accessing and products egressing the active site), all resulting in changes in reaction kinetics.

The immobilized enzyme is often present within porous carriers. In such cases, the substrate must also diffuse through the porous medium to reach the enzyme. Thus, the diffusion resistance of the particles must be considered together with the external mass transfer resistance, which can be altered by changing the flow rate.

10.8 Although there are variations in many implementations, size, and geometry of the reactors applicable for biotransformations in continuous flow mode, there exist two basic theoretical types of continuous flow reactors: continuous stirred tank reactor (CSTR) and continuous plug-flow reactor (CPFR).

A systematic study of reactor selection for continuous biocatalytic production of pharmaceuticals with a focus on residence time distribution and on the unique mass balance affected by enzyme kinetics for each reactor type revealed that CPFR should generally be the system of choice. In various microfluidics platforms, mostly the CPFR-type microreactors are used in different setups.

However, there are cases, such as the necessity of pH control, where they may need to be coupled with a CSTR or replaced entirely by a series of CSTRs, which can approximate the plug-flow behavior. In this respect, continuous production in enzyme membrane reactors (EMR) – such as in a cascade of continuously operated EMRs – can be considered theoretically as CSTRs. In practice, no reactor behaves ideally, instead it falls within the mixing limits of the ideal CSTR and CPFR.

Index

https://doi.org/10.1515/9783110693690-013

Bei Fragen zur Produktsicherheit wenden Sie sich bitte an:
If you have any questions regarding product safety, please contact:

Walter de Gruyter
Genthiner Straße 13
10785 Berlin
productsafety@degruyterbrill.com